Understanding Ventilation

Published by:
The Healthy House Institute
430 N. Sewell Road
Bloomington, IN 47408

Book and cover design by Lynn Bower

Copyright © 1995 by John Bower

All rights reserved. No part of this book may be reproduced or transmitted in any form, electronic or mechanical, including photocopying, recording, or by any information storage and retrieval system, without the written permission of the author, except for brief passages in a magazine or newspaper review.

10 9 8 7 6 5 4 3 2 1

Publisher's Cataloging-in-Publication Data.
Bower, John
Understanding Ventilation: How to design, select, and install residential ventilation systems
Includes index, bibliographical references, glossary, and list of sources.
 1. Ventilation—Handbooks, manuals, etc. 2. Housing and Health. 3. Dwellings—Environmental engineering.
 I. Bower, John. II. Title.
TH7678.B69   1995
697.9312

Library of Congress Catalog Card Number: 94–096635
ISBN 0–9637156–5–8 $31.95 Hardcover.

Understanding Ventilation

How to design, select, and install residential ventilation systems

Text and illustrations by
John Bower

THE HEALTHY HOUSE INSTITUTE

by John Bower

The Healthy House
How to buy one, How to build one, How to cure a sick one

Healthy House Building
A design & construction guide

Understanding Ventilation
How to design, select, and install residential ventilation systems

Your House, Your Health
A non-toxic building guide (VHS Video)

John Bower has appeared on dozens of television and radio programs, including ABC's "Home Show." He has also written scores of articles for magazines ranging from *Custom Builder*, and the *Journal of Light Construction*, to *Mother Earth News*, and *The Human Ecologist*. In addition, he has given presentations at conferences sponsored by the National Association of Home Builders, the U.S. Environmental Protection Agency, the American Institute of Architects, and the Energy-Efficient Building Association. In 1991, he was presented with a Professional Achievement Award by *Professional Builder and Remodeler* magazine for "healthy house advocacy." He lives with his wife, Lynn, in Bloomington, Indiana.

Acknowledgments

I am particularly indebted to six individuals in the building-science community for whom I have a great deal of respect—both personally and professionally. They were kind enough to review a draft version of this book and offer their insightful comments. I sincerely thank David Hill of Eneready Products Ltd. in Burnaby, British Columbia; Arnie Katz of North Carolina Alternative Energy Corp. in Research Triangle Park, NC; Joseph Lstiburek of Building Science Corp. in Chestnut Hill, MA; Gary Nelson of The Energy Conservatory in Minneapolis, MN; Marc Rosenbaum of Energysmiths in Meriden, NH; and Mike Uniacke of Residential Energy Consulting in Prescott, AZ.

I would also like to express my appreciation to the many ventilation-equipment manufacturers who supplied product literature, operating manuals, photography, and artwork; to the many individuals who provided me with documentation or answered technical questions over the telephone; and to Bob Baird for an excellent job of copy editing.

Contents at a glance

Part 1 Air quality & ventilation basics
Chapter 1 Air and air pollution 23
Chapter 2 How air moves 35
Chapter 3 Strategies for reducing indoor air pollution 41
Chapter 4 Ventilation is for people 47

Part 2 How air enters and leaves a house
Chapter 5 Natural ventilation 61
Chapter 6 Accidental ventilation 69
Chapter 7 Controlled ventilation 81

Part 3 Possible adverse effects of ventilation
Chapter 8 Ventilation-induced moisture problems 101
Chapter 9 Pollutant entry due to depressurization 117
Chapter 10 Backdrafting and spillage 129

Part 4 Design considerations
Chapter 11 House tightness 151
Chapter 12 Capacity of a ventilation system 161
Chapter 13 The cost of ventilation 175
Chapter 14 Controlling a ventilation system 189
Chapter 15 Air-filtration equipment 205
Chapter 16 Fresh-air inlets and stale-air outlets 229
Chapter 17 Distribution: moving air around a house 243
Chapter 18 Miscellaneous considerations 267

Part 5 Mechanical ventilation systems
Chapter 19 Central-exhaust ventilation systems 277
Chapter 20 Local-exhaust ventilation systems 299
Chapter 21 Central-supply ventilation systems 313
Chapter 22 Balanced ventilation systems 323
Chapter 23 Balanced ventilation with heat recovery 335
Chapter 24 Passive ventilation systems 361
Chapter 25 A nine-step design and installation process 369
Chapter 26 Typical central ventilation systems 373

Part 6 Resources
Appendix A Abbreviations 383
Appendix B Glossary 385
Appendix C Bibliography 395
Appendix D Organizations 409
Appendix E Equipment sources 411

Table of contents

Acknowledgments 5
Contents at a glance 6
Table of contents 7
About this book 17
Introduction 19
Preface 21

Part 1 Air quality & ventilation basics 23

Chapter 1 Air and air pollution 25
    The atmosphere 25
    Man-made pollutants 26
    Indoor pollutant categories 27
        Biological pollutants 27
        Pesticides 28
        Gases 29
        Metals 30
        Minerals 31
        Radiation 31
        Smoking 31

Understanding Ventilation

    How pollutants get into houses 32
        Pollutants can be released from materials in the house 33
        Pollutants can be sucked indoors by air-pressure differences 33
        Pollutants can be released by human and animal metabolism ... 34
    The bottom line 34

Chapter 2 How air moves 35
    The physics of air movement 35
        Holes in houses 36
        Pressures in houses 37
    The relationship between pressure and holes 37
        Changing the volume of air passing through a house 37
        Changing the tightness of a house 38
        Changing the pressure in a house 38
    Measuring house pressures 39

Chapter 3 Strategies for reducing indoor air pollution 41
    Source control 41
    Separation 42
    Filtration 43
    Ventilation 43
        Two types of controlled ventilation systems 44
            General ventilation 44
            Local ventilation 44
        Infiltration 45
    People need fresh air 47

Chapter 4 Ventilation is for people 47
    Sick-building syndrome 48
    Sensitive occupants 48
    Ventilation is for people 49
    Reasons for ventilating houses 50
        Diluting pollutants released by materials inside the house 51
        Diluting pollutants sucked indoors by pressure differences 52
        Diluting pollutants released by metabolic processes 53
        Diluting high moisture levels 53
            External moisture sources 54
            Internal moisture sources 55
            Controlling moisture levels in houses 56
    Are there reasons not to ventilate? 57

Part 2 How air enters and leaves a house 59

Chapter 5 Natural ventilation 61
    Natural pressures 61
        Wind 62
        Stack effect 63
        Diffusion 66
        Diffusion through solid surfaces 67
    Do natural pressures supply us with enough air? 67

Chapter 6 Accidental ventilation 69
    Clothes dryers 69
    Central vacuum cleaners 72
    Chimneys 73

Table of contents

Leaky heating/air-conditioning ducts 74
Closed doors 78
Does accidental ventilation supply us with enough air? 80

Chapter 7 Controlled ventilation 81
Drawbacks to uncontrolled air movement 82
    Too little air exchange 82
    Too much air exchange 82
Exhaust ventilation systems 83
    Local-exhaust ventilation 84
    Central-exhaust ventilation 85
Supply ventilation systems 85
Balanced ventilation systems 85
Heat-recovery ventilation 87
Passive ventilation 88
Partial pressurization & partial depressurization 89
Interactions 89
    Another way of looking at interactions 94
    What does it all mean? 96
The bottom line 96

Part 3 Possible adverse effects of ventilation 99

Chapter 8 Ventilation-induced moisture problems 101
Relative humidity & temperature 102
Measuring relative humidity 103
Predicting RH at different temperatures 104
Hidden moisture problems 107
    Hidden winter moisture problems 108
    Hidden summer moisture problems 109
    What about mixed climates? 111
        Wetting potential and drying potential 111
        Solutions for mixed climates 112
When does decay occur? 113
How ventilation affects the indoor relative humidity 113
Humidity control with HRVs 114
What is the correct amount of humidity to have indoors? 115

Chapter 9 Pollutant entry due to depressurization 117
Below-ground pollution sources 118
    Radon 120
    Pesticides and herbicides 121
    Ground moisture 122
    Other soil gases 122
Above-ground pollution sources 123
    Pollution sources in the outdoor air 123
    Pollution sources in building cavities 123
Preventing pollutants from being pulled into the living space 124
    Tightening the house 124
    Manipulating pressures 125
    What works the best? 125
Already-contaminated houses 126

Chapter 10 Backdrafting and spillage 129
- Combustion gases 132
- How house tightness affects backdrafting and spillage 134
- Other causes of backdrafting and spillage 135
  - Chimney function in the winter 136
  - Chimney function in the summer 137
- Solutions to backdrafting and spillage 137
  - Avoid unvented fuel-burning appliances 138
  - Controlling negative house pressures 138
  - Appliances that are immune from backdrafting and spillage ... 139
    - Fuel-burning appliances 139
    - Non-fuel-burning appliances 141
  - Remedial measures for natural-draft appliances 142
  - Outdoor furnaces and boilers 144
- Evaluating a house for backdrafting and spillage 145

Part 4 Design considerations 149

Chapter 11 House tightness 151
- How to measure tightness 152
  - Tracer-gas testing 152
  - Blower-door testing 152
    - Understanding blower-door data 154
    - Limitations of natural-infiltration estimates 155
    - Evaluating a house without a blower door or tracer gas . 156
- How tight is too tight? 156
- Ventilation for loose houses 157
- Tightening a house 158
  - Where to tighten 158
  - For more information 159

Chapter 12 Capacity of a ventilation system 161
- Calculating cfm and ACH 161
- How much capacity is enough? 162
- Understanding ASHRAE's recommendations 163
- Ventilation for formaldehyde removal 165
- Should a ventilation system be based on cfm per person or ACH? 165
- Continuous operation vs. intermittent operation 166
- Dilution is always a factor 169
- If a little ventilation is good, is more better? 169
- Factoring in infiltration 170
- Ventilating fans 172
  - Fan capacity 172
  - Motor and fan efficiency 173
- The bottom line 174

Chapter 13 The cost of ventilation 175
- Capital costs 176
  - Equipment cost 176
  - Incidental costs 176
  - Labor costs 177
  - Adding up the installation costs 177
- Operating costs 177
  - Maintenance costs 178
  - Cost of electricity to run fans 178

Table of contents

Cost of tempering incoming air 179
    Calculating the volume of air entering the house 180
    Calculating heating and cooling energy requirements 181
    Calculating heating and cooling energy costs 182
    Calculating the cost of dehumidifying incoming air 183

Is an HRV cost-effective? 184
Is ventilation worth the cost? 186

Chapter 14 Controlling a ventilation system 189

Manual controls 190
    On/off switch 191
    Manual timers 192
    Speed controls 194

Automatic controls 194
    No control 196
    Automatic timers 197
    Motion sensors 197
    Humidity sensors 199
    Carbon-dioxide sensors 200
    Mixed-gas sensors 201
    Air-pressure controllers 201

Combined control strategies 202
Controlling a ventilation fan and a heating/cooling fan simultaneously 202

Chapter 15 Air-filtration equipment 205

Efficiency and resistance 206
    Efficiency 206
        Weight-arrestance test 207
        Atmospheric-spot-dust test 207
        DOP-smoke-penetration test 208
    Resistance to airflow 208

Where to install a filter 209
Particulate filters 210
    Medium-efficiency filters 212
    HEPA filters 213
    Electrostatic precipitators 215
    Electrostatic air filters 216

Gaseous filtration (adsorption) 217
    Activated carbon 219
    Activated alumina 221

Miscellaneous air-cleaning strategies 222
    Negative-ion generators 222
    Ozone generators 223
    House plants 224

Packaged combination filters 225
The cost of filtration 227
Filtration for sensitive occupants 228

Chapter 16 Fresh-air inlets and stale-air outlets 229

Deliberate openings vs. random holes 230
Inlets and outlets are simply holes in a house 231
    What happens if the ventilation system is shut off? 231
    What if the ventilation system is running? 232
    Minimizing unwanted air movement 232

Inlet and outlet locations 233
    Fresh-air inlet locations 234
    Stale-air outlet locations 235
    Other considerations 235

Tempering the incoming air 236
    Through-the-wall vents 237
    Tempering with an HRV 239
    Electric duct heaters 239
    Hydronic duct heaters 240
    Mixing with warm air 240
    Permeable walls 241
    Solar preheaters 241
    Earth tubes 241

Chapter 17 Distribution: moving air around a house 243

Ventilation effectiveness 244
    Category A rooms and category B rooms 244
    Fresh-air supply grille locations 245
    Stale-air exhaust grille locations 245
    Short-circuiting of air within a room 247
    Effective ventilation layouts 247
    Displacement ventilation 248
Grilles 249
Dampers 252
Ducts 252
    Duct insulation 253
        Insulating for energy efficiency 253
        Insulating to prevent sweating 253
        Insulating materials 254
    Duct locations 255
    Duct sealing 256
    Using heating/air-conditioning ducts for ventilation air 257
Sizing and laying out a duct system 258
Adjusting the airflows 262
    Balancing a ventilation system 262
    Adjusting the airflows to individual rooms 263
Duct cleaning 265

Chapter 18 Miscellaneous considerations 267

Transparency 267
    Out of sight 267
    Can't be felt 268
    Quiet 269
    Odor-free 270
Taking responsibility 270
The KISS principle 271
Accessibility and maintenance 271
Occupant education 272
Future occupancy 273

Part 5 Mechanical ventilation systems 275

Chapter 19 Central-exhaust ventilation systems 277

Central-exhaust ventilation strategies 278
Advantages of central-exhaust ventilation 280
    Low installation cost 280
    Easy to maintain 280
    Minimizes hidden moisture problems in heating climates 280

Table of contents

- Disadvantages of central-exhaust ventilation 281
  - High operating cost 281
  - Unwanted pollutant entry 281
  - Cause of hidden moisture problems in cooling climates 281
  - Cause of backdrafting or spillage 282
  - Minimizing the disadvantages of depressurization 282
- Central-exhaust ventilation equipment 283
  - Individual fans 284
  - Multi-port systems 287
  - Central-exhaust heat-pump water heater 290
  - Through-the-wall vents 291
- Coupling a fresh-air duct with a forced-air furnace/air conditioner 294
  - Fresh-air duct 296
  - Adjustable damper 296
  - Motorized damper 296
  - Central-exhaust fan 297
  - System control 297
  - Air-handler fan 297
- Summary 297

Chapter 20 Local-exhaust ventilation systems 299

- Advantages of local-exhaust ventilation 300
  - Low installation cost 300
  - Minimizes hidden moisture problems in heating climates 300
  - Eliminates pollutants quickly 300
- Disadvantages of local-exhaust ventilation 300
  - High operating cost 301
  - Brings in unwanted pollutants 301
  - Cause of hidden moisture problems in cooling climates 301
  - Cause of backdrafting or spillage 301
- Avoiding depressurization when using a powerful local-exhaust fan ... 302
  - Extra through-the-wall vents 302
  - One large through-the-wall vent 302
  - Open a window 303
  - Add a supply fan 303
- Minimizing the number of exhaust fans in a house 304
- Local-exhaust ventilation fans 304
  - Kitchen-range exhausts 305
    - Recirculating range hoods 305
    - Overhead range hoods 305
    - Low-profile and microwave hoods 306
    - Surface-mounted downdraft exhausts 306
    - Pop-up downdraft exhausts 306
    - Remote-mounted kitchen fans 307
  - Bath fans 308
  - Quiet fans 309
  - Remote-mounted local-exhaust fans 310
  - Miscellaneous local-exhaust fans 310
- Summary 311

Chapter 21 Central-supply ventilation systems 313

- Central-supply ventilation strategies 313
- Advantages of central-supply ventilation 314
  - Low installation cost 314
  - Easy to maintain 314
  - Minimizes hidden moisture problems in cooling climates 315
  - Keeps out unwanted pollutants 315
  - Minimizes backdrafting or spillage 315

Understanding Ventilation

Disadvantages of central-supply ventilation 316
    High operating cost 316
    Cause of hidden moisture problems in cold climates 316
Central-supply ventilation equipment 317
    Individual fans 317
    Through-the-wall vents 318
Coupling with a forced-air furnace/air conditioner 318
    The old way 319
    A better way 320
    Equipment for fresh-air duct installations 321
Summary 322

Chapter 22 Balanced ventilation systems 323

Balanced ventilation strategies 324
Advantages of balanced ventilation 326
    Minimizes hidden moisture problems 326
    Keeps out unwanted pollutants 326
    Minimizes backdrafting and spillage 326
    Better control 326
    Filtering capability 327
Disadvantages of balanced ventilation 327
    High installation cost 327
    High operating cost 327
    Greater complexity 328
Balanced ventilation equipment 329
Coupling with a forced-air furnace/air conditioner 329
    Attaching only the fresh-air supply duct to the return duct 331
    Attaching only the stale-air exhaust duct to the return duct ... 332
    Attaching the stale- and fresh-air ducts to the return duct 332
    What if the furnace/air-conditioner fan is off? 333
Summary 333

Chapter 23 Balanced ventilation with heat recovery 335

Balanced heat-recovery ventilation strategies 338
Advantages of an AAHX 338
    Low operating cost 338
    Tempering of incoming air 338
    Minimizes hidden moisture problems 339
    Keeps out unwanted pollutants 339
    Minimizes backdrafting and spillage 339
    Better control 340
    Filtering capability 340
Disadvantages of an AAHX 340
    High installation cost 340
    High operating cost 341
    Greater complexity 341
Not all AAHXs are created equal 341
    AAHX core types 342
        Flat-plate cores 343
        Rotary cores 343
        Heat-pipe cores 344
    Moisture control 344
        Moisture control in air-conditioning climates 345
        Moisture control in cold climates 346
        Getting rid of the condensate 346
        Sweating on the outside of the cabinet 347
        Defrosting 347

Energy efficiency 350
    Apparent sensible effectiveness (ASEF) 353
    Sensible recovery efficiency (SRE) 354
    Total recovery efficiency (TRE) 354
    How important is it to consider latent heat? 355
    Fan wattage 356
Maximizing energy efficiency 356
  Balanced heat-recovery ventilation equipment 356
  Coupling an AAHX with a forced-air furnace/air conditioner 358
  Summary 359

Chapter 24 Passive ventilation systems 361

  Passive ventilation strategies 362
  Advantages of passive ventilation 363
  Disadvantages of passive ventilation 364
    High operating cost 364
    Minimal control 364
    Minimal predictability 365
  Passive ventilation equipment 365
  Summary 367

Chapter 25 A nine-step design and installation process 369

  Step 1 Fix any problems with the house 370
  Step 2 Determine the capacity 370
  Step 3 Select a strategy 370
  Step 4 Select the equipment 370
  Step 5 Select a control 371
  Step 6 Plan the general flow of air 371
  Step 7 Design the ducts and grilles 371
  Step 8 Install the equipment 372
  Step 9 Balance and measure the airflows 372

Chapter 26 Typical central ventilation systems 373

  Upgraded bathroom exhaust fan 375
  Multi-point central-exhaust fan 376
  Multi-point central-exhaust fan coupled
      with a forced-air furnace/air conditioner 377
  Fresh-air duct coupled with a forced-air furnace/air conditioner 378
  Dual-fan, independently ducted system 379
  Dual-fan system coupled with a forced-air furnace/air conditioner 380

Part 6 Resources 381

Appendix A Abbreviations 383
Appendix B Glossary 385
Appendix C Bibliography 395
Appendix D Organizations 409
Appendix E Equipment sources 411
Index 421

About this book

"...fond as our people are of improvement, the greatest possible improvement in a dwelling house—ventilation—is as yet a thing almost unknown in this country..." This quote is from a book titled *The Architecture of Country Houses* that was written in 1850 by the noted architect and prolific writer A.J. Downing.(Downing) You would think we would have come a long way since then. After all, ventilation is generally regarded as being more important in today's tighter houses than it was in the loosely built houses of the last century. Yet, in many ways, ventilation is almost as unknown today as it was a century and a half ago.

Without a doubt, we all could benefit from better ventilation in our houses, but many builders, architects, designers, and homeowners are unaware of precisely why it is needed, how much is necessary, what options are available, how to select the right equipment, and the installation requirements. When asked, we all say we want—indeed, we expect—our houses to be filled with clean, fresh, healthy air. This book explains how to obtain it.

Understanding Ventilation deals with all aspects of ventilating houses—everything from ventilation theory to the specifics of different pieces of equipment. But I have striven not to write in the technical jargon of engineers and scientists—but rather in a down-to-earth style that builders, designers, and homeowners can relate to and appreciate. Though you will find some technical terms and concepts are presented, they are discussed in easy-to-understand language.

As it turns out, there are many different ways to ventilate houses, but there is no single, unique solution that works well in all situations. The fact that there isn't one straightforward answer scares many people into doing nothing. This is unfortunate because understanding ventilation isn't difficult. Yet, a knowledge of some basic building-science concepts and a few simple laws of physics are essential in order to properly select, design, and install a ventilation system. Some of the concepts presented in *Understanding Ventilation* may conflict with what you have already read or heard in the past. This is because knowledge about how houses work has evolved considerably in the building-science community over the past few years. We now know that a house is more than the sum of its parts—it is a dynamic, ever-changing system.

When a house is viewed as a system, aspects of moisture control, infiltration, safe chimney operation, pollutant sources, duct leakage, pressure imbalances, and ventilation all become interconnected. In order to show how they relate to one another, you will learn not only about ventilation *per se*, but also the various side issues that affect indoor air quality. For example, you will learn how a fireplace can either ventilate a house (accidentally) or asphyxiate the occupants; that operating a range hood to rid the kitchen of pollutants can cause radon or other contaminants to be pulled indoors; that a particular ventilation system may be ideal in one climate but can result in disastrous moisture damage in another.

If these notions are unfamiliar to you, reading *Understanding Ventilation* from cover to cover will lead you logically through these interrelationships. If your knowledge of building-science is up-to-date, you may want to skim from chapter to chapter.

Part 1 of *Understanding Ventilation* covers the basics: what ventilation is, what it isn't, and why we need it. Part 2 discusses the differences between natural ventilation, accidental ventilation, and controlled ventilation (ventilation "on purpose"). Part 3 delves into the ways ventilation sometimes causes moisture and pollution problems. Part 4 explores a variety of factors involved in selecting, designing, and installing a ventilation system: capacity, cost, controls, filters, distribution, etc. Part 5 contains in-depth information about the different ventilation strategies (*e.g.* central and local exhaust, central supply, and balanced) as well as schematic drawings of several popular ventilation-system designs. And Part 6 is chock-full of appendices of resource information.

The acronym HVAC is used extensively in the construction industry. It stands for **H**eating, **V**entilating, and **A**ir **C**onditioning. There are HVAC contractors, HVAC designers, HVAC systems, etc. As a result of widespread use of this term, many people think that ventilation must be automatically a part of any heating and air conditioning system. Nothing could be further from the truth! You certainly can have a combined system, but in many instances the ventilation system is not coupled to the furnace or air conditioner. In fact, there are ventilation contractors and designers who don't deal with heating or air conditioning at all. *Understanding Ventilation* discusses all types of ventilation systems—those combined with a furnace or air conditioner and those that are solely ventilation systems. To avoid any misunderstandings, this book does not use the term HVAC to describe any of them.

In order to help readers locate organizations of interest, or specific manufacturers, **bold type** is used occasionally throughout the book. Whenever an organization that actively deals with residential ventilation issues is mentioned in the text, it is shown in ***bold italic type*** and its address and telephone number is listed in Appendix D. Dozens of manufacturers of specific ventilation products are also shown in **plain bold type** whose addresses and telephone numbers are listed in Appendix E. In addition, particularly helpful books and magazine articles are flagged with the author's name in parentheses(like this) so they can be located easily in the extensive bibliography (Appendix C).

Introduction

Which house would you rather have: Mary's energy-efficient, comfortable, draft-free, odor-free, healthy house, or Jim's drafty, moldy, smelly, energy-hog-of-a-house, where runny noses and headaches are common?

No doubt about it, we all want houses that are energy efficient, affordable, comfortable, and healthy. Yet, many houses feel stuffy due to lingering odors and too much humidity; condensation on windows leads to mold growth and decay; and many of us suffer health effects due to polluted indoor air. All these issues interact in a variety of ways, and ventilation is a major piece of the puzzle.

There are two primary reasons for ventilating houses: 1) to provide the necessary fresh air for the occupants to breathe and 2) to dilute indoor air pollutants and excess moisture. Complaints of stuffiness, unpleasant odors, and illness are common in houses that contain too little fresh air. Outdoor air pollution is bad enough, but we are now learning that indoor air pollution is almost always considerably worse—and this polluted air can make us sick. Reports of "sick building syndrome" are being reported with increasing frequency all across the country.

To feel comfortable and healthy, people simply need clean air. While breathing is something few of us consciously think about, each of us breathes every minute of every hour of every day of every year of our lives. We breathe primarily to take the oxygen into our bodies that we need to survive, but we also exhale moisture, carbon dioxide (CO_2) and a variety of other by-products of metabolism. We take air in, we use it, then we expel it, and even though we do this unconsciously most of the time, our bodies are specially designed to provide this service to us continuously.

But, don't houses breathe also? Yes, but in many cases they don't do it very well. A ventilation system should be as reliable and predictable as our own lungs. Houses without ventilation systems get their air quite

Understanding Ventilation

by chance—the air moving into and out of them is totally uncontrolled. For example, air may infiltrate indoors only when the wind is blowing. We deserve to have control over the air we breathe, we deserve more than "ventilation-by-chance," we deserve "ventilation-on-purpose."

When air enters a leaky house through the cracks, it may contain radon or other pollutants, and thus, not be very clean or healthy. We deserve a ventilation system designed to bring in clean air. We deserve a system as efficient as our lungs. Our automobiles typically have better ventilation systems than our houses. Even though some houses get enough fresh air by chance some of the time, research is showing us that most newer houses simply don't have enough fresh air on a regular basis.

What about the fact that most houses have never had ventilation systems? Why do we need them now? Part of the answer lies in the fact that we expect more out of our houses than our grandparents did. We aren't willing to accept drafty, uncomfortable, unhealthy, energy-wasting houses. At the same time, considering all of the various synthetic materials and cleaning products we bring indoors, our houses have more and different sources of indoor pollution today. We spend more of our time indoors than our parents or grandparents did, and we generate more indoor moisture with steaming-hot showers, washing machines, dishwashers, and hot tubs. Today's houses are also more energy-efficient, are more tightly constructed, and have smaller furnaces—all of which lead to less "natural" ventilation.

Why not just build looser houses? People really don't want to live in loose houses—they are drafty, uncomfortable, and consume too much energy. We can't afford to waste energy heating and cooling our houses any more than we can continue to waste it in gas-guzzling automobiles. And, just because a house is loosely constructed doesn't mean the occupants are getting enough fresh air—loose houses can be unhealthy, too.

What about air filters? Filters are one of several ways of improving the air quality in houses. But filters can't supply us with oxygen and they can't remove moisture from the air. So, if you use an air filter in your house, you still need ventilation—although you might not need quite as much. As it turns out, all the different strategies for reducing air pollution should be used *in addition to* ventilation, not instead of it.

If ventilation is so important, why don't building codes require it? Some codes do require ventilation. In fact, Canada, Sweden, and France have specific ventilation requirements in their national building codes. In the U.S., the state of Washington has incorporated ventilation into its building code. But most local, regional, and national building codes in the U.S. have only token ventilation provisions, such as requiring either a small exhaust fan or an operable window in the bathroom. However, building codes are continually being revised, and as code officials learn more about the issues involved, ventilation will, no doubt, be a requirement throughout the U.S. in the near future.

How much does ventilation cost? Houses are expensive enough already and now we have to spend more on a ventilation system? Yes, ventilation costs money. But then, so does everything else. Bathtubs cost money, as does furniture, newspapers, television sets, and almost everything! Having clean air to breathe is far more important to our health and well-being than any of these "things," yet ventilation is often rejected because of its cost. Surprisingly, as you will learn, the cost is often quite reasonable. And besides, can a price tag be placed on fresh, healthy air?

Without a doubt, each and every one of us could benefit from a ventilation system in our home. In fact, a ventilation system should be as important as air conditioning, electric lighting, closets, kitchen cabinets, and indoor plumbing. Ventilation is not a mysterious high-tech process; it is very simple once you understand the basics. In fact, it is no more complicated than the electrical, plumbing, and heating/cooling systems we already take for granted in our homes. But, ventilation is frequently misunderstood, so it is a subject that is often ignored. We tend to spend far more time selecting the color of curtains wallpaper, paint, and cabinets than we do on ventilation. Yet, in comparison, proper ventilation is vital.

Fresh air is not something to be left to mere chance. We simply cannot continue to pollute our indoor environment, then expect Mother Nature to keep it healthy for us. Natural ventilation just doesn't assure that the air in a house will be kept clean—especially if that house is filled with polluting materials. The bottom line is this: ventilation isn't an option; it is fundamental to a well-built house.

Preface

Words such as *infiltration*, *air exchange*, *combustion air*, and *ventilation* are often used incorrectly. This definitely leads to a certain amount of confusion. Therefore, the following brief discussion is offered to give you an understanding of what various ventilation-related terms mean. The concepts will be expanded upon in the chapters that follow.

For air to move from one place to another, there must be an *air-pressure difference* to push it or pull it. Air-pressure differences can be placed in three categories: *natural pressures* resulting from natural phenomena such as the wind; *accidental pressures* caused by mechanical devices such as clothes dryers that move air into or out of a house for a purpose other than improving air quality; and *controlled pressures* caused by ventilating fans.

When the air pressure indoors is greater than that outdoors, a house is said to be *pressurized*, or experiencing a *positive* pressure. When the air pressure indoors is less than the air pressure outdoors a house is *depressurized*, or experiencing a *negative* pressure. Sometimes, part of a house is pressurized and part of it depressurized. Pressurization and depressurization are neither good nor bad, but they can occasionally result in adverse effects. For example, radon can get sucked into a depressurized house.

Exhaust air is air that is leaving a house. It is often called *stale air* because it has been contaminated by people, activities, or materials inside the house. The *outdoor air* that enters a house is either called *make-up air* (because it makes up for what was exhausted), or *intake air*. It is also often called *fresh air* even though it may be contaminated with outdoor air pollutants.

Air that enters or leaves a house for the purpose of improving the air quality is called *ventilation air* or *supply air*. A *general ventilation system* is designed to improve the air in the whole house. A *local ventilation system* is designed to improve the air in one part of a

Understanding Ventilation

house (e.g. a bathroom). There are three basic ventilation strategies. *Exhaust ventilation* blows stale air outdoors, causing a house to be *depressurized*. *Supply ventilation* blows fresh air into a house, causing the house to be *pressurized*. *Balanced ventilation* uses two fans to blow fresh air indoors and stale air outdoors simultaneously, so the house experiences a *neutral pressure*. *Heat-recovery ventilation* is a special form of balanced ventilation that is more energy-conserving. (Actually, there is one heat-recovery ventilator on the market that is an exhaust ventilator.)

When air moves between the indoors and the outdoors (or between the outdoors and the indoors) an *exchange* of air results. The speed at which the exchange takes place is called the *exchange rate*. When air moves within a room, or from one room to another room, it is not being exchanged—it is being *circulated*. Forced-air furnaces and air conditioners are primarily designed to circulate (or recirculate) air in a house, as well as heat or cool it. (If a forced-air furnace or air conditioner has leaky ducts, an exchange of air might result, but this isn't usually a very desirable way to exchange air.) Ventilation systems are specifically designed to exchange the air in a house—as well as circulate it.

Exchanging the air in a house is important in order to *dilute* the concentration of pollutants found in the indoor air. If indoor-pollutant concentrations are too high, they can negatively affect the health of occupants. Pollutant concentrations can also be reduced through *source control*, *separation*, and *filtration*. These strategies can be used effectively in addition to ventilation, but they are not a substitute for it because they cannot supply oxygen to living persons in the living space.

When a fan causes air to move directly, it is called *active air movement*. When air moves indirectly because of a fan, or because of something else causing an air-pressure difference, it is called *passive air movement*. For example, when a window fan *actively* blows air outdoors through a window, an equal volume of air *passively* enters through another window. At any given moment, the total amount of air entering a house always equals the total amount of air leaving the house.

For an air-pressure difference to cause an exchange of air in a house, the house must have holes in it. The holes can be *deliberate* (cut through a wall on purpose), or *random* (miscellaneous holes in the structure that are often hidden or too small to be visible). Air moves through random holes passively when a house is either pressurized or depressurized. If a fan is connected to a deliberate hole, air will pass through that hole actively when the fan is operating. If the fan is not operating, but something else applies pressure to the house, air can move through a deliberate hole passively.

Infiltration is air that passively enters a house through the random gaps and holes in the building. *Exfiltration* is air that passively leaves a house through the random gaps and holes in the building. Infiltration and exfiltration can be caused by anything that results in an air-pressure difference between the indoors and the outdoors.

In a house with combustion appliances (*e.g.* gas or oil furnaces, water heaters, etc.), *combustion air* passes into the combustion chamber of the appliance where it mixes with the fuel and burns. Combustion air may come directly from the outdoors or it may come from the living space. If a combustion appliance is connected to a conventional chimney, the *combustion by-products* rise up through the chimney because warm air rises. When this happens, a negative pressure is created inside the chimney called a *draft*. Besides the combustion by-products, a certain amount of *dilution air* also leaves the house through a conventional chimney. Even though a fan isn't used, conventional chimneys are considered *active* exhausts. If a conventional chimney is not in use, it is *inactive*, and air can move through it *passively* in either direction. If a negative pressure in a house is stronger than the draft in an active chimney, the chimney may not function correctly. Some combustion appliances use a fan to expel combustion by-products outdoors, rather than relying on chimney draft.

Part 1

Air quality & ventilation basics

Chapter 1 Air and air pollution
Chapter 2 How air moves
Chapter 3 Strategies for reducing indoor air pollution
Chapter 4 Ventilation is for people

Chapter 1

Air and air pollution

Fact: the air inside our houses is making us sick. According to a Special Legislative Commission of the Commonwealth of Massachusetts, "Indoor air pollution is a growing problem in the United States and accounts for 50% of all illness. Health care costs are estimated at $100 billion per year."(Special) It is becoming very clear that poor indoor air quality is something that must be dealt with. In fact, indoor air pollution is often ranked as the number one environmental problem facing us. It is responsible for symptoms ranging from sinus congestion to cancer, from depression to immune-system damage. While many houses don't have life-threatening pollution problems, it isn't unusual for everyday symptoms such as headache, drowsiness, runny nose, lethargy and inability to concentrate to be related to poor indoor air quality.

The good news is that it is possible to build houses that don't make us sick. Actually, there are several ways to improve the air quality indoors—but all the solutions ideally should to be coupled with better ventilation. And that is what this book is all about—*ventilation*—supplying houses with fresh, clean air and removing the stale, polluted air. But before discussing how to ventilate, it is important to understand what air is composed of, how air gets polluted, and why polluted air is not good to breathe. In other words, the problem needs to be understood before an attempt is made to solve it. This chapter will discuss what kinds of pollutants we are exposed to and where they come from.

The atmosphere

Like all living creatures, human beings require oxygen to survive. The atmosphere that all of us have been breathing for millions of years contains about 21% of this vital element. High-school science textbooks tell us that the remaining 79% of the air surrounding our planet is almost entirely nitrogen, with small amounts of gases such as argon, carbon dioxide, helium, krypton, neon, and xenon. Air also contains a certain amount of water vapor. This is basically what we have been breathing during our evolution from cave-dwelling creatures to modern man.

A wide variety of naturally occurring air pollutants also float around us waiting to be breathed. For example, volcanoes spew out dust and sulfur compounds. Forest fires contaminate the air with soot and smoke. Radon is a naturally occurring radioactive gas found in small quantities in the air virtually everywhere. Swamp gas, sea salt, sand particles, and petroleum leaking to the surface of the earth can also "naturally" affect air quality. Living creatures pollute the air with their waste products and the by-products of their metabolism—carbon dioxide, ammonia, methane gas, water vapor, etc.—and when living creatures die, their decomposition releases additional pollutants into the air.

"Natural" pollutants also include pollen and mold spores—tiny particles that can aggravate a variety of symptoms in allergic or asthmatic individuals. The characteristic odor of mold or mildew is something many people find irritating or offensive. Disease-causing viruses and bacteria are also considered air pollutants.

Ozone is a gaseous pollutant that has been around for millennia. It is a powerful oxidizing agent—meaning it can react with a variety of substances, including human tissue. As a result, it is very irritating to the respiratory tract. Ozone consists of three oxygen atoms bound together, while the oxygen we need to breathe is found in atomic pairs. This may seem like a minor difference, but our bodies were not designed to metabolize the three oxygen atoms of ozone. In the lower levels of the atmosphere, ozone is often a component of smog, where it can cause burning of the eyes and breathing difficulties. Paradoxically, in the upper atmosphere—where there are no people—ozone is vital to our survival. High above the earth's surface, a thin layer of ozone blocks a portion of the ultraviolet light from the sun. This is the ozone layer in which scientists have recently discovered a "hole." Where the ozone layer in the upper atmosphere is thin or missing, excess ultraviolet light can pass through the atmosphere and reach the earth's surface. Too much ultraviolet light increases incidences of skin cancer.

Man-made pollutants

In addition to all the naturally occurring air pollutants, many man-made contaminants exist as well. One of the earliest examples of man-made pollution was a campfire. When the fire was brought inside a cave, it was responsible for the first case of *indoor* air pollution. Over the centuries, human beings have discovered many more ways to contaminate the air. Today, primarily as a result of the industrial revolution and "better living through chemistry," the air we are breathing contains hundreds of new pollutants, some of which are difficult to pronounce. We have all heard of formaldehyde, but other compounds such as limonene, 4-Phenylcyclohexene, and ethylbenzene rarely pop up in daily conversation.

The air in a typical house is filled with hundreds of gaseous pollutants that have only been synthesized in the last few decades. Many of these contaminants originate in paints, caulking, adhesives, cleaning supplies, building materials, home furnishings, printing ink, and dozens of other products that we come in contact with daily. While these substances may not have an immediate and noticeable effect on many of us, people who have an illness called multiple chemical sensitivity (MCS) often exhibit severe symptoms when exposed to these common sources of indoor pollution. MCS is discussed in more depth in Chapter 4.

Since the advent of the industrial revolution in the late 1800s, everyone on the planet has been exposed to ever-increasing amounts of combustion by-products. While emissions from factory smokestacks and automobiles are now strictly regulated—at least in industrialized countries—we still breathe far more pollution from combustion processes than is good for us. As a result, some parts of the U.S. have restrictions on the use of wood stoves and fireplaces, but for the most part, pollution from these devices is very poorly regulated. Gas or oil furnaces and water heaters, space heaters, and gas ranges are also common sources of combustion gases. While their contribution to atmospheric pollution is certainly a serious concern, the greatest immediate danger to human health is when they directly pollute the indoor air—something that, unfortunately, happens on a regular basis.

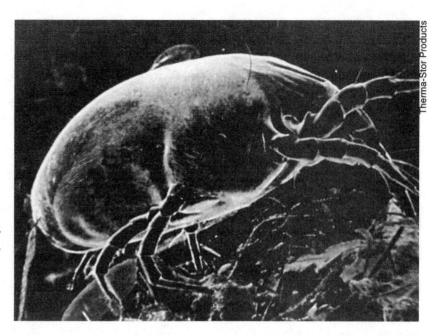

Figure 1–1. Dust mites and their fecal pellets often bother allergy sufferers.

Indoor pollutant categories

Literally thousands of possible air pollutants can be found inside a house. They range from lead paint to asbestos, radon to carbon monoxide, formaldehyde to dust mites. With new chemical compounds being developed daily, the list continues to grow. Following is a brief rundown of what types of materials contaminate the air we breathe indoors.

Biological pollutants

Biological pollutants are sometimes appropriately called "bio-nasties" in indoor pollution circles. They include such agents as dust mites (Figure 1–1), mold (Figure 1–2), mildew, *Legionella pneumophila* bacteria, decay-causing fungi, viruses, pollen, etc. These are all living organisms that can negatively affect health.[Spengler] There are currently some 41 million people in the United States who have conventional allergies, so there is a good chance that someone in a majority of households will end up with a runny nose, itchy eyes, sinus congestion, or a headache if these pollutants are allowed to proliferate indoors. When someone suffers from an allergic reaction, their body is weakened, making them more susceptible to many other air pollutants.

Of course, even more devastating health effects than allergies are attributed directly to biological pollutants. Legionnaire's disease, a vivid example, has certainly taken its toll in human life. Some molds can be equally deadly and an asthmatic attack can be quite frightening. There were 4,580 asthma-related deaths in the U.S. in 1988.[Williams] And the incidence is rising—today, asthma affects some 3.7 million children, up from 2.4 million in 1980.[Childhood]

One of the most common indoor biological pollutants is the dust mite.[American College] These microscopic creatures find carpeting to be a comfortable home where they enjoy their favorite meal—tiny flakes and particles of dead skin we all shed every day. If someone is allergic to dust mites, he or she typically reacts to the extremely small pellets of mite feces that can be stirred up simply by walking across carpeting. Once airborne, mite fecal pellets remain aloft for extended periods of time and are easily inhaled.

Mold and mildew are also biological pollutants prevalent in homes, especially where water or high humidity is present. Basements and bathrooms are particularly vulnerable. If a moisture problem is long-lasting, it can progress to decay, resulting in struc-

Understanding Ventilation

Figure 1–2. Under a microscope mold looks surreal, but mold spores are a common air pollutant.

tural damage to the house itself. Even in a house that seems dry, areas of high humidity—microclimates—always exist near cold surfaces. Because of this, the center of a room can be fairly dry, while mold may be growing on a cold window frame or in the corner of a closet. A variety of moisture sources affect the humidity in a house: bathing, laundering, washing dishes, exhaled breath, defective gutters, plumbing leaks, inadequate dampproofing of basements or crawl spaces, etc. Humidity and water vapor will be discussed in much greater detail in both Chapters 4 and 8.

Histoplasmosis is a serious infectious disease caused by a fungus that can be associated with pigeon droppings in attics. It should be of concern if you plan on doing any demolition during a major remodeling project. Pets are another common source of allergic reactions because of their urine, dander, shedding, and saliva. More people are probably allergic to cats than to any other pet, but sensitivities to pets ranging from birds to dogs are well-documented. And allergies to insects, such as cockroaches, isn't unusual.

Pesticides

Sometimes, biological pollutants themselves don't have a direct effect on human health, but the toxic pest-control chemicals used to eradicate them cause problems. A good example involves termites—small creatures we seldom see. The termite workers that actually do the damage to the wood structure of a house can pass through a crack as small as $1/32"$, so it is quite easy for them to remain hidden from view. The chemicals that have typically been used to control termites have destroyed the health of many people around the world. These pest-control chemicals are literally formulated to kill, and most don't differentiate between termites and human beings—they affect all forms of life. Because termites are so small, they are easily killed with these toxic solutions. It takes considerably more to kill a human being, or a pet dog—simply because of their larger size—but small amounts of termiticides have caused a variety of negative health effects ranging from immune-system damage, headaches, nausea, and dizziness to muscle spasms, confusion, and multiple chemical sensitivity (MCS).

In *Safety at Home,* the National Coalition Against the Misuse of Pesticides (NCAMP) reported that consumers in the U.S. buy and use a phenomenal 285 million pounds of toxic pesticides every year. The Coalition says these chemicals "are nerve poisons, can cause cancer, respiratory problems, birth defects, genetic damage, injure wildlife, and pollute the environment and drinking water."(National Coalition) When used indoors, pesticides can be serious indoor air pollutants. Fortunately, less-toxic methods of pest control are available.(Olkowski)

Gases

Gases are another major category of air pollutants. While it is obvious that an automobile's exhaust pipe is emitting noxious combustion by-products, many houses contain unvented furnaces and space heaters that expel combustion gases directly into the air we breathe. Natural gas and propane kitchen ranges can emit carbon monoxide, carbon dioxide, and nitrogen dioxide. Even furnaces and fireplaces that are connected to a chimney can be problematic because it is not uncommon for the combustion gases to backdraft, that is, flow back down the chimney. This will be covered in greater detail in Chapter 10 because it is something that can actually be caused by an improperly designed ventilation system.

Volatile organic compounds (VOCs) are gases that are emitted, or outgassed, from a wide variety of modern materials (Figure 1–3). The word *volatile* means these gases evaporate easily—usually below room temperature—and the word *organic* means they contain one or more atoms of carbon. Many VOCs are difficult to spell and pronounce: cyclopentadine-ethenyl-2-ethylene, hexamethylene triamine, tetrachloroethylene, 4-phenylcyclohexene, etc. Formaldehyde is probably the best known. It is colorless, only has an odor at high concentrations, and is a probable human carcinogen as well as a sensitizer.(Godish 1986) Exposure to formaldehyde can sensitize you so that your body will react to very tiny amounts of it—amounts that were previously not a problem. Once sensitized, you may begin to react to a wide range of other VOCs as well, and a typical house could have a hundred or more different VOCs

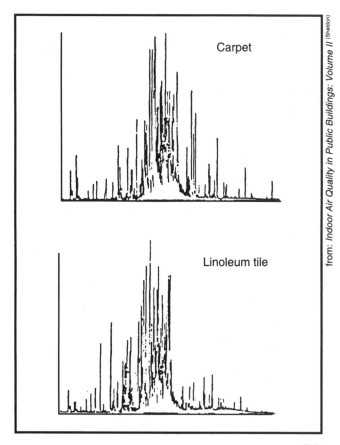

Figure 1–3. Gas chromatography/ mass spectrometry can be used to analyze the outgassing of building materials. In these chromatograms, each vertical line represents a different VOC.

floating around in the air. These include compounds that are neurotoxins (toxic to the nervous system, spinal cord, brain, etc.), carcinogens (cancer-causing), mutagens (alter chromosomes, genes, etc.), teratogens (interfere with fetal development), and irritants (cause part of the body to become overly sensitive). The characteristic odor of a new car's interior or a new vinyl sofa is composed of dozens of different VOCs.

Most VOCs have not been studied for their precise health effects, but an evaluation of 52 compounds released from common building materials found that 25% were known or suspected carcinogens and 82% were known or suspected irritants.(Molhave) To learn how VOCs act in combination with each other (synergism) is virtually impossible because the concentrations and specific compounds vary so much from house to house. What is known is that, in general, it is wise to limit our exposure to VOCs.

VOCs from products such as fresh paint will dissipate relatively quickly, often in a matter of days, but other compounds can linger for weeks or months. Building materials such as particleboard and medium-density fiberboard will outgas formaldehyde into the air for years. The outgassing rate for VOCs changes with the seasons: it increases as the temperature and humidity go up. In other words, VOCs outgas faster during a hot, humid summer than in a cold, dry winter.

People and animals also release several different gases. These are the pollutants we all give off just because we are alive—they are normal by-products of metabolism, but they are pollutants nevertheless. They include such gases as acetaldehyde, acetone, ammonia, carbon dioxide, hydrogen sulfide, methane, toluene, and water vapor. Sometimes we release more pollutants than normal, for example, after heavy exercise or eating certain foods.

Metals

Metals can also pollute our houses. Although it has not been sold since the early 1980s, lead paint can still be found on the walls of tens of millions of houses. A window painted with lead paint can, as it is raised and lowered over the years, result in a considerable amount of powdery lead dust on the window sill. Children looking out such a window will invariably put their hands on the sill, then later put their fingers in their mouths. Lead paint applied to the outside of a house often chalks off, contaminating the soil with lead dust. This then gets tracked indoors on shoes and builds up in the carpeting. Children playing on contaminated carpet will get lead dust on their hands and, again, end up putting their fingers in their mouths. Lead ingested as a result of this normal, hand-to-mouth activity can result in permanent brain damage.(Center 1991) It is often surprising to realize that children generally ingest more lead by coming in contact with lead dust that has settled on window sills and in carpeting, than by chewing on lead-painted woodwork.

In 1989, a four-year-old Michigan child developed a rare form of mercury poisoning after his family painted the interior of their home with what they believed was benign latex paint.(Center 1990) Mercury had been used in the paint as a fungicide and the toxic metal evaporated into the air of the house as the paint dried. At that time, approximately 25% of the interior latex paint sold in the U.S. contained mercury. Because of swift action by the U.S. Environmental Protection Agency (USEPA), mercury can no longer be used in interior latex paints. However, it is still possible to buy exterior paint containing mercury, and other heavy metals such as cadmium are occasionally used in interior formulations.

The metal arsenic is a common component of the chemically treated lumber used for wooden decks, porches, and railings. This lumber has a greenish tint and is often described as being "salt-treated." It isn't table salt that is used, however, but rather an arsenic salt. Several lawsuits have resulted from individuals being made ill after working with treated lumber. One worker, whose job it was to build picnic tables, was so affected that he vomited seven to eight units of blood—nearly half of his total body supply—before he was able to get to a hospital for treatment.(Arsenic) When tested, he and a co-worker had arsenic levels in their hair and nails that were hundreds of times higher than normal. Although manufacturers warn against using chemically treated lumber on eating surfaces, this material is often found on picnic tables. Most building codes prohibit the use of treated lumber that is directly exposed to the living space of a house, but it is occasionally unknowingly used indoors.

Minerals

Asbestos is a naturally occurring, fibrous mineral that can cause a variety of lung diseases including lung cancer.(USEPA 1985) While it is no longer used in new construction, asbestos can still be found in many older houses as a component of insulation, vinyl flooring, gaskets on the doors of furnaces and wood stoves, siding, and drywall joint compound. If these materials remain intact, and do not release any asbestos fibers, they are often considered safe. But if they become damaged or start to deteriorate, the asbestos can easily become airborne and be inhaled.

Fiberglass is believed by some experts to also cause lung disease, especially if the fibers are similar in size to asbestos.(Man-Made) Fiberglass is currently classified as a man-made mineral and a possible carcinogen. It is widely used as house insulation hidden within the walls of a house. Fiberglass is also being used as a soundproofing material inside heating, air conditioning, and ventilation ducts. In this application, it can easily pollute the air passing through the ducts. Fiberglass was recently implicated by researchers at Cornell University as a cause of sick-building syndrome.(Hedge)

Minerals, such as calcium, that are dissolved in tap water can also become air pollutants because, when the water is used in humidifiers, the minerals can be spewed into the air. Their particle size is so small that they can be inhaled deeply into the lungs. As a result, in December 1988 the U.S. Consumer Product Safety Commission issued an alert recommending that you use demineralized or distilled water in ultrasonic and impeller-type humidifiers.(U.S. Consumer)

Radiation

Radiation is another well-known cause of illness. In houses, the main source of radiation is radon. Radon has been widely discussed in the media, yet many people do not fully understand its danger. It is a radioactive gas that is released during the natural decay of the mineral radium which is found in small amounts in the soil virtually everywhere on the planet. You are probably breathing some radon as you read this, and while it isn't good to breathe any, low levels can't be avoided totally. Problems occur when radon seeps into houses and builds up to dangerous concentrations. Breathing large amounts of radon can lead to lung cancer. The USEPA has estimated that up to 20,000 lung cancer deaths a year can be attributed to radon,(USEPA 1986) although some experts doubt the validity of the statistics.

Electromagnetic radiation (EMR) may be even more pervasive than radon, but its seriousness is the subject of much debate. It is the invisible electrical and magnetic energy that surrounds electrical wiring and appliances. Most sources of EMR are relatively weak, but there are usually some areas indoors where it can be strong enough to be of concern, and it is possible to measure high levels of EMR in a house if it is located near an electrical power station or high-voltage power lines.(Nair)

A few readily available consumer products actually contain radioactive material. For example, some types of compact fluorescent lights and some smoke detectors contain tiny amounts of what is usually considered hazardous radioactive waste. These products are claimed to be safe by the manufacturers, but many people question the wisdom of supporting an industry that sells products that require radioactive material to operate—especially when safer alternatives are available.(Mayell)

Smoking

Smoking indoors is probably the largest source of combustion by-products inside a house. It contaminates the air with nicotine, tars, acetaldehyde, nitrogen dioxide, aerolin, carbon dioxide, carbon monoxide, etc. Actually, there have been thousands of different compounds identified in tobacco smoke that affect both smokers and nonsmokers who happen to be living in the same house with a smoker. Negative health effects range from decreased attention span, headache, nausea, and drowsiness, to death. It is estimated that 53,000 nonsmokers die every year from secondhand smoke, that nonsmoking wives have a 30% increased risk of lung cancer if their spouses smoke, and as many as 300,000 infants suffer from bronchitis, pneumonia, and other infections related to secondhand smoke.(Lawson)

The most effective way of reducing the danger associated with smoking is to ban smoking indoors. To combat the negative effects of smoking with ventilation is difficult because massive amounts of ventilation are typically needed to do an adequate job. As a result, many governmental and commercial buildings now have smoking bans rather than "designated smoking areas."

If smoking is to be permitted in a house, it should be confined to a single room that has an exhaust fan. For example, if a smoker only smokes in a bathroom while the exhaust fan is running, the rest of the house can often be kept reasonably smoke free. However, smoke will cling to smokers' hair and clothing while they are smoking and then is released when they return to other rooms in the house.

How pollutants get into houses

Because air pollutants originate from a wide variety of sources, a control strategy effective with one particular contaminant may not be very effective with another; in fact, some control strategies can reduce the concentration of one pollutant but increase another.

Ventilation is often considered a universal method of controlling indoor pollution. However, ventilation can only reduce the concentration of *airborne* pollutants, and not all forms of pollution are airborne. For example, electromagnetic pollution from a house's electrical system is impossible to cure with ventilation. This is because EMR is composed of energy—something that can't be diluted with fresh air.

Sometimes particulate pollutants (as opposed to gaseous pollutants) such as lead dust, asbestos fibers, or mold spores will be airborne for a while, then will settle out of the air when the air is still. When they aren't floating in the air, you can't breathe them—but you can still get them on your fingers by touching a contaminated surface. If a ventilation system stirs up these pollutants and causes them to become airborne, two things can happen: 1) your exposure will increase (while they are airborne) and 2) the ventilation system will help to blow them out of the house, thus reducing the duration of your exposure. If a forced-air heating/cooling system stirs up these pollutants, your exposure can also increase, but instead of blowing them out of the house, it can circulate them to other rooms. Of course, some forced-air heating/cooling systems contain a filter that can capture airborne pollutants.

In order to determine the best indoor-pollution control strategy, it is helpful to place airborne indoor contaminants into three categories: those that are released from materials inside the house, those that can be sucked into the house by air pressure differences, and those that are released by people (Figure 1–4). Pol-

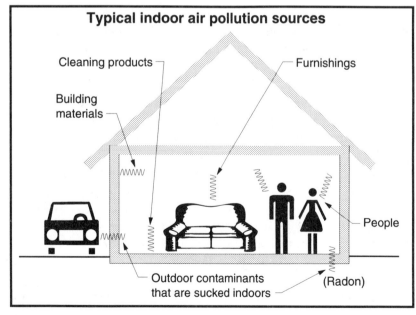

Figure 1–4. Pollutants enter the air of a house in three primary ways: they are released from materials inside the house, they are sucked indoors by pressure differences, and they are released by the metabolism of people.

lutants can also diffuse from outside the house through solid building materials into the living space, but this is a far less significant route of entry. (Diffusion is discussed in depth in Chapter 5.)

Pollutants can be released from materials in the house

Many building materials, cleaning products, and household furnishings release contaminants directly into the indoor air. Formaldehyde is given off by kitchen cabinets. Wallpaper is treated with fungicides. The odor associated with new carpet consists of over a hundred different VOCs. Disinfectant and pesticide aerosol sprays often contain hazardous ingredients. Bothersome chemical treatments are found on upholstery fabric. And on and on. When analyze everything found inside a typical house, it quickly becomes apparent that we are surrounding ourselves with unhealthy materials. The good news is that there are many alternative products on the market that are much more benign that can be used to build, furnish, and maintain our houses. If we used more of these healthier products, our indoor air quality would be much better. When indoor air quality is fairly good to begin with, a small-capacity ventilation system is often enough to keep it clean.

Sometimes, pollutants released from materials in a house are described as being either avoidable or unavoidable. In reality, all the pollutants in this category are avoidable because there are low-tox choices on the market for virtually everything—from less polluting building materials and furnishings to healthier cleaning products. However, from a practical standpoint, it can be expensive to remove all the polluting materials from an existing house and replace them with healthier alternatives. So, even though many pollution sources are technically avoidable, in the real world, some are more avoidable than others. For instance, because books, newspapers, and magazines release VOCs as the ink ages, they could be avoided by banning them from the house, but most people would be unwilling to do so. Even though avoidance is still generally more effective, for pollutants that are less avoidable, a ventilation system that blows them outdoors can minimize their negative effect.

Biological pollutants can grow on surfaces inside the living space and be released into the air. For example, mold can grow on the walls of a room with a high relative humidity and dust mites can grow in carpeting. In these instances, the pollutants are released from living creatures that grow on materials in the house.

Pollutants can be sucked indoors by air-pressure differences

Some air pollutants originate outdoors but get sucked indoors by air-pressure differences. The pressure differences we are talking about are quite small—usually too small to even feel—but they are quite common. For example, when you turn on a clothes dryer, it blows a certain amount of air out of the house. This creates a slight negative pressure in the house, and an equal volume of air gets sucked in (infiltrates) from the outdoors through small hidden gaps and cracks in the house. This infiltrating air can bring pollutants with it.

Actually, it is quite common for infiltrating air to bring pollutants indoors. The prime example is radon. It is found in the soil but it gets pulled through cracks in the foundation into the basement if the air pressure in the basement is less than the air pressure outdoors. This can also happen with other pollutants found in the soil—things such as lawn chemicals, termiticides, and biological pollutants such as mold. Air-pressure differences can cause all -these pollutants to be pulled from the ground, through a crawl space, into the living space, or from the ground directly indoors through cracks in a basement wall or concrete-slab floor. Water vapor can also be pulled indoors from the soil when a house is depressurized.

Combustion gases often migrate into the living space from a furnace, water heater, or wood stove, even though they are supposed to be expelled through a chimney. If the air pressure indoors is less than that outdoors, the gases will have difficulty going up the chimney and can remain in the house. This will be covered in more depth in Chapter 10.

Particles or gases from insulation can also be sucked indoors by air-pressure differences. This is more of a problem with very potent insulating materials such as the rarely used urea-formaldehyde foam insulation

Understanding Ventilation

(UFFI) than with today's commonly used insulations. Fiberglass, cellulose, and the various foam boards—though there are negative health effects associated with most of them—are more benign by comparison.

With an understanding of how and where these pollutants enter a house, it is possible to control them by either manipulating the air pressures, or by blocking entry points. With proper attention to detail during the design and construction of a house, this category of pollutant can be kept out of the living space in the first place. Therefore, the ventilation system won't need to dilute them, so it can have a smaller capacity.

Some ventilation strategies can cause positive or negative pressure differences in a house. Typical problems will be discussed in Chapters 9 and 10.

Pollutants can be released by human and animal metabolism

All living creatures release a variety of pollutants as a part of their metabolic processes. Oxygen, water, and food are consumed, and by-products of the life process are released. For example, the odor associated with mold is given off by the tiny creatures as they live and grow. We are all familiar with the unpleasant odor of a sweating body, or the smell that occasionally permeates a bathroom. Human beings (and animals) give off a wide variety of pollutants. Our exhaled breath contains dozens of chemical compounds. These are normal by-products of our metabolism, and they all contribute to indoor air pollution.

The pollutants given off by people are usually much less noxious that the pollutants given off by building materials or cleaning products, or the contaminants that are sucked indoors by air-pressure differences. Even though it is possible to eliminate all the toxic materials in a house and prevent pollutants from entering a house because of pressure differences, the only way to counteract the pollutants given off by people is to either eliminate the people (usually not a viable solution!) or dilute the pollutants with ventilation air. The best way of dealing with the metabolic by-products from mold is to prevent the mold from growing by controlling the moisture it needs for survival. People's need for fresh air is discussed further in Chapter 4.

The concentration of "people pollutants" in a house depends on several factors. The number of people inside a house is very important, as is the size of the house. Six people in a small house will contaminate the indoor air faster than six people in a large house. Behavior patterns are also significant. Someone who exercises daily will release more by-products of metabolism than a couch potato. A household with several teenagers may contain more moisture as a result of frequent showers than a household with young children, so the age of the occupants is also a consideration. Because each of us is biologically unique, each of us has a different comfort level and degree of sensitivity, so we tend to perceive different levels of people pollutants as being comfortable or tolerable. Therefore, the amount of ventilation air that we need indoors to feel comfortable can vary slightly from person to person.

People also bring pollutants indoors attached to their bodies. We are all familiar with how tobacco smoke can cling to our clothing and hair. Our lungs also take up the pollutants as we inhale. In fact, many air pollutants—VOCs, perfume, exhaust gases, etc.—can be carried indoors on our bodies. Once contaminated clothing and bodies are indoors, the pollutants will be released slowly, contributing to indoor pollution. Hypersensitive people sometimes have trouble being close to family members whose bodies have become contaminated with heavy doses of pollutants. People can also track pollutants indoors on their shoes (*e.g.* lawn chemicals, animal waste, road dust containing asbestos, lead, rubber, etc.), and deposit those pollutants in carpeting.

The bottom line

We are all surrounded by air pollutants every day, both indoors and outdoors. While a human body is capable of tolerating a certain amount of contaminated air, evidence from a variety of sources tells us that we are being exposed to more pollution than our metabolism can adequately process—especially when we are indoors—and it is making us sick. But there is no reason for this trend to continue. After all, a variety of strategies can be used to build houses with minimal indoor air pollution. These low-pollution "healthy houses" have one thing in common: they all have ventilation systems.

Chapter 2

How air moves

Various forces of nature have caused air to move through houses for as long as there have been houses. In a given house, at a given time of the year, there may be a great deal of air moving from the outdoors to the indoors and vice versa. Such a house will have a high *exchange rate*. At another time of the year, there may be almost no air exchange in that same house, but there may be plenty of air being exchanged in a similar house in a different geographic location. None of this occurs by magic. In fact, there are some basic laws of physics that describe what causes air to move into and out of houses.

While there is no sorcery or magic involved, houses can be very complex as far as air movement is concerned. So it can sometimes be difficult to predict exactly how much air will be entering and leaving, and precisely where it is entering and leaving. Because ventilation involves exchanging the air in houses, it is important to have an understanding of the basic principles involved.

This chapter discusses what factors cause air to be exchanged in houses—and why there may not be very much air being exchanged in a particular house.

The physics of air movement

In order for air to move into or out of a house, two basic requirements are necessary. First, there must be a path through which the air to travels, and second, there must be an *air-pressure difference* (wind often causes an air-pressure difference) to push the air molecules through the pathway. In other words, in order for air to move from the outdoors to the indoors (or indoors to outdoors), there must be an opening in the house (the pathway or *hole*), and the air pressure indoors must be different than the atmospheric pressure outdoors.

Air can't move into or out of a house with only openings—or with only an air-pressure difference—both are necessary. Therefore, to cut down on air movement, you have a choice: You can either close up the holes or reduce the air-pressure difference. To increase air movement, you either add more holes or increase the air-pressure difference.

Furthermore, at least two holes in a house are necessary in order for air to move through it—an entry hole and an exit hole. To understand why, try to blow air into a bottle, or suck air out of it. You can't do it. The best you can do is to pressurize or depressurize the air inside the bottle—but you can't blow air through it. With a house, you can't have a certain amount of air entering unless an equal volume of air is also leaving. If one hundred cubic feet of air enters a house, one hundred cubic feet of air must leave the house somewhere else. Scientists sometimes call this *conservation of mass*—the mass of the air going in must equal the mass of the air coming out. While you can't blow air through a bottle, it is easy to blow air through a piece of tubing because it has two holes—one at each end.

While most houses have quite a few holes in them, they may not be distributed evenly around the house so there may not be much air movement through the holes. For example, if all of the holes are on one side of a house, and wind blows on that wall, it will try to push air into all the holes at once. With no holes anywhere else in the house, there will be no exit holes, so it will be like trying to blow air through a bottle. The result will be very little air movement through the house. A house with holes on only one side probably isn't very likely, but mobile homes usually have many more holes through the floor than they do anywhere else, and this has an effect on how much air enters and leaves.

A tight house is one that either doesn't have very many random holes, or the random holes it does have are very small. In either case, it takes quite a bit of pressure to blow very much air through a tight house. This is why tight houses often have poor air quality in them—there aren't enough holes for the available pressures (such as the wind) to blow air through. As a result, tight houses are often underventilated. This has led many people to recommend building looser houses. Actually, there are some excellent reasons to build a house as tightly as possible—tight houses are less drafty, more comfortable, and more energy-efficiency than loosely built houses. The real answer is not loose houses—it is tight, energy-efficient, comfortable houses *with mechanical ventilation systems*.

Holes in houses

One of the largest "holes" in a house is an open window, but an open window can be an invitation to a burglar, so people rarely leave their windows open for extended periods. Even when all the windows are closed, there are still holes in houses—they just aren't as noticeable. There are hidden gaps between the window or door frames and the 2x4s holding up the wall. There are even narrower gaps between the floor and the walls, and even smaller gaps around electrical outlets. There are also many hidden holes inside the structure that were cut through studs, floor joists, and rafters by plumbers, electricians, or heating/cooling contractors.

Sometimes, when you start looking around in an attic or basement, it is possible to locate holes in the structure that are quite large—large enough to poke your leg into. It is through these big holes that most of the air enters and leaves a house. Weatherization contractors have learned in recent years that when you caulk around windows and doors, you really aren't tightening up a house very much. To do an effective job of weatherizing a house, you usually need to get into the attic and basement and plug up the big holes.

If it were possible to combine all of the small holes and gaps in a very tight house into one single hole, you would end up with an opening several square inches in size. In a very loose house you might end up with an opening several square feet in size. So, just because you can't see any holes doesn't mean there aren't any. Most houses have quite a few.

The way house construction has evolved over the years has had an effect on house tightness. For example, earlier in this century, when houses were constructed with solid-wood 1x8 boards, there were gaps between each of the boards. When builders switched to using sheet goods such as plywood, there were still gaps between the sheets. However, because of the large size of the sheets, the number and size of the gaps were less and, thus, houses became tighter. The holes we are talking about are called *random holes*, because they are randomly found throughout a house and they weren't created for the purpose of supplying the occupants with fresh air. If a hole is created on purpose, specifically to provide a pathway for air to travel through, it is called a *deliberate hole*. The installation of a controlled ventilation system requires one or more deliberate holes.

Today, some energy-efficient builders are purposefully using special techniques to build houses that are almost hermetically sealed. Their goal is to create an airtight house with no random holes. When this is done, no amount of naturally occurring pressure can possibly provide enough air to supply the needs of the occupants—*unless you install a ventilation system.*

Pressures in houses

Houses can be pressurized or depressurized in a variety of ways. For example, "naturally" occurring pressures—the result of Mother Nature—are quite common. They result in *natural* ventilation, something that varies considerably from day to day. This is primarily the result of pressures caused by the wind and by temperature differences (warm air exerts a small upward pressure as it rises up into cooler air). Natural ventilation is discussed in more depth in Chapter 5.

Pressures in houses can also be caused by mechanical equipment. Some mechanical devices aren't specifically designed to cause air to move into and out of a house for the purpose of ventilating the house, but they do so anyway. For instance, the fan in a clothes dryer blows air out of a house, so it has an effect on the air pressure in a house. A clothes dryer's main purpose isn't to ventilate—it is to dry clothes—so it contributes to *accidental* ventilation. Accidental pressures only cause air movement when the mechanical device is operating. Because these pressures are rarely continuous, they are sporadic in causing air movement through a house. Thus a number of different factors cause accidental ventilation; they are covered in Chapter 6.

Houses also contain mechanical devices, such as window fans, that are deliberately designed to exchange air in a building for the purpose of supplying fresh air or expelling stale air. This is what ventilation is really all about. This is called *controlled* ventilation—ventilation that is created "on purpose." See Chapter 7 for the basic controlled-ventilation strategies.

So, the air pressures that push air through the holes (either random or deliberate holes) in houses fall into three categories: *natural*, *accidental*, and *controlled*. The direction the air moves (outdoors to indoors, or indoors to outdoors) depends on which way the pressure pushes. Air will always move from an area of high pressure, through a hole, to an area of low pressure.

If air movement is directly caused by a fan, it is said to be *active*. If air movement isn't directly caused by a fan, it is said to be *passive*. For example, a window fan *actively* blows a certain amount of air out of a window, but in doing so an equal volume of air will enter the house *passively* somewhere else.

If something causes the air pressure inside a house to be higher than the atmospheric pressure outdoors, the house is said to be *pressurized*, or experiencing a *positive pressure,* and indoor air will passively leak toward the outdoors through any random holes. If something causes the air pressure inside a house to be lower than the atmospheric pressure outdoors, the house is said to be *depressurized*, or experiencing a *negative pressure,* and outdoor air will passively leak toward the indoors through any random holes. It isn't unusual for part of a house to be pressurized and another part of the same house to be depressurized.

The relationship between pressure and holes

There is a close interrelationship between 1) the pressures in a house, 2) the number and size of the holes (house tightness), and 3) the volume of air moving through the holes. If you change any one of these three items, you affect the other two. This is an important concept because pressurization/depressurization, house tightness, and air movement are very basic to an understanding of how ventilation systems work—and what problems can be caused by them.

Changing the volume of air passing through a house

If you have a house with a certain number of random holes, and you depressurize the house by blowing air out of it (*e.g.* with a window fan), then a certain

Understanding Ventilation

amount of air will be sucked into the house through the random holes in the structure. If you increase the volume of air leaving (*e.g.* by turning a window fan up to high speed), more air will be sucked in. But the house will also sense slightly more depressurization because it will be more difficult for the increased amount of air to enter through the limited number of random holes in the structure. If you reduce the volume of moving air (*e.g.* turn the fan down to low speed), then less air will move through the holes and there will be slightly less depressurization.

This same relationship holds true if you turn the fan around and blow air into the house and pressurize it. If you increase the volume of air, you will also increase the pressurization slightly. If you decrease the volume of air going in, you will decrease the amount of pressurization slightly. In this example, the number of holes remains the same, so the tightness of the house itself is unchanged.

The tighter the house, the easier it is to change the pressure in it. For example, in a tight house (one with few random holes) you may only need to speed up the window fan a little bit to measure a big change in pressure, but in a loosely built house (one with many random holes), you may need to speed up the fan considerably to get the same change in pressure.

Changing the tightness of a house

If you weatherize or tighten up a house to make it more energy-efficient, you are reducing the number of random holes in it, so you will affect the amount of air being exchanged, as well as the pressures the house experiences. A loosely built house will have quite a few random holes in it. If a window fan blows air into such a house, there will be many pathways for the air to escape, so the house will only be slightly pressurized. If you weatherize and tighten up the house, that same window fan will try to blow the same amount of air through, but because there are fewer random holes for it to escape through, the house will become more pressurized. So, when you tighten a house, a fan will move less air through the house, but you will also be able to measure an increase in pressure. The same thing happens if you turn the fan around and blow air out (depressurizing the house) and air gets sucked in through the random holes—you will move less air with the same fan and have more depressurization.

Actually, when you tighten an average house, the amount of air flowing and the pressure will change very little. For the pressure and the airflow to change significantly, you must have an extremely tight house. Years ago, when most houses were very leaky, pressures weren't a significant concern. Pressures are very important today because houses are tighter and many modern fans are quite powerful. Potential pressure-related problems are discussed in Chapters 8, 9, and 10.

As was said earlier, tightening a house is often a very good idea. But *a tight house must have a controlled mechanical ventilation system* to supply it with fresh air and to remove stale air. This is because natural and accidental pressures are not very effective at moving air through a tight house.

Changing the pressure in a house

If the air-pressure difference between the indoors and the outdoors (positive or negative) changes, only two things could have caused it—a change in the volume of air passing through the house (*e.g.* a different-sized fan, or a change in the naturally- or accidentally-induced pressures) or a change in the tightness of the house. Sometimes we don't realize it, but by caulking around windows, closing a fireplace damper, or accidentally painting shut a dryer vent, we are affecting the tightness of the house, and as a result, the air pressure indoors.

If a house is tightly built in the first place—and many modern houses are—there can be significant pressure changes due to an increase in the volume of air movement. In other words, turning on a fan can create big pressures, positive or negative, in a house. It is more difficult to significantly change the pressure in a house of average tightness with a fan because there are so many random leakage points.

Indirectly, the size of the house is also a factor in pressurization (and depressurization). Consider a small house that is built of average tightness. It will have an average number of random holes in its structure—say there is on average 1 square inch of leakage area for every linear foot of exterior wall. Compare this to a large house of similar tightness which also has 1 square inch of leakage area for every linear foot of exterior wall. The large house will have more holes in it simply because it has more linear feet of exterior walls. Therefore, if all else is equal, it will be easier to pressurize (or depressurize) a small house (or an apartment) than a large one.

Measuring house pressures

You can measure the difference in air pressure between the indoors and the outdoors with a length of clear flexible tubing containing some water. Carefully close a door or window on the tubing, without smashing it, so one end of the tube is indoors and the other is outdoors. This isn't always easy to do but often the tube, if it is small enough, can be slipped around a window or door near a corner without crushing it. If the house is pressurized (the air pressure indoors is greater than the pressure outdoors), the level of the water in the tube will be unequal. It will be lower inside the house, as in Figure 2–1 because the higher pressure indoors pushes down on the water. In Figure 2–1, the difference in height of the water is 1", thus the pressure difference is said to be 1" of water. This is abbreviated 1" w.g. (w.g. stands for water gauge). (Actually, it would be extremely unusual for a house to experience as much as 1" of water pressure. In fact, the pressures typically experienced by houses are too small to measure with a tube partially filled with water—but this is a valid way of explaining the concept.) The pressure difference between the indoors and the outdoors is often different from one side of a house to another, or at the top of a wall compared to the bottom of a wall.

Air pressures can also be measured in inches of mercury (inches hg. or " hg.) or in pounds per square inch (psi). The pressure in an automobile tire might be about 30 psi. This is equal to 830" w.g. or 61" hg. The

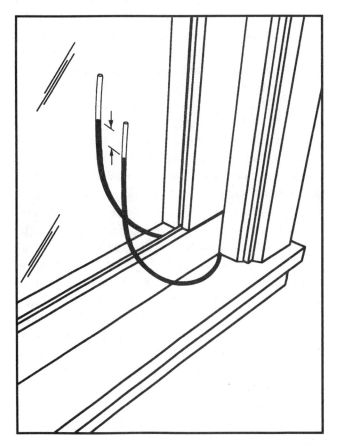

Figure 2–1. A length of clear tubing partially filled with water can be used to measure the pressure difference between the indoors and the outdoors. If there is a pressure difference, the water level will be unequal. (While this is a useful way to explain the concept of positive and negative pressures in a house, in reality house pressures are too small to measure accurately in this way.)

Understanding Ventilation

Figure 2-2. Magnahelic gauges are often used to accurately measure pressures in houses.

air pressure in houses is usually measured in metric units called Pascals (Pa.). A Pascal is actually a fairly small unit of pressure. One pound per square inch is equal to about 7,000 Pa., so an automobile tire inflated to 30 psi would contain over 200,000 Pa. of pressure. To understand how little pressure a Pascal is, if you stand in a 20 mph wind, you will feel a pressure of about 50 Pa. on your face.

The pressures measured in houses are usually less than 50 Pa., and they are often less than 10 Pa. This may not seem like very much pressure—and it really isn't—but it is enough to cause air to move through the random holes in a house. It is also enough pressure to cause some serious problems. For example, 3 Pa. of negative pressure in the vicinity of a chimney is often enough to cause backdrafting (combustion gases flowing down a chimney), something that will be discussed in Chapter 10.

Different building trades tend to use different units of measurement. For example, heating/cooling contractors generally use inches of water and weatherization contractors usually use Pascals.

To convert between different units of pressure measurement, use the following formulas:

1 Pascal = 0.004" w.g. = 0.000295" hg. = 0.000145 psi
1" w.g. = 249.18 Pa. = 0.07355" hg. = 0.03613 psi
1" hg. = 3,386.54 Pa. = 13.59" w.g. = 0.4912 psi
1 psi = 6,895 Pa. = 27.67" w.g. = 2.036" hg.

Pressure gauges are available in a range of styles and prices. Inclined manometers can be very accurate, but they contain a liquid that is easily spilled. Digital manometers are also quite accurate, but they tend to be fairly expensive. Magnahelic gauges are an accurate, reasonably priced alternative that are easy to use. (Figure 2-2) Pressure gauges are manufactured by **Dwyer Instruments, Inc.** and **Modus Instruments, Inc.** and are available from **AirPro, Inc.**, **The Energy Conservatory**, **Infiltec**, and many other suppliers of ventilation equipment.

Chapter 3

Strategies for reducing indoor air pollution

Ventilation is one of four basic ways of improving the air quality in houses. The other three are: *source control*, *separation*, and *filtration*. Of the four, source control is usually considered the most important, but even if you use source control, separation, and filtration effectively, ventilation is still necessary. This is because source control, separation, and filtration cannot supply oxygen to a house, they cannot dilute the carbon dioxide we exhale, and they cannot remove excess humidity. (Of course, ventilation can't always solve a high humidity problem either—sometimes you must rely on a dehumidifier or an air conditioner to control humidity in a house.)

It is entirely possible to rely solely on ventilation (without source control, separation, or filtration) to clean up the air in a house, but it is rarely very cost-effective to do so because you would need a very large capacity (expensive) ventilation system—a system so large that it would simply not be practical. It always makes the most sense to practice source control and separation first; then a relatively small ventilation system can be used to dilute the remaining pollutants. In other words, ventilation is best reserved for those pollutants that can't be minimized by any other means. In many cases, when source control, separation, and ventilation are implemented effectively, filtration is not even necessary. However, since it can be difficult to practice source control and separation perfectly, a modest amount of filtration is often combined with ventilation to maintain good indoor air quality.

Source control

When you eliminate the source of a pollutant, you are practicing source control. For example, if you build a house without products containing formaldehyde, you will have no formaldehyde to contend with

Understanding Ventilation

Figure 3–1. *There are several methods of dealing with indoor air pollution—but some are more practical than others.*

(at least not from the building materials). If you use a floor covering other than synthetic wall-to-wall carpeting, you will have less outgassing and fewer dust mites. If you use low-tox maintenance products, you won't add pollutants to the indoor air every time you clean the house. Source control is often the strategy of choice for dealing with lead paint—just get rid of it.

When you practice source control, you don't have to rely on extra ventilation to dilute the polluted air after-the-fact. Source control should always be the first step in improving indoor air quality because it is generally the most effective, the most reliable, and the most cost-effective solution. To practice source control, you first need to identify the pollution sources, then either eliminate them completely, or substitute a more benign product.

Books such as *The Healthy House*[Bower 1989a], *Healthy House Building*[Bower 1993b], *Your Home, Your Health and Well-Being*[Rousseau], *Why Your House May Endanger Your Health*[Zamm], and several others listed in Appendix C contain a great deal of information about source control—and separation.

Separation

While there are low-tox substitutes for nearly all building materials, cleaning products, and furnishings, sometimes the alternatives can be expensive. That is when the strategy of separation comes in handy. Separation means building a barrier between the occupied part of the house and the pollution source. If a pollutant can't reach you, it can't harm you. For example, if you have a jar of formaldehyde and the lid is screwed on tight, the formaldehyde can't get to you, so it can't affect your health. Of course, polluting materials that are to be separated from the occupied part of the house may cause harm to the workers who install them, so the workers may need to use protective equipment such as respirators or goggles.

Lead paint is sometimes separated from the occupants by encapsulating it—sealing it behind drywall, or behind a new protective coating of paint. Because there are negative health effects associated with common insulating materials such as fiberglass and cellulose, it is often less costly to separate the insulation from the living space by using airtight construction techniques than by using a more inert (and more expensive) insulating material. An airtight house is one without very many hidden random gaps and holes. Such a house won't have many pathways for the insulation to get sucked indoors by air-pressure differences. So, an airtight house can be a good way to separate pollutants such as insulation from the occupants. *But an airtight house must have a ventilation system so its occupants can be supplied with fresh air.*

Filtration

Filtration is another method of reducing indoor air pollution, but the other strategies are often more effective—and cheaper. Air filters can be costly. If you practice source control and separation, and you provide some mechanical ventilation, then you might be able to use a small filter to clean up any remaining minor pollutants—in other words, to polish the air. This is sometimes helpful for very sensitive people. However, if you intend to rely solely on filtration to deal with highly polluted air, you will need a very large (and expensive) filter. Keep in mind that filters do nothing to reduce high indoor humidity levels, nor do they supply oxygen, so they aren't a substitute for ventilation. Filters will be covered in detail in Chapter 15.

Ventilation

Ventilation is often defined as the *controlled* movement of air into and out of the living space of a building. In other words, ventilation is a process that exchanges specific amounts of air between the indoors and the outdoors by moving it through deliberate holes. To accurately control the amount of air, you generally need to use a fan and to create some deliberate holes for the air to move through. Ventilation is more than just air circulation between rooms—although that is a part of ventilation. With ventilation, there must be an *exchange* of air between the indoors and the outdoors.

If we stick with the above definition, it is easy to see that we sometimes use the word ventilation incorrectly. To talk about *natural* ventilation or *accidental* ventilation isn't quite accurate because they aren't controlled. However, because natural and accidental pressures do cause an air exchange in buildings, in this book we will stray a bit from the "official" definition. But, in the remaining chapters, it will become apparent that the best and most efficient way to exchange the air in a house is to rely on a controlled ventilation system, and not on natural or accidental pressures.

Attic and crawl-space ventilation aren't really ventilation either—unless a fan of some type is involved. Of course, ventilation in the attic or crawl space shouldn't do much for improving the air in the living space—although it is certainly possible for a change of air pressure in an attic or crawl space to affect the air pressure in a house. While the popular paddle-type ceiling fans do move air, they don't exchange the indoor and outdoor air, they only circulate air within a single room. Because they don't cause air to enter and leave the building, they shouldn't be considered ventilating fans. They may feel comfortable, and they may help keep the temperatures in a room evenly distributed, but they don't affect indoor air quality. Just because air is moving doesn't mean there is an exchange between the indoor and outdoor air.

Air fresheners don't help improve air quality either because they don't bring in fresh air, they don't create oxygen, and they don't remove pollutants. Instead, they add fragrance to the air to cover up odors. The pollutants are still there to be inhaled; they just aren't as noticeable. Actually, most air-quality experts consider air fresheners to be *pollutant* sources because they add fragrance molecules to the air.

Filtration is not the same thing as ventilation either. Filters come in a variety of types, depending on the kind of air pollutant to be removed. But filters do not bring in fresh air, they do not remove excess humidity, and they cannot create oxygen; therefore, they should not be relied on alone to improve air quality. They can, however, be combined with a ventilation system. For example, a filtration system can be used to clean the outdoor air before a controlled ventilation system brings the air indoors. Or a portable room-sized filter can be used to clean the air after it is brought indoors. Some ventilation systems have built-in filters, but many aren't very efficient.

Two types of controlled ventilation systems

As it turns out, even if a house is constructed, maintained, and furnished with inert materials, and moisture and pollutants are prevented from being sucked indoors by air-pressure differences, to achieve optimum efficiency most houses can benefit from two types of controlled ventilation systems. A high-volume system is useful for quickly removing pollutants or moisture where they are generated (*e.g.* a kitchen range hood or bathroom exhaust fan), and a low-volume system is necessary to remove pollutants and moisture that are generated throughout the house more gradually.

A low-volume *general ventilation* system is good for dealing with the average "background" pollution in a house. A high-volume *local ventilation* system is good for dealing with the occasional peak pollution levels or excess moisture generated periodically in specific rooms. Local ventilation is desirable in bathrooms, kitchens, and other service rooms that regularly have high levels of moisture or pollution. In some cases, it is possible to use the same system for both purposes.

General ventilation

General ventilation is *whole-house* ventilation because it is for every room in the house. It is often called *central* ventilation. Ideally, general ventilation should provide fresh air to all rooms and remove stale air from all rooms. In a house built and maintained with very toxic materials, the general ventilation system may need to be large and powerful to dilute the pollutants sufficiently, but in a more inert house, the general ventilation system can be much smaller. In a perfectly inert house, general ventilation may only need to dilute the by-products of metabolism of the occupants themselves. A low-volume system is desirable because it is less expensive to buy, its smaller fan is cheaper to operate, it is quieter, and it will bring in less outdoor air so the furnace or air conditioner won't have to work as hard to heat or cool the incoming air.

General ventilation is important because people rarely spend all their time in one room of their house; they move from room to room for different activities. They eat in the dining room, cook in the kitchen, sleep in the bedroom, watch TV in the family room, and visit with friends in the living room. We need fresh air wherever we are, so we need ventilation in each room of the house. It would make little sense to have fresh air only in the bedroom, and not worry about ventilating the living room.

Some people call general ventilation *primary* or *basic* ventilation because its purpose is to provide the air the occupants need on a continual basis. After all, human health is a primary or basic consideration.

Local ventilation

Local ventilation is used to capture and remove pollutants quickly at their source. In all houses, large quantities of moisture or pollutants are occasionally generated in certain rooms from specific activities such as cooking, showering, laundering, or using the toilet. A high-volume exhaust system (local ventilation) can be used to clear these rooms of contaminates quickly. Local ventilation is also often useful for workshops, home offices, or hobby rooms, if they regularly contain concentrated pollution sources.

When a fairly powerful local ventilation fan removes high levels of pollutants or excess moisture from near the source quickly, the contaminants won't have a chance to dissipate into the rest of the house. This is the purpose of local ventilation—to exhaust air from one particular locale. If large quantities of moisture and pollutants are allowed to dissipate from one room into all the other rooms in the house, then the general ventilation system will take longer to remove them. Ventilating one highly contaminated room quickly is more efficient, uses less energy, and results in less exposure to the occupants than ventilating an entire house slowly. Both methods will work, but a powerful local ventilation system can get the job done much more effectively. It is only necessary to run a local ventilation fan periodically; for example, only when pollutants or moisture are generated in the bathroom. However, a general ventilation fan is most effective if it runs continuously.

Sometimes local ventilation is called *spot* ventilation because it ventilates one spot in a house. Local ventilation is also sometimes called *secondary* ventilation because it isn't as important as general (primary) ventilation. It is still important, but it is secondary to the occupants basic needs.

Infiltration

Infiltration is sometimes referred to as being the opposite of ventilation, but that isn't quite accurate. Infiltration is also occasionally called natural ventilation, but that too, isn't quite right. Ventilation is the controlled movement of air *into* or *out of* a house through *deliberate* holes (inlets and outlets). Infiltration is the movement of air *into* a house through the *random* gaps and holes in the structure. The pressure that pushes the air through the random holes can be either naturally occurring, accidental, or the result of a controlled ventilating fan. *Exfiltration* (air moving *out of* a house through the random holes) is the opposite of infiltration and they often occur simultaneously (but through different random holes).

Infiltration occurs whenever a house having random holes experiences a negative pressure. Exfiltration occurs whenever a positive pressure is applied to a house having random holes. In a house without a controlled ventilation system, all the air moves through the building by infiltrating in and exfiltrating out. In a house with a controlled ventilation system, some of the air will be entering and leaving through the deliberate holes in the house that are a part of the ventilation system, but some air will also be infiltrating and exfiltrating through the random holes in the structure. Controlled ventilation doesn't eliminate infiltration and exfiltration. Because natural and accidental pressures are almost always present to some extent, the only way to eliminate air movement through random holes is to eliminate the random holes in the structure—to tighten the house.

There are several fundamental drawbacks to infiltration and exfiltration: random air movement is unpredictable (it is hard to tell when and where it will be occurring), random air movement is uncontrolled (you can't turn it on and off), random air movement is hit-or-miss (it occurs when and where it wants to), random air movement can bring pollutants indoors (such as pollen, radon, or termiticides), random air movement can cause moisture-related problems (sometimes hidden within the walls or roof), and random air movement rarely provides the correct amount of air (either too much or too little). Of course, a poorly designed mechanical ventilation system can also have some of these drawbacks.

Infiltration is often uncomfortable. For example, if a house has a great deal of infiltration near the floor (many houses have gaps along the baseboard molding—between the floor and wall—or in the floor itself, around registers), there will be cold drafts on your feet in the winter. Cold drafts can often be felt around window and door frames. Loosely built houses can have a great deal of infiltration and exfiltration, and as a result, are often very uncomfortable.

In most cases, infiltration is not a very efficient way of providing the occupants of a house with fresh outside air. This is because air often infiltrates and exfiltrates through hidden pathways within the wall, floor, or roof cavities. When this happens, it doesn't provide any air to the occupants in the living space, but it does rob heat from the house. Some loosely built houses have 2–5 times as much air flowing between the indoors and the outdoors (mostly through building cavities) as is necessary for the health and comfort of the occupants, yet very little of the air actually reaches the occupants—it just makes the house uncomfortable. A mechanical ventilation system that removes polluted and moisture-laden air from near the ceiling of the kitchen and bathrooms, then delivers fresh outdoor air to the living room and bedrooms is a much more efficient way to exchange the air in a house.

Infiltration and exfiltration are perhaps best described by the word erratic. Unfortunately, this is how many houses get most of their fresh air—haphazardly through the various hidden cracks and gaps in the structure depending on whatever pressures just happen to occur. While infiltration and exfiltration are certainly better than nothing, fresh air is vital for all of—and we deserve better.

It may seem that when you rely on infiltration and exfiltration to move air through a house, you are getting something for free, namely fresh air. But as with most things in life, you get what you pay for. When you

Understanding Ventilation

rely on infiltration to provide your house with fresh air and exfiltration to remove the stale air, you simply don't get very much for your money because you can't control how much air is being exchanged. The process is simply unreliable. And it really isn't free because it will increase your heating bill in the winter and your cooling bill in the summer. In fact, you often have more infiltration and exfiltration than you need during the coldest months of the winter, when it has the greatest impact on your heating bill.

Chapter 4

Ventilation is for people

This chapter emphasizes the two primary reasons for ventilating houses: 1) because the people who live in them require clean air to breathe, and 2) because the people who live in them are pollution sources themselves. In other words, we should ventilate houses *for* people and *because* of people. While there are other reasons why houses should be ventilated, they can often be remedied more effectively by other means, such as source control or separation.

People need fresh air

Evolution is a very slow process, often taking many centuries, yet we are asking our bodies to adapt to higher levels of hundreds of different air pollutants, many of which have never existed before in human history. Some of these new contaminants are carcinogenic, some cause birth defects, some damage our cells, and some can cause our bodies to become hypersensitive to tiny amounts of air pollution. Some experts believe we are asking our bodies to tolerate too much.

Outdoor air pollution can certainly be severe, especially in large, industrialized areas. But the really bad news is the fact that indoor air pollution is almost always worse—in fact, it is usually at least 5–10 times worse than outdoor pollution, and in some cases it is over a hundred times worse. This is because so many building materials, cleaning products, furnishings, etc. introduce dozens and dozens of pollutants directly into the indoor air, and our houses are often so poorly ventilated that the contaminants can't easily escape.

Many rules and regulations deal with outdoor air quality, but very few laws address indoor pollution. Our homes are our castles, so the government is reluctant to invade our privacy and regulate the air that we are at greatest risk from breathing. Yet, Americans spend between 80–90% of their time indoors where we are exposed to poor-quality air on a daily basis. We should be very concerned about contaminated indoor air, because those who are at greatest risk are children, the elderly, and the sick—categories we all fall into sooner or later. We all need fresh air.

Sick-building syndrome

The words "sick-building syndrome" imply that a building is ill, when in fact they refer to conditions which make the *occupants* ill—generally as a result of poor air quality. Sick-building syndrome is different from conventional maladies such as the flu, diabetes, heart disease, or broken bones because the symptoms often disappear when an affected person leaves the sick building. Complaints may be reported after a building is remodeled or, in commercial buildings, after maintenance crews have reduced the amount of fresh air entering the building (ventilation rates are often reduced to save energy). Pollutants released from construction products or cleaning solutions, microbial contamination of wet materials, and inadequate ventilation are often-cited causes of sick-building syndrome.

Typical symptoms caused by poor air quality in sick buildings include drowsiness, eye irritation, nose or throat irritation, sinus congestion, etc. However, virtually any complaint can be related to poor air quality: aches, chest pains, circulatory disturbance, depression, anxiety, diarrhea, digestive problems, fatigue, itchy skin, menstrual irregularity, rash, breathing difficulties, tension, weakness, etc.

Incidences of sick-building syndrome are on the rise. As more and more people grasp that poor indoor air quality might be the cause of their illnesses, they are cleaning up their indoor environment and improving their ventilation.

Sensitive occupants

Whether a particular house will make its occupants sick depends on two factors: 1) how contaminated the air is indoors and 2) how susceptible the occupants' metabolisms are to the polluted air. Throughout this book, sensitive or hypersensitive occupants are occasionally mentioned. These are people who react to very low levels of indoor air pollutants—often levels that are considered safe by many experts. Sensitive people require air that is much cleaner than normal. Their condition is most often referred to as multiple chemical sensitivity (MCS).

We have all heard of people with extremely strong constitutions—they smoke four packs of cigarettes and drink a quart of hard liquor every day and yet live to a ripe old age. Sensitive people exist at the opposite end of the spectrum; their bodies react to very low levels of a wide variety of pollutants. Most of us fall somewhere between these two extremes.

Individuals with MCS generally exhibit a wide variety of symptoms. Eye and respiratory irritation are common, but headaches are also widely reported. Other symptoms can range from joint and muscle pain to seizures. Because the brain is very sensitive to some pollutants, MCS patients often report cerebral symptoms such as depression and anxiety that are related to poor indoor air quality. Sometimes, because these people are reacting to such low levels of air pollution, their symptoms seem unrelated and fit no set pattern.(Ashford)

MCS was first described in the early 1960s.(Randolph, 1962) At that time the condition was called environmental illness (EI). In the intervening years it has been referred to by a number of names: total allergy syndrome, toxic response syndrome, twentieth century disease, etc., but today MCS seems to be the most widely accepted term. Individuals with MCS often react to artificial fragrances, very low levels of combustion by-products, synthetic clothing, printing ink, and a wide variety of commonly encountered consumer products. Some estimates place the number of hypersensitive people at over one million. Perhaps as many as 25% of us are sensitive to a less debilitating degree—yet sensitive enough to derive a substantial benefit from breathing cleaner air.

Sensitive people tend to react to building materials such as carpeting, adhesives, particleboard, paints, cabinets, etc. For them, a house must be constructed of materials that are as inert as possible.(Bower 1989a, 1993b) In addition, the air entering their house through a ventilation system must be as clean as possible. A house located near a busy highway will generally have too much outdoor pollution to be a suitable location—unless the outdoor air is adequately filtered before it is brought into the living space.

Occasionally, sensitive people will decide to install a general ventilation system in order to make their polluted house tolerable. This often doesn't work because most general ventilation systems on the market have too little capacity to substantially improve the air quality in a problem house. If source control and separation aren't practiced first, the general ventilation system may need to be 4–10 times (or more) as powerful as a general ventilation system in a house that is fairly unpolluted to begin with. This could be equivalent to a wind blowing through the house. Even with a very energy-efficient ventilation system, this is a great deal of air exchange, and it will place a considerable burden on the furnace or air conditioner. When a ventilation system is this powerful, it is very difficult to properly humidify the incoming air in the winter and dehumidify it in the summer.

So, for sensitive occupants, it is imperative to clean up the indoor air through source control and separation, then you can use a ventilation system to keep the air clean.

Ventilation is for people

When you think about it, ventilation really is a *people* consideration. If there are no people in a house, there's no need to ventilate it. If you aren't interested in the health and comfort of the people in the house, then you probably don't need to ventilate. (There are times when it is a good idea to ventilate an unoccupied house to prevent polluting materials from contaminating inert materials.) Actually, pets need fresh air too. Because we are all interested in health and comfort—for ourselves and our pets—we all need fresh air. In some loosely built houses, the needs of the occupants can be

Figure 4–1. Ventilation is for people—to provide them with fresh air to breathe, and to remove by-products of their metabolism such as water vapor (H_2O), carbon dioxide (CO_2), and various unpleasant bodily odors.

partially met by infiltration and exfiltration, but as houses are built tighter and tighter, mechanical ventilation systems are much more effective.

Credit should go to ventilation specialists David Hill and Yvonne Kerr of **Eneready Products Ltd.** for coining the phrase *Ventilation is for people*. To illustrate this concept, they have created a life-size mannequin called Charlie (Figure 4–1). Using slides and a video tape depicting Charlie in various situations, they give presentations showing when and why people need ventilation. When Charlie is at rest, he releases about 1 quart of water a day and 1 *olf* of body odor. (An olf is a unit of measuring odor. One person releases 1 olf per day when performing normal activities.) Charlie's exhaled breath contains about 20 cubic feet of pure carbon dioxide per day. When Charlie gets on his exercise bike, he releases additional moisture through his breath and pores, extra carbon dioxide, and a little more odor.

By taking a shower, cooking, and doing the laundry, Charlie adds another quart of water to the air every day. And he releases odor at a rate of 3 olfs when he uses the toilet. To make sure that Charlie and his activities don't create an indoor pollution problem, Hill and Kerr suggest two ventilation systems. A general low-level ventilation system that runs whenever Charlie is at home will remove the moisture, body odor, and bad breath that he gives off during the day. A higher-capacity local ventilation system is used occasionally in the bathroom to remove moisture from showering or those extra odors when he uses the toilet. Another local ventilation system is also used in the kitchen when Charlie is cooking. Charlie isn't much of a cook, so he uses a two-speed exhaust fan in the kitchen to clear out the smoke quickly when he burns his toast.

In an ideal situation, a general ventilation system will only need to deal with the types of pollutants Charlie generates, not pollutants from outside the house, or from building materials, furnishings, or cleaning products. However, most houses aren't built of, furnished with, or maintained using totally inert products, and some pollutants can get sucked indoors by air-pressure differences. In an imperfect house, it is often a good idea to operate the general ventilation system even when the house is unoccupied, but perhaps at a lower-than-normal capacity. This will prevent the pollutants from building up to high levels and overexposing the occupants when they return home. Building materials often absorb moisture quickly, then release it slowly, so a continuously running general ventilation system will help keep the indoor humidity at a more constant level. If you do have an ideal house—one built, furnished, and maintained with inert materials, and one where pollutants aren't sucked in by pressure differences—the general ventilation system can be shut off during the day when everyone is at work or school.

Reasons for ventilating houses

The purpose of ventilation is to provide occupants with healthy and comfortable air. In Chapter 1 we pointed out that pollutants contaminate the indoor air in three different ways: 1) they are released by materials inside the living space, 2) they are sucked into the living space from somewhere else, or 3) they are released by metabolic processes. Therefore, it would seem that ventilation can be used to 1) dilute the pollutants released by materials found indoors, 2) dilute the pollutants that are sucked indoors by pressure differences, and 3) dilute the pollutants released by metabolic processes. An aspect related to this third point is that a ventilation system is needed to supply occupants with oxygen. A fourth important reason to ventilate is to dilute high indoor humidity levels to prevent moisture-related problems such as mold growth and rot, both of which involve metabolic processes. Controlling excess moisture protects both the health of occupants and the health of the house.

While air infiltrating into a house will dilute the concentration of indoor contaminants, infiltration is unreliable when compared to mechanical ventilation. In fact, it has been determined that houses with mechanical ventilation systems generally have better indoor air quality, and use less energy than houses that rely on infiltration.(Feustel, 1986)

An old adage states that "Dilution is the solution to pollution." Because all the above four reasons for ventilating houses use the word dilution, many people believe that this adage is true. However, source control and separation are often much more effective than dilution. Also, one of the drawbacks to relying solely on dilution to reduce indoor pollution levels is the fact that it suffers from the law of diminishing returns. As you ventilate at greater and greater rates, you get less and less benefit. To understand the limits of ventilation, let's look at each of the four reasons for ventilating houses a little more closely.

Diluting pollutants released by materials inside the house

If a house isn't built, furnished, or maintained in a perfectly inert manner—and most aren't—ventilation can be used to dilute the indoor contaminants that are released from everything inside the house. In order to determine how much ventilation air is needed to dilute the pollutants to a safe level, you must know precisely what pollutants are in the air and what constitutes a safe level. Because every house is built differently, cleaned and maintained differently, and because no one has studied all the possible indoor air pollutants to determine how much an average human body (let alone a hypersensitive body) can safely tolerate, this is a formidable task indeed. From a practical point of view, it is far easier to build, furnish, and maintain a house with inert materials than it is to try to figure out how much to dilute the effects of those materials after-the-fact with a ventilation system.

However, any amount of fresh air will help. Yet there is no way of knowing what the optimum amount of fresh air is for any particular house. So, if you are interested in ensuring that the concentration of indoor pollutants is reduced to a healthy level, you have two choices: you can guess what the correct amount of ventilation would be, or you could build and maintain the house with inert materials.

If you decide to guess what an appropriate ventilation rate would be, it will be helpful to actually measure some of the indoor pollutants to determine just what you are dealing with. It is certainly possible to have a laboratory analyze everything in the air, but that can be impractical and expensive. Such an analysis would cost thousands of dollars, and you would be given a list of over a hundred compounds, most of which have unknown effects on the body. Rather than measure everything, you might decide to focus on only a few key pollutants for which some scientific information is known. For example, formaldehyde and carbon dioxide (CO_2) are very common indoor pollutants and they have been studied extensively.

Perhaps you decide to measure the indoor formaldehyde or CO_2 concentration and so you design a ventilation system that is powerful enough to keep them in check. Of course, you still have no idea if other unknown pollutants are even worse for you than formaldehyde or CO_2, but you could assume that since their level is reduced, then everything else will be, too. And, in fact, the levels of all the other pollutants will be lower, but it is still merely a guess that they will be at a healthy level. Actually, some guidelines can be used to determine how much capacity a ventilation system needs, so it becomes possible to make an "educated" guess. These guidelines are often based on CO_2 levels. See Chapter 12 for a discussion.

Suppose you are a designer planning a new house. You can't measure formaldehyde levels or any other pollutant indoors because the house isn't built yet. So you must guess at a ventilation rate, and your guess should be based on the worst pollutant—whatever that might be. Or, you can specify that the house be built with inert construction products, and explain to your clients that the best way to furnish and maintain their new home will be to use inert cleaning products and furniture. For low-tox furnishings see *The Healthy Household*[Bower 1995]; for low-tox construction practices see *The Healthy House*[Bower 1989]; *Healthy House Building*[Bower 1993]; and *Your Home, Your Health and Well-Being*[Rousseau]; all are listed in Appendix C.

Without a doubt, a ventilation system will dilute the concentration of any air pollutant found indoors, but it is also very clear that the best way to deal with any kind of pollutants is to use healthy products in the first place by practicing source control, or by using the principle of separation. Sometimes, contaminants are released by materials indoors that simply can't be eliminated or separated from the living space; for example, cooking odors. While the aroma of baking bread is pleasant, there are a wide variety of air pollutants released by cooking, such as smoke and grease; that must be removed from the house by a ventilation system. Or someone may have a hobby that uses paint or glue whose odors need to be exhausted to the outdoors. Because it isn't exactly practical to banish cooking and hobbies to

Understanding Ventilation

the outdoors, these and other pollutants usually must be removed from the air in a house with a local ventilation system.

Diluting pollutants sucked indoors by pressure differences

All houses experience air pressures from a variety of sources. These pressures will be discussed in depth in Chapters 5–7. When the air pressure indoors is less than that outdoors, we say the house is depressurized. When this happens, air from outside the occupied space will be sucked indoors in an attempt to balance the pressures and relieve the depressurization. If a house has random holes—and most do—air will pass through those holes. In doing so, it can bring in pollutants that originated outside the house. Because it is actually quite common for houses to become depressurized, pollutants are often sucked indoors.

Consider a basement. Many basements have a few cracks in the walls or floor. They also have small gaps where plumbing and gas lines or electrical wires pass through the walls from the outdoors. Many basements also have an open sump pump pit or a floor drain that is connected to a gravel bed under the house. All these openings are simply holes through which air can move. When a basement gets depressurized, air will get sucked into it through these random holes. Because most soils are porous to a certain extent, the air will move from the atmosphere, through the porous soil, through the random holes, and into the basement. As the air moves through the soil, it can pick up pollutants such as herbicides applied to the lawn, termiticides injected around the house, radon gas, sewer gas from a cracked drain, or ground moisture. These contaminants will all be pulled indoors.

If a furnace room gets depressurized, air can enter the room from the outdoors by coming down the chimney, picking up bits of soot and gases clinging to the inside of the chimney. This is potentially very dangerous because it can be difficult for the combustion gases from the furnace to rise up the chimney and escape from the house.

If an entire house gets depressurized, outdoor air will infiltrate wherever it can. As air infiltrates through the random gaps and cracks in a wall, it can bring with it odors from the insulation or particles of the insulation itself. Infiltrating air can also pick up contaminants from the outdoors (say, from a nearby automobile's exhaust) and bring them indoors.

In order to predict how much pollution will infiltrate into a house due to depressurization, you must know the concentration of the pollutants, how big the random holes are, and how much pressure is applied to the house. Then, if you can determine what a healthy level of these contaminants would be, you can calculate how much ventilation air will be needed to dilute the pollutants to a healthy level. From a practical standpoint, this is virtually impossible to do accurately.

There are two solutions to dealing with pollutants sucked indoors by air-pressure differences that are far more effective than trying to dilute those pollutants with a ventilation system. First, you can manipulate the pressures so that a house doesn't get depressurized (you can at least minimize depressurization) or, second, you can plug up the random holes. Remember, you must have two things before infiltration can take place—holes and pressure—and if you eliminate either one, you eliminate the infiltration.

Air pressures in a house can often be manipulated by selecting a certain type of ventilation system, perhaps one that creates a positive pressure in the house. The random holes can be plugged up with standard weatherization techniques. The amount of radon in a basement can often be reduced by sealing cracks and putting a tight fitting cover over the sump pump pit because those are two places where radon is often sucked indoors. Chapter 11 covers house tightening in more detail.

When you tighten a house, keep in mind that there is a relationship between pressure, tightness, and airflow. If a mechanical device, say a clothes dryer, blows air outdoors, the house will become depressurized as soon as the dryer starts running and an equal volume of air will find its way indoors. If you tighten the house to prevent radon from being sucked indoors, the clothes dryer will try to blow the same amount of air outdoors, but there will be fewer holes for the equal volume of air to enter the house. The tighter house will be depressurized more than the loose house. Depressurization itself isn't necessarily bad, but it can have some negative consequences. For example, depressurization can adversely affect the operation of a chimney. While there may appear to be several simple solutions to a particular problem, there are often other consequences that should also be considered.

The bottom line is this: it is generally more effective to tighten up a house (seal the holes), or manipulate the air pressures (pressurize the house) to deal with pollutants that get sucked indoors by pressure differences, than to attempt to dilute them with a ventilation system. In other words, it is more effective and easier to prevent them from coming indoors in the first place than to try to dilute them after-the-fact. But if you select a tightening strategy, make sure that you don't create other problems, especially problems affecting chimney operation. While pressurization may seem to be a viable strategy, it can cause hidden moisture problems in cold climates (see Chapter 8).

Diluting pollutants released by metabolic processes

All living creatures (including people and animals) give off a variety of pollutants as a normal part of the life process. While most by-products of metabolism are not toxic (at least not at the levels found in houses), if they are allowed to build up indoors, the occupants will begin to complain about dampness, drowsiness, stuffiness, and poor air quality. To keep the contaminants released by people and pets to a comfortable and tolerable level, they should be diluted with a ventilation system. The only other way of dealing with these pollutants is to remove the source of pollution: ban the people and pets from the house, usually an unlikely proposition. To keep the by-products of metabolism that are released by mold to a comfortable and tolerable level, you should eliminate the mold by controlling excess moisture—see *Moisture control* below.

Considerable of research is available on how much ventilation is needed to dilute the by-products of human metabolism in order for people to feel comfortable. The **American Society of Heating, Refrigerating, and Air-Conditioning Engineers (ASHRAE)** is one of the most active professional organizations in the country dealing with ventilation in buildings; it has based its ventilation recommendations on how *comfortable* people feel (see Chapter 12)—usually involving only a modest amount of ventilation. Which brings us back to the title of this chapter, *Ventilation is for people*. Pollutants released from materials inside the house, those sucked indoors, and those released by mold—while they can be dealt with by ventilation—are more effectively dealt with by other means.

When we inhale air, our bodies absorb the oxygen molecules into our lungs. The oxygen then travels through our blood and is processed throughout the body. If a person were in a sealed room, he or she would eventually consume all the oxygen in the air and suffocate to death. Thus, one purpose of a ventilation system is to supply a house with enough oxygen so the occupants won't die. In reality, most houses, even tightly constructed ones, have enough infiltration, or occasionally opened doors or windows, to supply them with sufficient oxygen. (People only need about $1/2$ cubic foot of fresh air a minute to have enough oxygen.) In a very tight house with no ventilation system, people will do something about stuffiness and odors long before they suffer from lack of oxygen. Because of the way our bodies function, and the concentrations of oxygen and CO_2 in the atmosphere, it is possible to calculate that when a person's exhaled breath raises the concentration of CO_2 in a room by 300%, the oxygen level in that room will only have decreased by 0.5%. So there will be a large increase in CO_2 (and other pollutants) long before there is a significant reduction in oxygen in a house. The real problem in houses has nothing to do with oxygen deprivation but rather too many pollutants.

Diluting high moisture levels

Although moisture itself isn't a pollutant, with a build-up and accumulation over several days, it can contribute to a variety of pollution problems in a house. If there is excess moisture in a house in the form of high humidity, there can easily be a proliferation of biological pollutants. While mold, mildew, or dust mites will start to thrive as the humidity rises, and the occupants may start experiencing allergy or asthma symptoms, the culprits may not be visible, at least at first. This is because the organisms often live in carpeting or upholstered furniture (especially in cooler parts of the

Understanding Ventilation

house—unheated rooms, near exterior walls, or in closets) or hidden within building cavities. If the indoor relative humidity continues to rise, the problem will become more visible with mold or mildew growing on walls or ceilings. A bad situation can even progress to decay. Once mold starts growing when the relative humidity is high, it can continue to thrive at lower humidities because one of the by-products of mold's metabolism is water vapor. In other words, mold actually creates some of the water it needs to survive.

If high moisture levels continue unchecked, decay organisms can attack the structure of a house and rot can result. The hidden, wooden structure of a house (studs, floor joists, rafters, etc.) can begin to rot unnoticed by the occupants because of unseen moisture damage. A damp structure is also more susceptible to termite or carpenter-ant attack because those insects prefer moist wood. Sometimes the heads of the nails holding the drywall in place or other steel components of a house will begin to rust if the indoor relative humidity is high enough. And even if biological pollutants aren't in evidence, high humidity will cause VOCs, such as formaldehyde, to outgas at a faster rate.

Before discussing how ventilation can be used to reduce high indoor humidity, let us look at the different ways moisture builds up in houses. Moisture can either be generated inside the living space (an internal source) or it can migrate into the living space from somewhere else (an external source).

External moisture sources

When excess moisture enters a house through a leaking roof, dripping water will generally be noticed quickly by the occupants who will take steps to repair the leak. Equally large amounts of water can migrate up into a house from a damp basement or crawl space. However, this moisture may not be noticed because, rather than being a dripping liquid, the moisture may be in the form of an invisible vapor.

Crawl spaces and basements become damp for a variety of reasons. There could be a high water table in the ground around a house, allowing water to seep in. There could be defective gutters or downspouts, allowing rainwater to saturate the ground around the foundation. Normal soil dampness can release moisture from the dirt floor of a crawl space. Moisture can wick up by capillary action through foundation walls and evaporate into basements and crawl spaces. This is sometimes called rising damp. For a complete discussion of basement moisture problems, and methods of solving them, see *Investigating, diagnosing, and treating your damp basement*[Canada 1992] This book also contains information applicable to crawl spaces.

Moisture problems in the upper floors of a house are often due to excessive moisture in a crawl space or basement. The moisture generally finds its way from the crawl space or basement into the living space by one of two methods. It can diffuse through the floor system: the water vapor molecules can migrate from an area of higher concentration (the basement or crawl space), through the floor system itself, into an area of lower concentration (the upper level of the house). More significantly, moisture-laden air can be pushed or pulled into the living space by air-pressure differences. If the crawl space or basement is pressurized and the main floor of the house is depressurized, the humid basement or crawl space air will move through the random holes in the floor (around plumbing lines, electrical wires, etc.) up into the living space. A forced-air heating/air-conditioning system can often create significant air-pressure differences from one part of a house to another (see Chapter 6).

In a climate where air conditioning is highly desirable, the humid outdoor air can be loaded with moisture. If this humid air gets sucked indoors when a house becomes depressurized (or if outdoor air enters through a controlled ventilation system), it can easily contribute to a high humidity indoors.

Whenever moisture originates outside the house and passes through the structure into the living space, two potential problems can occur: the moisture can trigger biological growth hidden within the structure, or it can trigger biological growth inside the living space. If mold starts growing hidden within the structure, it won't be readily noticed. Yet it could be releasing spores, some of which will find their way into the living space (pulled there by air-pressure differences). If a moisture problem continues over an extended period of time, it can progress to rot and threaten the integrity of the building. When moisture triggers biological growth within the living space, susceptible occupants will probably be affected quickly, and discoloration or minor damage

Chapter 4 Ventilation is for people

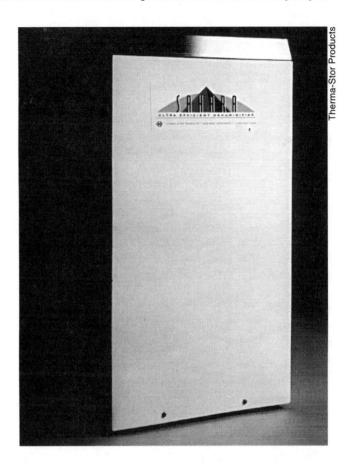

Figure 4–2. High-efficiency dehumidifiers can be used to lower the relative humidity in a house.

will generally prompt a search for a solution before serious structural damage has occurred.

Internal moisture sources

Inside the living space, moisture can be released from both people and their activities. For example, breathing and perspiration release moisture into the indoor air. A person at rest will release about a quart of water a day—half from the lungs in exhaled breath and half from perspiration through the pores. This rate of moisture release doubles or triples when someone is physically active.

Moisture is also released from activities such as washing clothes, drying clothes on an indoor clothesline, transpiration from house plants, using humidifiers, and washing dishes. It wouldn't be unusual for 2 gallons of water to evaporate from an uncovered spa in a single hour. Green firewood stored indoors will also release moisture into the indoor air, as will an uncovered aquarium. Unvented combustion appliances and gas ranges—or combustion appliances that are backdrafting—can introduce moisture into the indoor air because water vapor is a by-product of combustion. It is estimated that an average family of four in an average house puts 2–3 gallons of water into the air every day from all sources.

In a new house, a great deal of moisture can be released from construction materials. This is due to the high moisture content of framing lumber and excess moisture released from new concrete, plaster, drywall joint compound, and paint. It has been estimated that during its first year, the building materials in a new house will release 20 liters (approximately 5 gallons) of water per day.(Product) Usually, 75–90% of the moisture in new building materials will be released during a house's first year of life. So, new houses can have higher indoor humidities than older houses, but those high levels will taper off considerably after the first year. The rate at which moisture is released by building materials depends on the climate; when there is a short heating season, it may take several years before most of the moisture is released from the construction materials.

Plumbing leaks can also create significant moisture problems in houses. If the leak is very severe, it will most likely be noticed quickly and repaired. However, a small plumbing leak inside a wall cavity can go unnoticed for some time, contributing to a hidden mold or decay problem.

Controlling moisture levels in houses

It is almost always a good idea to prevent moisture from entering the living space from an external source. Roof or plumbing leaks should be repaired promptly (source control works for moisture as well as air pollutants). If moisture is kept out of basements and crawl spaces (separation), then it won't be able to migrate into the living space. It is also almost always a good idea to prevent moisture from any source (whether internal or external) from getting into hidden parts of a house in order to minimize the chances of hidden condensation, mold growth, or decay in unseen locations. The proper control of moisture in houses involves the correct use of moisture barriers, dampproofing, etc. This topic is covered in depth in the *Moisture Control Handbook*.(Lstiburek and Carmody)

Two common ways of dealing with the moisture generated from internal sources are dehumidification and ventilation. Air conditioning is a form of dehumidification; not only do air conditioners cool the air, they also dehumidify it. Air conditioning can be a viable way of lowering indoor relative humidity levels, but it is often only feasible in the summer and only then at lowering the general level of a humidity throughout an entire house. In the winter, commercial buildings sometimes use a cooling coil to remove excess humidity from the air, then a heating coil to warm the air back up. This can be an expensive way to control humidity compared to using a local exhaust fan, and so is rarely an option in residences.

Though it is possible, air conditioning isn't a very useful method of reducing high humidity levels in a single room, such as a bathroom. In any case, a bathroom would still need a spot ventilation fan to remove odors. In residences, portable dehumidifiers are available from most department stores that can be used when air conditioning isn't practical. They are often good choices for lowering the humidity in a damp basement when it is too difficult or expensive to prevent moisture from entering from the soil.

Therma-Stor Products manufactures a high-efficiency Sahara dehumidifier (and a Hi-E Dry commercial unit) that can be used as either a stand-alone portable dehumidifier, as a central ducted unit, or in conjunction with a central forced-air heating/cooling system(Gehring 1991, 1992, 1993, 1994) (Figure 4–2). When it is 80°F at 60% RH, a Sahara can remove 100 pints (12 gallons) of water per day. This unit has been used successfully to lower the humidity in houses where the occupants are highly allergic to mold, dust mites, and other microorganisms.

Where air conditioners or dehumidifiers aren't an option, ventilation is the only way to expel excess moisture from a house. In fact, in many houses, some form of ventilation is the best way to lower the humidity indoors, but it depends on the climate. The climate is important because the process of ventilating a house can either raise or lower the indoor relative humidity, depending on the specific indoor and outdoor temperatures and humidities involved. Therefore, ventilation will be more effective at reducing indoor relative humidity some times of the year rather than others. For example, in the winter, outdoor air contains very little moisture, so if it is brought indoors and heated up, it can dramatically dry out the indoor air. On the other hand, ventilating in the summer can cause humid outdoor air to enter a house, thus raising the indoor humidity. However, it depends on the situation; a local exhaust generally blows more moisture outdoors from a steaming shower than is sucked in with the make-up air, even in the summer.

Some times of the year, when outdoor and indoor air contain similar amounts of moisture, ventilation may have little effect on the indoor humidity. Because of indoor and outdoor temperature and humidity variables, it is often difficult to maintain a constant indoor humidity level throughout the year. Often, the structure of a house and its furnishings will tend to absorb and release moisture seasonally. Materials will take up moisture during a humid summer and then give it up during the fall. For a more complete discussion of relative humidity and how ventilation systems affect the humidity indoors, see Chapter 8.

Are there reasons not to ventilate?

Several reasons have been given not to ventilate. It has been suggested that if the outdoor air is badly polluted, a ventilation system will bring poor-quality air into the house. That may be true, but because indoor air quality is usually worse, ventilation will often help improve indoor air—even in a polluted city. But the best a ventilation system can do is make the indoor air as good as the outdoor air; it can't make it any better than outdoor air. In a situation where the outdoor air is especially polluted, it may be important to couple the ventilation system with a filtration system. Commercial buildings are often required to do this in heavily polluted cities.

It is also pointed out that an improperly designed ventilation system can cause a furnace's chimney to backdraft. When this happens, combustion gases flow down the chimney, rather than up as they are supposed to, and enter the living space. The answer is to correctly design the ventilation system in the first place so it won't cause a problem, rather than not ventilating. Actually, leaky forced-air heating/air-conditioning ducts are often a more significant cause of backdrafting than ventilation systems, but backdrafting is a very serious issue, and ventilation systems can definitely be a factor, especially in the summer when the draft in a chimney is weak. Backdrafting will be discussed in more detail in Chapter 10.

Ventilation can sometimes result in indoor air that is too dry in the winter or too humid in the summer, but there are ways to minimize this. Critics of ventilation systems also point out that they use energy (to run the fans), must be maintained (periodic cleaning or lubrication), can be noisy (sometimes), and can be complex (although not always). Some builders don't install ventilation systems because homeowners don't ask for them, and some homeowners who do ask have difficulty finding a builder or mechanical contractor who has enough understanding to properly design and install a system.

Certainly, ventilation is not a cure-all. It will do nothing to reduce electromagnetic pollution, because it is only effective at removing or diluting airborne pollutants. A ventilation system won't prevent children from getting lead dust on their fingers that has settled into the carpeting. Ventilation is also ineffective with some pesticides that have an oily base and don't evaporate very quickly, yet can be absorbed through the skin. Children and pets are especially vulnerable if they get pesticide residues on their hands or paws.

Actually, the most often cited reason not to ventilate houses is to save money. Unfortunately, very little in life is free, and ventilation does have a price. There is a cost in buying ventilation equipment, maintaining it, and operating it. However, the cost of ventilation should be considered just as basic as a house's foundation, walls, or cabinets. It is a cost we must learn to live with if we are to provide for our health and comfort. We would never consider building a house without a bathtub simply because they cost money. Nor should we build a house without a ventilation system just because there is a cost associated with it.

Part 2

How air enters and leaves a house

Chapter 5 Natural ventilation
Chapter 6 Accidental ventilation
Chapter 7 Controlled ventilation

Chapter 5

Natural ventilation

Because Mother Nature is at least partially responsible for the air-exchange rate in all houses, it is important that we understand how she does her work. People have been arguing for many years whether houses should receive their air randomly, by means of naturally induced infiltration offered by Mother Nature, or mechanically, through a ventilation system. In the past, there was no such thing as mechanical ventilation, and Mother Nature seemed to supply houses with plenty of fresh air—or did she? As it turns out, Mother Nature sometimes does an adequate job of ventilating houses but sometimes she doesn't.

This chapter covers the naturally occurring pressures that cause an exchange of air in houses. Keep in mind that positive and negative pressures themselves are neither good nor bad, but they can have good or bad effects. For example, they can cause fresh air to enter a house (a good effect) or they can cause pollution- or moisture-related problems (a bad effect). Potential ventilation-related problems will be covered in depth in Chapters 8–10.

Natural pressures

To review, for air to enter or leave a house, there must be holes for it to pass through and there must be a an air-pressure difference to push the air through those holes. There are three pressures in nature that can cause air to move through the holes in a house. One pressure, diffusion, is actually too weak to be of much significance. The other two natural pressures—wind and stack effect—can easily push air through the random holes in a house, even when those holes are so small they are unnoticed by the occupants. Mechanical devices, such as clothes dryers, central vacuums, forced-air heating/cooling systems, and exhaust fans, can also apply pressures to a building, resulting in air movement between

Understanding Ventilation

the indoors and the outdoors. These will be discussed in Chapters 6 and 7.

Wind

The easiest source of natural pressure for people to visualize is the wind. When wind blows on the side of a house, it pushes air through any hole it can find. This is infiltration. On the opposite side of the house, air will be exfiltrating, or leaving the house. When a certain amount of air enters a house somewhere, an equal amount of air must exit somewhere else. (While the *total* amount of air entering a house always equals the *total* amount of air leaving a house, the air flows due to two different pressures aren't always additive. For a discussion, see Chapter 7.)

A strong wind obviously will move more air through a house than a weak one, and bigger holes in a loosely built house will allow more air to pass through than smaller holes in a tightly constructed house.

Sometimes the effect of wind is very visible, as when it blows through a leaky window frame and rattles the window sash, rustles the curtains, or blows papers off a desk. A cold infiltrating winter wind can often blanket the floor with cold uncomfortable air—and it can drive up heating bills.

Figure 5-1 shows a simplified diagram of how the wind can apply pressure to a building. Infiltration takes place on the windward side and exfiltration takes place on the leeward side. In reality, the pressures of the wind can be quite complicated. This is because the wind rarely strikes a wall at a precise 90° angle. Also, because turbulence is created at various locations, wind can result in unusual pressures as it passes around corners or over a roof. Figure 5-2 shows a more realistic diagram of what wind pressures look like on houses. The positive pressures are pushing inward, the negative pressures are pushing outward.

It is important to realize that if a house is very tightly built (one with few holes), the wind will have a hard time blowing air through it. On the other hand, if a house is full of holes (loosely built), and the wind isn't blowing, there still won't be any air passing through the house (unless some other pressure is at work).

How much air will be flowing through a house because of the wind depends on the speed of the wind, how many random holes there are in the house (its tightness), and where the holes are located. (There will be a different amount of air flowing through a house that has all of its holes on one side compared to a house having holes on all sides.) In most cases, the indoor/outdoor pressure difference due to wind is less than 7

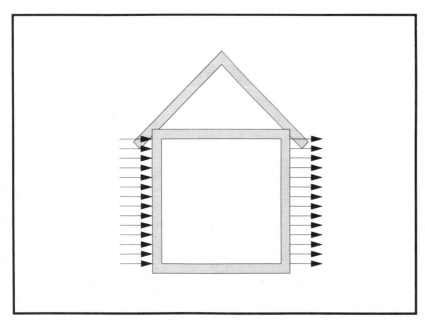

Figure 5-1. *In this simplified diagram of wind pressures, there is an inward pressure on one side of the building and an outward pressure on the opposite side.*

Pa. This isn't very much pressure, but it can be enough to move a certain amount of air through random holes in a house. Of course, higher-strength winds do occur; a hurricane or tornado can create a strong enough pressure difference to cause windows to either explode (when the house experiences a strong positive pressure) or implode (when the house experiences a strong negative pressure).

Stack effect

Stack effect is just as common a pressure in causing air to move through houses as the wind, but it isn't as well recognized. Stack effect is sometimes called chimney effect and is based on the principle that warm air rises.

We don't often think of air as having weight, but it does, and warm air weighs less than cool air. To clarify, as air is warmed, individual air molecules vibrate faster and bounce off each other. As the air molecules run into one another, they push against each other, forcing each other further apart so each molecule can have more room to maneuver. As a result, warm air has fewer molecules per cubic foot than cool air. Thus, warm air is less dense (lighter) than cool air, so warm air rises up above cool air. Although air is in fact very light (compared to solid or liquid matter), it is surprising to many people that the air in a typical 1,200 sq. ft. house weighs about 750 pounds.

When weathermen talk about cold air moving into an area, the incoming cold-air mass will slide under the existing warm-air mass. Because the warm air is lighter and more buoyant, it floats on top of the cold air. This same principle explains how a hot air balloon rises. If the air within the balloon is warmer than the air outside the balloon, the balloon will be buoyant and it will start to rise up above the cooler air. Warm air rises because it is light and buoyant, but it will not rise up into air of the same temperature; it will only rise up into air that is cooler and denser. Thus, it is not simply the fact that the air is warm that makes it rise. It will only rise if there is some cool air above it, so there must be a *temperature difference*, not just warm air.

In the winter, when a house is filled with buoyant warm air, that warm air will rise and try to escape out the top of the house through any random holes it can find there (exfiltration). At the same time, cold outdoor air will enter, or infiltrate, somewhere else. Like a weatherman's cold air mass, it will enter the house underneath the warm air. If you feel a draft of cold air rising up from an open basement door in the winter, it is no doubt caused by stack effect.

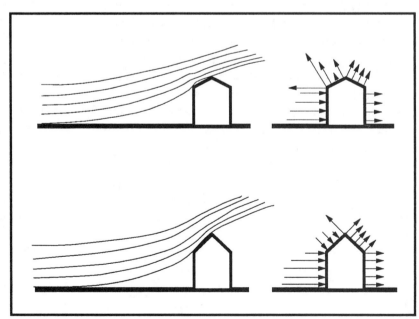

Figure 5–2. *The actual pressures exerted on a building by the wind depend on the precise angle the wind strikes the building and the shape of the building.*

Figure 5–3. In the winter, stack effect causes air to enter at the bottom of a house and escape at the top.

Figure 5–3 illustrates how the pressures from stack effect work in the winter. The pressures cause air to enter at the bottom of the house and exit at the top. The entire house acts like a chimney, or stack. If there are pesticides or radon gas in the soil around the house, stack effect can cause them to enter through cracks or gaps in the lower portion of the building. The positive and negative pressures resulting from stack effect are greatest at the highest and lowest points in the house, respectively. Figure 5–4 shows a more detailed diagram of how these pressures might look. Note that in the middle of the wall there is neither inward or outward pressure. This is an area of neutral pressure or the *neutral pressure plane*. If you used a pressure-sensing instrument to measure the pressure difference between the indoors and the outdoors at the neutral pressure plane, no pressure difference would be found.

Because stack effect only occurs when warm air rises up into cooler air, there will be no stack effect if the indoor and outdoor temperatures are the same. There must be an indoor/outdoor temperature difference for there to be any pressure because of stack effect, and the greater the temperature difference, the higher the pressure difference. Height is also a factor, so a taller building will experience greater pressures than a short building when the temperature difference is the same. So, a two-story house will be under greater pressures due to stack effect than a single-story house, and both will experience higher pressures as the outdoor temperature drops in the winter.

Stack effect works in reverse in the summer (if a house is air conditioned) because the temperatures are reversed. When it is cool indoors, the heavy cool air falls and escapes (exfiltrates) through holes in the lower part of the house. Then the warm outdoor air is pulled in (infiltrates) through holes at the top of the house. Figure 5–5 shows stack-effect pressures in the summer.

Because the indoor/outdoor temperature difference usually isn't as great in the summer as it is in the winter, stack effect won't be as significant as in the winter in a cold climate. To understand why, let's consider some typical temperatures. When it is 90°F outdoors and 70°F indoors during the summer, the temperature difference is 20 degrees. In the winter it might be 10°F outdoors and 70°F indoors, in which case the temperature difference would be 60 degrees. The greater the temperature difference between the indoors and the outdoors, the greater the pressures due to stack effect. If a house is airtight (no random holes), then it doesn't matter how much pressure there is because of stack effect—there will be no air movement between the indoors and outdoors.

Stack-effect pressures are generally fairly small when compared to wind pressures—usually less than ±4 Pa. in the winter, and less than ±2 Pa. in the summer. But if there is a temperature difference for 24 hours per day, 7 days a week, the pressure will be continuous. As a result, stack effect often accounts for more infiltration (and exfiltration) during the summer and winter than wind because the wind isn't always blowing. On the other hand, there is very little stack effect in the spring and fall when the indoor and outdoor temperatures are similar, so wind has more influence on house pressures in those seasons.

Even when stack effect isn't a significant factor, small temperature differences within the living space can cause air to circulate around the inside of a house. Because the temperature of the air in a room is almost never uniform (*e.g.* it is warmer near the ceiling than near the floor), there will generally be very small air currents that result in a continuous mixing of air in a room. These are called convective air currents because they are driven by convective heat transfer. (While a discussion of heat transfer is beyond the scope of this book, heat is transferred by three basic mechanisms: conduction, convection, and/or radiation.) Because of convective air currents, the air in a house is in constant slow motion. When there is stale or polluted air in one

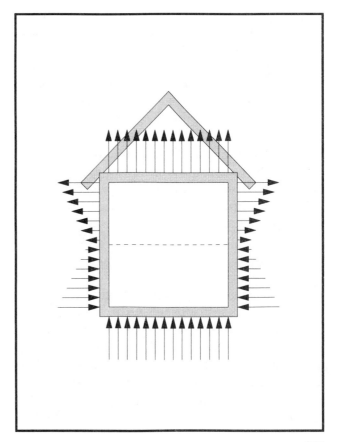

Figure 5–4. The actual pressures due to stack effect vary at different heights in the house. In the winter, the greatest outward pressure is at the ceiling and the greatest inward pressure is at the floor.

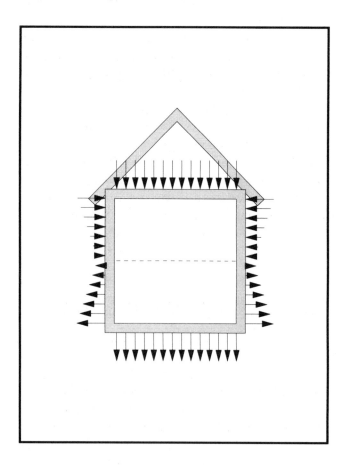

Figure 5–5. In an air-conditioned house in the summer, stack effect will result in air entering at the top and escaping at the bottom.

part of a house, convective air currents will cause it to eventually disperse throughout the house, becoming more and more diluted along the way. This explains why a bedroom is more likely to get stuffy if the door is closed—the door blocks the convective air currents between the bedroom and the rest of the house. As a result, the pollutants released by the people in the bedroom can't disperse into the rest of the house, so they build up.

Diffusion

When you place a drop of food coloring into a bowl of water, the color gradually moves through the entire bowl until it is evenly distributed. This is because nature tries to balance everything out by moving liquids from an area of high concentration to one of low concentration. The same thing occurs with gases: a high concentration of a particular gas will move toward an area of lower concentration. The process is called *diffusion*. Diffusion is the third natural pressure that causes air (actually, individual molecules of the components of air) to move.

For diffusion to occur, there must be a difference of concentration between gas or vapor molecules in one area compared to another area. The greater the difference in concentration, the faster the diffusion. (The pressure in this case is called *vapor pressure*.) If you release a small amount of perfume in a room, eventually the fragrant molecules will diffuse throughout the entire room. This may take a while, but sooner or later the concentration of perfume molecules will be evenly distributed in every cubic foot of air in the room.

If there is a high concentration of automobile exhaust gases in the outdoor air, and a window is open, the exhaust gas will diffuse from the area of high concentration (outdoors), through the window, to areas of low concentration (indoors). This happens even if there is no stack effect or wind. However, diffusion is such a slow process that it usually isn't worth considering.

While diffusion can cause air pollutants to disperse throughout a house, in reality the concentration of air pollutants can be dispersed much more quickly and effectively by using air-pressure differences (perhaps, using a fan) to circulate the air through the house, than by diffusion. In fact, convective air currents also account for significantly more air circulation between the various rooms of a house than diffusion.

Diffusion through solid surfaces

Besides causing molecules to move through a room or through a hole such as an open window, diffusion can also account for movement through a solid wall. No open windows or holes are necessary for diffusion to do its work. Those exhaust gases outdoors can sense that the concentration is lower indoors and will start passing between the molecules in the solid wall until the indoor/outdoor concentrations are similar (Figure 5–6). The greater the difference in concentration, the faster this will happen.

Diffusion through a solid wall is, of course, much slower than diffusion though air, and it can be slowed down even further by constructing the wall out of a low-porosity material. Builders often use plastic sheeting in walls as a diffusion retarder to slow the movement of water vapor through the wall. Such a retarder will also slow down considerably the movement of all the different components of air—the oxygen, nitrogen, and various other gases.

It is important to realize that, no matter what the concentration of the various molecules, or what the wall is made of, diffusion through solid materials is an extremely slow process. In fact, wind and stack effect can move hundreds of times more air through the tiny, random holes in a house than diffusion can move through both the random holes and the solid materials combined. To repeat, in the vast majority of cases, diffusion is so slow that it is not even worth considering.

Do natural pressures supply us with enough air?

Wind and stack effect are the most significant natural pressures that cause air to move through houses. In contrast, diffusion is insignificant. Do wind and stack effect supply enough fresh air to the rooms where people work and play? Sometimes they do, sometimes they don't. Do wind and stack effect exhaust stale air effectively from the most polluted rooms? Sometimes they

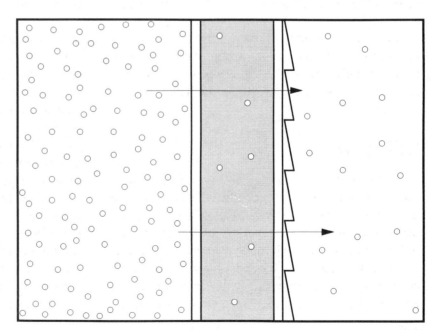

Figure 5–6. Diffusion can account for the movement of gases and vapors through a solid surface, but because it is an extremely slow process, diffusion is rarely the cause of air pollutants moving through a wall—but it can sometimes cause water vapor to travel through a wall.

Understanding Ventilation

do, sometimes not. Because natural pressures are not continuous or predictable, they can only be expected to move air through a house in an equally unpredictable manner.

In a tight house, there will be very little air infiltrating or exfiltrating because of wind or stack effect simply because there aren't enough random holes. A few builders have suggested constructing looser houses—ones with more holes—so that more air will pass through them. However, because the wind isn't always blowing, and stack effect isn't a factor for much of the year, the extra holes don't always help much.

More progressive builders advocate that houses be built purposefully airtight—for comfort, energy efficiency, and moisture control —and a mechanical ventilation system can supply fresh air. Actually, a mechanical ventilation system can be a good idea in all houses, tight or loose, because the pressures induced by wind and stack effect are very erratic and unpredictable. We need fresh air continually, not once a week when the wind picks up.

Still, many people like the concept of *natural* ventilation. To them, *mechanical* ventilation sounds unnatural. As a recent ventilation article asks, "Which sounds better to you: 'The *natural* taste of orange juice' or 'the *mechanical* taste of orange juice?'"(Hill and Kerr) With ventilation, however, mechanical is better than natural. Keep in mind that without *mechanical* refrigeration, most Americans wouldn't be able to enjoy the natural taste of orange juice throughout the year. Natural ventilation may sound good, but mechanical ventilation generally works more effectively.

Chapter 6

Accidental ventilation

Most modern houses contain a variety of mechanical devices—clothes dryers, central vacuums, etc.—that cause air pressures within the house to change. These devices were never really designed to help provide a house with fresh air, but by increasing or decreasing the indoor air pressure, they definitely result in more infiltration and exfiltration. These devices supplement the naturally induced pressures of wind and stack effect by supplying houses with additional air.

If, for any reason, a mechanical device causes the air pressure indoors to be greater than the air pressure outdoors, a house is considered to be pressurized. In other words, it is experiencing a positive pressure. If a mechanical device causes the indoor air pressure to be less than that outdoors, the house is depressurized, or under negative pressure.

As with natural pressures, these positive and negative *accidental* pressures themselves are neither good nor bad, but sometimes they can cause pollution- or moisture-related problems. The potential problems will be discussed in Chapters 8–10.

Clothes dryers

Most clothes dryers have their exhausts connected to the outdoors. An operating dryer is, therefore, blowing air outdoors at a rate up to 150 cubic feet per minute (cfm). This can slightly depressurize the entire house. If 150 cfm is actually leaving the house, an equal volume of air (150 cfm) will need to infiltrate into the house from the outdoors to balance the indoor/outdoor pressures (Figure 6–1).

If a house is loosely built, it will have plenty of random holes, so 150 cfm of air will infiltrate easily whenever the dryer is running and very little, if any, depressurization of the house will result. The tighter the house, the more the depressurization. A clothes dryer

Understanding Ventilation

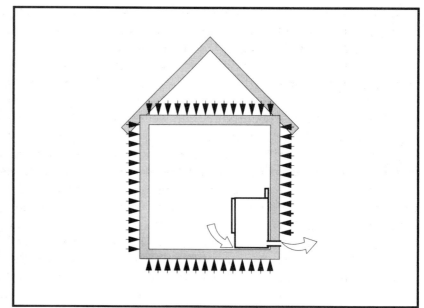

Figure 6–1. When a clothes dryer blows air out of a house, the house will become depressurized.

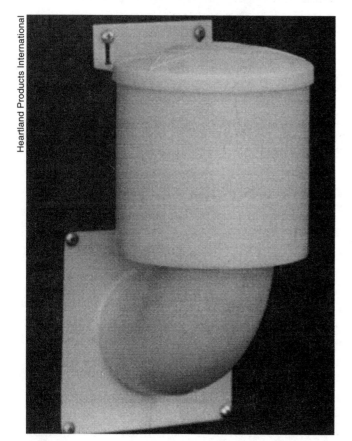

Figure 6–2. A tight-fitting dryer vent closure will prevent unwanted infiltration when the dryer isn't running.

might cause as much as 10 Pa. of depressurization in a very tight house. When running for an hour, a clothes dryer will blow about 9,000 cubic feet of air outdoors (150 cfm x 60 minutes). This is enough air movement to exchange $^3/_4$ of the air in a typical 1,500-sq.-ft. house.

Occasionally, to conserve energy in the winter, some homeowners have disconnected a dryer's exhaust and directed it into the interior of the house. This is no longer recommended. Even though some energy is saved, a dryer's exhaust contains a great deal of moisture, and when all that moisture is blown into the indoor air, it can easily result in problems such as mold and mildew growth. If allowed to continue for an extended period of time, it can lead to decay. A high relative humidity can also result in increased outgassing of formaldehyde from manufactured wood products such as wall paneling and cabinets.

Besides moisture, a dryer's exhaust contains a variety of pollutants such as lint, detergent residue, or odors from fabric softeners. If it is a gas dryer, the exhaust will contain combustion by-products that definitely aren't good to breathe. So, a clothes dryer should always be vented to the outdoors.

Sometimes, clothes dryers are located a considerable distance from an exterior wall, despite the fact that dryer manufacturers recommend they be connected to the outdoors with a short exhaust duct. Since an extra long exhaust duct will have more resistance to airflow than a short exhaust duct, it can be difficult for the dryer to blow air outdoors efficiently. A simple solution is to install a larger-diameter exhaust duct that has less resistance to air flow. For a more expensive solution, **Reversomatic Htg. & Mfg. Ltd.** and **Fantech, Inc.** both offer "dryer exhaust fans" that can be used to boost the flow of air from a clothes dryer to the outdoors. These fans are specially designed to prevent lint buildup.

When a clothes dryer is not operating, outdoor air can enter the house through the exhaust duct if the house gets depressurized because of something other than the dryer. This can be prevented if the dryer's outlet vent is fitted with a backflow prevention device that will only allow air to move in one direction—out of the house. Aluminum or plastic flapper closures are the most common, but they often fail to close properly because they are poorly made, warped during installation, and are rarely cleaned of the inevitable lint buildup. **Heartland Products International** manufacturers a dryer vent closure that forms a much better seal that is less likely to become clogged with lint (Figure 6–2).

If a house gets pressurized for some reason, and the dryer isn't operating, air will be able to leak out of

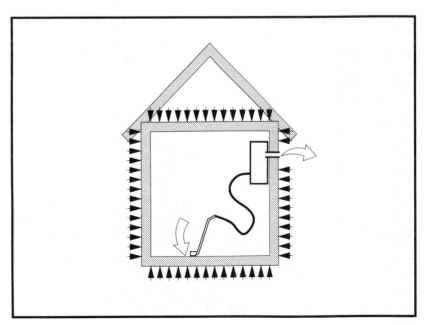

Figure 6–3. A central vacuum blows air outdoors, resulting in house depressurization as make-up air tries to enter.

the house through the backflow preventer because it only prevents flow in one direction. However, this may not be a significant consideration; air leaking in is usually more of a concern because it can bring lint or odors into the living space.

Central vacuum cleaners

Most central vacuum systems are also exhausted to the outdoors. In such instances, approximately 100 cfm of air will leave the house whenever the vacuum is running. When 100 cfm of air is leaving a house, another 100 cfm of air will need to infiltrate from somewhere else. As with clothes dryers, this can result in whole-house depressurization (Figure 6–3). The tighter the house, the greater the negative pressure indoors.

Central vacuums that aren't exhausted to the outdoors won't depressurize a house, but they will blow a great deal of very fine house dust into the room in which the vacuum motor and canister are located. This is because most vacuum cleaners (including portable vacuums) have relatively inefficient filters. The tiniest particles of house dust (consisting of dust mite feces, mold spores, and who knows what else) pass through the filter and are blown back into the living space. If the canister is in the basement, this dust will be distributed throughout the basement. If it is in a utility room, the dust will contaminate the clothing and linens stored there. Therefore, the best central-vacuum installations have their exhaust vented to the outdoors.

Air entering a house through a central vacuum system that is turned off isn't a problem because snug-fitting covers over hose inlets block this pathway into the house.

Sometimes a central vacuum without an outdoor exhaust will be placed in an attached garage. This will still depressurize the house because the vacuum pulls

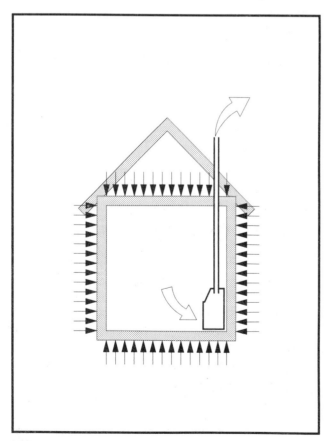

Figure 6–4. Chimneys can expel enough air to significantly depressurize a house. Open wood-burning fireplaces expel more air than efficient gas furnaces.

Chapter 6 Accidental ventilation

Figure 6–5. An inflatable fireplace pillow can be used to semi-permanently seal an unused chimney to prevent infiltrating air from entering a house by passing down the chimney.

air out of the living space. And this type of installation isn't immune from polluting the living space because very few garages are sealed well enough to keep 100% of the pollutants out of the house; some pollutants may leak back indoors because of house depressurization.

Chimneys

Until recently, most houses had chimneys that operated on a daily basis during the heating season. Today, many houses are kept warm by electric heaters or with heat pumps, neither of which require a chimney. While some electrically heated houses also have fireplaces connected to a chimney, many such fireplaces are only operated occasionally.

During normal operation, combustion by-products travel up a chimney and out of the house. The negative pressure induced inside a chimney by the rising warm air is called a draft. The process works because the combustion gases are warm, and warm air rises. This is based on the same principle of stack effect discussed in Chapter 5. In fact, chimneys are sometimes called stacks.

If warm air leaves a house by way of a chimney, it is possible for the entire house to become depressurized (Figure 6–4). The tighter the house, the more depressurization can result. If the house is too tight, there won't be enough holes for make-up air to enter (make-up air is air that enters a house to replace air that is leaving), so there may not be enough air leaving the house through the chimney, and some of the combustion gases will spill into the living space around the base of the chimney. This is called spillage and it is discussed more fully in Chapter 10.

When a fireplace, wood stove, or furnace is operating and combustion by-products move up the chimney as they are supposed to, air is being expelled from the house. Therefore, an equal amount of air will need to infiltrate wherever it can. The draft in a chimney from a roaring fire in a large open fireplace can exhaust enough air from a house that a breeze can be felt in the room. There can be as much as 25–50 Pa. of pressure inside a chimney due to a roaring fire in a fireplace, with as much as 400 cfm of air leaving the house. Chairs with winged-backs and skirting were developed to divert this breeze around people sitting in front of a such a fireplace. A 100,000 Btu natural-draft gas furnace might expel 40–50 cfm up the chimney while an induced-draft gas furnace would expel a little less—perhaps 25–30 cfm. A chimney has the potential to move air through a house whenever it is in operation. And

73

Understanding Ventilation

whatever volume of air goes up the chimney, an equal volume will enter the house through random holes.

Relying on a chimney to exchange the air in a house is considerably less efficient with smaller, more efficient furnaces because less air is expelled up the chimney. For example, an "airtight" wood stove might expel only 10 cfm from the house. If a natural-draft furnace is located in a basement, the incoming air may infiltrate through cracks or leaks in the basement walls, find its way to the chimney, then leave the house. In this example the chimney is causing the basement to be ventilated. However, people generally spend more time upstairs than they do in the basement, so upstairs is where the fresh air is needed.

Something important to keep in mind about a chimney is the fact that it is a hole in the house, and air can move through that hole in either direction. When a furnace or fireplace is operating, air is leaving the house through the chimney. If the furnace or fireplace aren't operating, and another device, say a clothes dryer, is blowing air out of the house, causing the building to be depressurized, then infiltrating air may enter the house by traveling down the chimney. This will be covered in more depth in Chapter 10.

To prevent unwanted air from entering a house through a chimney, the damper can be closed. This will certainly help, but many dampers are not very airtight. **Chim-A-Lator Co.** and **Lymance International** have dampers that mount on the top of a chimney to seal out both the weather and air leakage when the chimney is not in use. Both can be opened and closed from inside the building. For a semipermanent installation, **Enviro-Energy Marketing Services Ltd.** has an inflatable fireplace Draftstopper that can be used to seal up an unused chimney to prevent unwanted infiltration or exfiltration (Figure 6–5). It resembles a small air mattress and, when wedged above a firebox and inflated, seals a chimney quite well. A warning label hangs down to warn homeowners not to accidentally build a fire under it.

Leaky heating/air-conditioning ducts

Not all forced-air furnaces are connected to a chimney or flue. For example, electric furnaces and heat pumps have no chimney, nor do the newer gas furnaces with sealed combustion-chambers. But all forced-air furnaces, as well as central air-conditioning systems, can radically affect air pressures in a house, especially if their ducts are located outside the conditioned space,

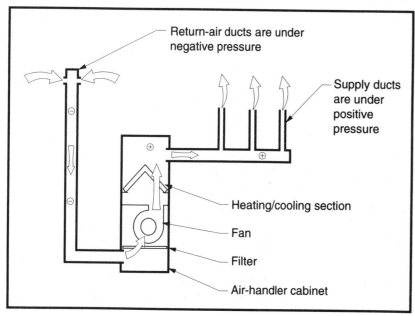

Figure 6–6. In a typical forced-air heating/air-conditioning system, air is pulled from the living space through return-air ducts that are under negative pressure and blown back into the living space through supply ducts that are under positive pressure.

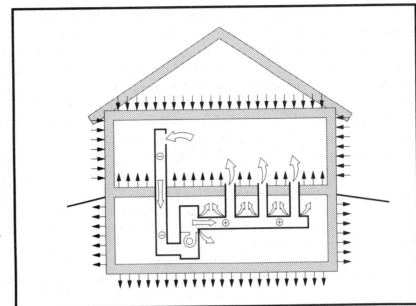

Figure 6–7. Leaky ducts can easily affect pressures in a house. In this example, leaky supply ducts pressurize a basement which results in the main floor becoming depressurized.

such as in the attic or crawl space. This happens because the ducts are often quite leaky, and the pressures caused by those leaky ducts can significantly affect the infiltration and exfiltration rates of a house. (Houses with electric baseboard heaters or electric radiant ceiling/floor heating have no chimney, nor do they have leaky ducts, so they usually have the least amount of accidental ventilation.)

The forced-air heating/air-conditioning system in a typical house often consists of a single return-air grille connected with ducts to one side (the return-air side) of a fan cabinet (usually called an air handler). Besides the fan, the air handler contains some method to heat or cool the air and often contains a filter. The other side of the air handler (the supply-air side) is typically connected with ducts to several supply registers throughout the house (Figure 6–6).

During operation, air is drawn from the living space into the return-air grille, and pulled into the air handler where it is filtered and heated or cooled. Then the air is blown out through the supply ducts and supply registers, and back into the living space. All the air within the ducts between the return-air grille and the air handler is under a negative pressure (this can be thought of as the suction side of the system). The air in the supply ducts (between the air handler and the supply registers) is under positive pressure because it is being pushed along by the fan.

Occasionally, forced-air heating/air-conditioning systems have at least one supply register and a return-air grille in each room. (Some people don't like return-air grilles in bedrooms because they are slightly noisier.) More often, however, there are supply registers in each room but only one or two return-air grilles in the entire house. A few systems are installed with no return-air grilles—instead, the air handler simply has an opening in its side for the return air to enter.

Furnace fans are quite powerful. As a result, they typically generate up to 50 Pa. (0.2" w.g.) of pressure within the ducts, and there can be well over 100 Pa. (0.4" w.g.) near the fan. This results in a great deal of leakage if the ducts aren't well-sealed. It isn't unusual for there to be a 15–20% leakage rate into or out of leaky ducts. At this rate an air handler capable of moving 1,800 cfm would lose 270–360 cfm because of duct leakage. And there are some cases where a 50% leakage rate has been measured.

Ducts are manufactured from a variety of different materials and there are several types of fittings (elbows, boots, take-offs, etc.) in the entire system. Between each fitting, there is a potential leak that can cause unplanned air loss from the system, and pressure

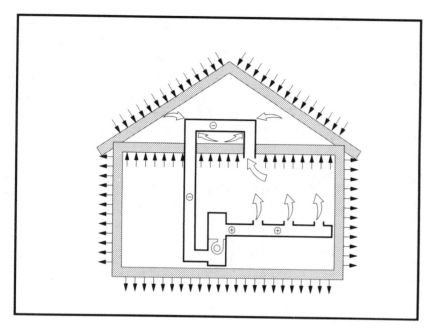

Figure 6–8. In this example, attic air is being sucked through leaks in return-air ducts. This results in a depressurized attic and a pressurized living space.

Figure 6–9. If leaky return-air ducts depressurize a mechanical-equipment room, the rest of the house can become pressurized. In addition, air can try to enter the mechanical room through a chimney.

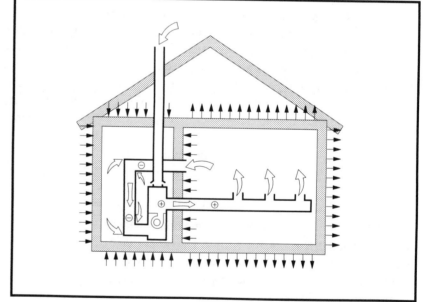

changes in buildings. Unless all of the joints are sealed with high-quality aluminum-foil duct tape or duct-sealing mastic, air will move through the various leaks. Air can also leak into or out of a poorly sealed air handler because of inherently leaky sheet-metal joints, holes drilled for piping or wiring, or loose-fitting, field-fabricated filter racks.

Sometimes building cavities inside floors and walls are used as ducts. For example, the hollow space inside a wall between the studs might be used as a return-air duct. The spaces between floor joists are often used as return-air ducts. When building cavities are used as ducts, they tend to be much leakier than if metal, plastic, or fiberglass is used, because they are rarely

Chapter 6 Accidental ventilation

sealed very well. In fact, building cavities often have a number of fairly large holes in them that were cut to run wiring or plumbing lines through. Most of the time, building cavities were not intentionally constructed to be airtight.

Because the air in supply ducts is under a positive pressure, air tends to flow out of their leaks. Conversely, air tends to flow into leaks in return ducts because they are under negative pressure. If supply ducts are located in a basement, and air leaks out of them, the basement will become pressurized—even if there are no supply registers in the basement. The basement will then be subject to exfiltration. When this happens, the return-air grille upstairs doesn't get enough air (because air is leaving the house through the basement walls), so the upper part of the house becomes depressurized and experiences infiltration (Figure 6–7).

If leaky return ducts are located in an attic, they will suck attic air into them. This will depressurize the attic and cause the rest of the house to be pressurized (Figure 6–8). After being pulled into the return ducts, the attic air, which is actually outdoor air, will need to be heated or cooled as it passes through the fan cabinet. This means an increase in the heating or cooling bill.

Sometimes, because of leaky ducts, the operation of a forced-air furnace or air conditioner can cause a very large negative pressure in the mechanical-equipment room. If there is a chimney in that room, outdoor air will flow down the chimney to relieve the negative pressure (Figure 6–9). When air is flowing down the chimney, combustion by-products from the furnace or water heater will have difficulty flowing up the chimney. This will be discussed more fully in Chapter 10.

To summarize, leaky ducts can account for very large pressure changes in houses. Depending on where the leaky ducts are located (inside or outside the conditioned space), this can result in either positive or negative pressures indoors, or a combination of positive pressure in one part of a house and negative pressure in another part. In any case, it isn't unusual for there to be a great deal of infiltration and exfiltration whenever the forced-air heating or cooling system is running.

Minimizing infiltration and exfiltration due to leaky ducts is simple in theory: Seal all the joints and gaps in all the supply and return ducts. But this is often easier said than done, especially in an existing house where it is difficult to get to ducts hidden inside building cavities. It is very important to keep in mind the fact that a partially effective duct-sealing job might result in more leakage through the unsealed portion of the system than originally existed. Contractors who specialize in duct sealing will pressurize the duct system

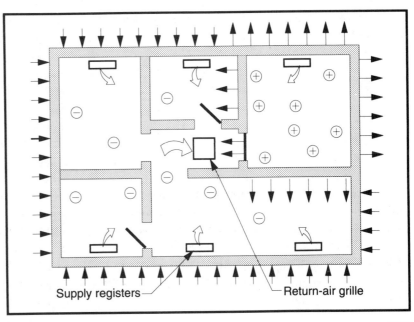

Figure 6–10. When doors between rooms are closed, it can be difficult for air leaving supply registers to get out of a room and find its way to a return air grille. The result is that rooms with closed doors become pressurized and other rooms become depressurized.

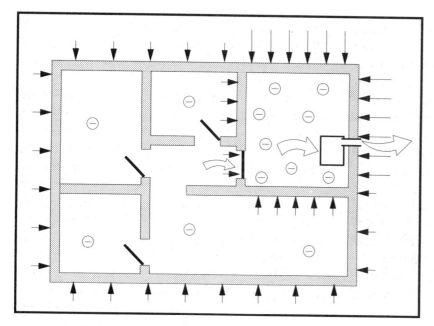

Figure 6–11. Any type of device that blows air outdoors (clothes dryer, central vacuum, exhaust fan, etc.) can depressurize a house, and the effect will be more pronounced if the exhausting device is located in a room with the door closed.

to determine if they have done an effective job without causing problems.

Duct-sealing mastics are available from a number of manufacturers including **Foster Products Corp., Hardcast, Inc., Mon-Eco Industries, Inc., RCD Corp., RectorSeal Corp.** and **United McGill Corp.** Water-based mastics are generally preferred over solvent-based mastics because they are less noxious. Duct-sealing mastics are generally longer-lasting that duct tape, do a better job, and are faster to apply. See also Chapter 17.

Closed doors

Simply closing the doors between rooms also can also change the air pressures in a house. For example, consider a forced-air heating/cooling system that utilizes a single return-air grille in the central hallway (or perhaps just an opening in the side of the furnace and no return-air grilles) but has supply registers located in every room. When the air handler is on, but the door to a bedroom is closed, 60–100 cfm of air tries to enter the room through each supply register, but it can't easily get out of the room because the door is closed. This will result in the entire bedroom being under a positive pressure. Since the air can't escape easily from the bedroom, the air entering the return-air grille in the hallway will be pulled from all the parts of the house with open doors. The result is less air entering the return-air grille and a negative pressure in that part of the house (Figure 6–10). The more doors that are closed, the more the pressures in the house get out of balance. Closing a bedroom door while a central forced-air furnace or air conditioner is running can cause that room to experience as much as 10–20 Pa. of positive pressure when compared to the rest of the house. Sometimes, the operation of a forced-air furnace/air conditioner will cause enough air movement to actually blow a door shut. Once shut, the room quickly becomes pressurized.

If a clothes dryer, central vacuum, or any other device is blowing air outdoors from a utility room with the door open, air will tend to infiltrate through random holes everywhere in the house. Such a house may sense only a small amount of overall depressurization. If the utility-room door is closed, there will be fewer paths for the air to enter that particular room, so that room will become more highly depressurized (Figure 6–11).

Exfiltration will be concentrated in any part of a house that is experiencing a positive pressure. In other words, air will try to leak out of a pressurized bedroom with a closed door. Infiltration, on the other hand, will occur wherever a house is under a negative pressure, such as a laundry room containing a clothes dryer behind a closed door. Sometimes a forced-air system will have a supply register in the basement, but the only return-air grille is located upstairs. If the basement door is closed, then the basement will be pressurized whenever the fan is running, and the upstairs will be depressurized. (Of course, there may be enough duct leaks to counteract the imbalances due to the closed door.)

In reality, there is a small gap beneath doors that will allow some air to escape from a pressurized or depressurized room when the door is shut. However, this gap is often not high enough to prevent a room from sensing some change in pressure when the door is closed. A solution would be to cut an extra inch or two off the bottom of all the doors. This will often relieve some of the pressure, but it can be unattractive. A one-inch gap is usually acceptable from an aesthetic standpoint. This is usually enough to relieve a significant amount of depressurization in a room having a single supply register, but it is usually not enough to relieve the pressure generated in a room having multiple supply registers.

Another solution to pressure changes in a house with a single return-air grille can be to redesign the system of ducts so that there is a return-air grille in every room that contains a supply register. This option is generally only exercised in more expensive houses because of the increased cost.

A third option to relieving room pressurization or depressurization—one that is easy to incorporate into an existing house—is to cut *pass-throughs* between rooms above the doors. A pass-through (sometimes called a *transfer grille*) is simply a hole in the wall with a grille on each side. Pass-throughs provide a path for air to move through—even when a door is shut—and they often are less obtrusive than a door with 2–3" cut off the bottom. Pass-throughs are usually not too con-

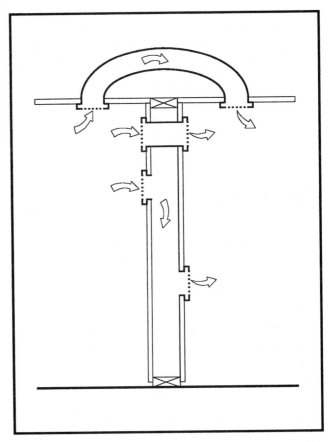

Figure 6–12. A jump duct connects two rooms via the ceiling to allow air to flow from one room into another when the door is closed. Pass-throughs can be installed in walls for the same purpose. A direct pass-through can be lined with sheet metal, making it easy to clean. An offset pass-through won't transmit light between rooms, but since it allows air to flow through a building cavity, it must be well-sealed to prevent air from being pulled from other cavities. Offset pass-throughs can be difficult to clean.

spicuous when placed above doors. In older houses, open transoms can easily function as pass-throughs. To minimize the amount of sound and light moving through a pass-through, the grilles can be offset on either side of a wall cavity. A *jump-duct* or *jump-over duct,* sometimes called a *transfer duct,* that connects between the ceilings of two rooms is another possibility. If a jump duct goes through the attic, it should be insulated and sealed (Figure 6–12).

If a pass-through is installed in a laundry room containing a clothes dryer, that room will no longer experience a large negative pressure when the appliance is operating and the door is closed, but the entire house may still sense a small negative pressure, depending on how tightly it is built. Louvered doors are a good solution to relieving pressures resulting from closed basement, closet, or utility-room doors, but louvered doors can compromise privacy when used in bathrooms and bedrooms.

Compared to forced-air heating/air-conditioning systems, most central ventilation systems move relatively small amounts of air between rooms, say 15–30 cfm. This amount of air entering a room with the door closed is generally too little to pressurize the room significantly, especially if the door has at least a $3/4$" gap at the bottom.

Does accidental ventilation supply us with enough air?

Accidental ventilation has the capacity to supply houses with more air than natural ventilation simply because the pressures are usually stronger. So, even in a semi-tight house (one having few random holes) accidental ventilation can exchange a considerable amount of air between the outdoors and the indoors. But some accidental ventilation is sporadic and inconsistent, so it won't supply air on a continuous basis. After all, clothes dryers and central vacuum cleaners are only operated a few hours per week.

On the other hand, a furnace that is expelling combustion gases continually up a chimney can cause a continual depressurization, which will result in continual infiltration. This can supply a house with a certain amount of air during the winter, but none during the summer when the furnace is off. Leaky ducts provide accidental ventilation whenever a forced-air furnace/air conditioner is running. However, it isn't unusual for the amount of infiltration due to leaky ducts to be excessive, resulting in high heating/cooling bills. Remember, the furnace runs the most when it is cold out, so you get the most accidental ventilation (due to the chimney and leaky ducts) when it is going to cost the most to warm up the incoming air.

One of the biggest drawbacks to accidental ventilation is the fact that it is unplanned. Therefore, air infiltrates and exfiltrates wherever it can find a random hole in the house. Sucking air in through cracks in a foundation wall or through fiberglass sidewall insulation isn't the best way to provide air for occupants to breathe. And the air probably isn't delivered to the occupants in a comfortable manner, or to the rooms they are located in. Does accidental ventilation supply us with enough air? It can, but whether the incoming air is healthy enough depends on where it enters the house and how sensitive the occupants are. Like natural ventilation, accidental ventilation has its drawbacks, but it is better than nothing—sometimes.

Chapter 7

Controlled ventilation

In addition to all the ways air enters and leaves a house naturally and accidentally (*e.g.* wind, stack effect, clothes dryers, chimneys, leaky ducts, etc.), there are mechanical systems specifically designed to ventilate houses using *controlled* pressures. This is *controlled ventilation*—in other words, ventilation *on purpose*—and it is the only way to exchange the air in houses that is consistent, reliable, and predictable. (Of course, consistency, reliability, and predictability with any mechanical system can be compromised by lack of maintenance or the occasional power failure.)

There are three basic controlled ventilation strategies—exhaust ventilation, supply ventilation, and balanced ventilation—each of which has some variations. Each differs in equipment cost, ease of installation, operating cost, maintenance, energy efficiency, effectiveness, and how pressures in houses are affected. No single approach is best for all situations. The decision to choose a particular ventilating strategy will be based on the climate, the heating system, the size and layout of the house, and the amount of money you are willing to spend. Some systems cost little to install, but have a high operating cost. Other systems are more expensive to install but offer a savings on operating expenses. It is important to select a ventilation strategy carefully because the wrong strategy can cause as many problems as it solves. Possible ventilation-induced troubles will be covered in Chapters 8–10.

With a little forethought, incorporating a controlled ventilation system into the design of a new house is relatively easy because provisions can be made for locating the equipment, ducts, controls, etc. during the design stage. However, if the ventilation system isn't considered early in the design process—if it is an afterthought, considered after the house is under construction—compromises will probably need to be made. In retrofits or remodeling situations, it can be more difficult to physically fit ventilation equipment into a house.

Understanding Ventilation

It may also be difficult to locate ducts in the optimum locations. Therefore, when adding a controlled ventilation system to an existing house, some compromises are often necessary. But, in most cases, a ventilation system can be added to any house to provide the occupants with better comfort and air quality.

Before we discuss controlled ventilation systems, let's quickly review the drawbacks of relying on natural and accidental ventilation.

Drawbacks to uncontrolled air movement

With the wide variety of natural and accidental pressures that cause air to enter and leave a house, it may not seem necessary to have a mechanical system specifically designed for ventilation. This may be true in some houses, but in more and more cases, mechanical ventilation is now being considered a necessity. One of the main reasons for this is that natural and accidental pressures rarely account for the *optimum* amount of air exchange between the indoors and the outdoors to satisfy the needs of occupants: There is either too little air infiltrating and exfiltrating, or too much. Too little air can mean health problems, too much can mean high utility bills.

Another significant problem with uncontrolled air exchange is the fact that you have no control over *where* it is supplied and exhausted. For example, there may be a great deal of air movement through various building cavities such as floor-joist spaces or chases around chimneys. This can lead to high utility bills, but it may not provide much air exchange within the living space itself.

Too little air exchange

Too little air exchange between the indoors and outdoors is a problem in all unventilated houses to some extent, but it is more of a concern in houses that are tightly constructed. And various studies tell us that houses are tighter than they used to be.

When you consider all the various pressures involved in moving air, the natural pressures of wind and stack effect are relatively weak when compared to the pressures generated by some mechanical devices. In order for the pressures typically generated by wind to move significant amounts of air movement into and out of a house, the house must usually be fairly leaky. In general, it will take a strong wind to generate enough pressure to push air through a tight house; this is only a factor when the wind is blowing.

Pressures due to stack effect are often weaker than those of the wind, but they can be continuous for several months at a time. Therefore, stack effect can cause air to be exchanged in a house whenever there is a temperature difference between the indoors and outdoors—perhaps all winter long. But the tighter the house, the less air exchange there will be. Because there can be weeks at a time when there is very little temperature difference, stack effect can be as unreliable as the wind over the course of a year. Overall, natural pressures simply aren't uniform enough in most climates to depend on them to supply air on a regular basis.

Mechanical devices such as clothes dryers and central vacuums can generate higher pressures than the natural pressures, so they can account for more air exchange, even in a tight house. However, they are only operated intermittently, so they aren't consistent at supplying fresh air that the occupants require every day.

Too much air exchange

While there is usually nothing wrong with too much fresh air strictly from a health standpoint, it is certainly possible to have so much air moving into and out of a house that it results in high heating or cooling bills. And too much incoming cold air in the winter can make a house feel drafty and uncomfortably dry. In southern climates, too much incoming humid air in the summer can make a house feel uncomfortably muggy.

When a combustion furnace, water heater, or fireplace is operating, the chimney draft can cause a certain amount of air to move through the house. In fact, there can be much more air than is actually necessary for the health and well-being of occupants. For example, a chimney connected to a fireplace will move so much air out of a house that there is more heat going up the chimney than the fireplace is radiating into the living space. In other words, because they expel so much warm air, fireplaces can waste more heat than they generate. While the occupants may benefit from the extra air (if it passes through the living space), it can mean high heating bills.

When a forced-air heating/air-conditioning system has leaky ducts, there can be some very large pressures (positive and/or negative) generated indoors. These pressures can result in a great deal of infiltration and exfiltration in a house. Again, this is often more air than the occupants actually need. There have been cases where so much outside air infiltrated indoors that the heating/air-conditioning system couldn't keep up with the extra load. Usually, the air infiltrates into the living space from wherever the leaky ducts are located—the basement, crawl space, or attic—and it is often contaminated with a variety of pollutants (*e.g.* insulation, radon, mold, etc.) that are not good to breathe.

Now, let's consider some controlled, mechanical ventilation systems.

Exhaust ventilation systems

Exhaust ventilation is the most common controlled ventilation strategy used in houses. An exhaust ventilation system utilizes a fan to blow air out of a house solely for the purpose of ventilating. If there are no other pressures acting on the house, this results in an equal volume of air passively infiltrating wherever it can to make up for the air being exhausted. During this process, a house can become uniformly depressurized (Figure 7–1). In other words, the air pressure indoors is less than that outdoors.

The air that infiltrates into a depressurized house is often called *make-up air* because it makes up for what was exhausted. In a loosely built house, there will be plenty of random holes through which the make-up air can enter. A tight house will have fewer random holes so the make-up air will have a harder time finding a way in. Because of this, it can be helpful to purposefully place some holes (*through-the-wall vents*) in the exterior walls so that make-up air can enter easily when the fan is running. At first, this may seem illogical—build a tight house, then poke holes in it! But it is often necessary if you want to control where the outdoor air enters the living space. (As it turns out, most houses are loosely built enough that through-the-wall vents aren't necessary. However, in small, tight houses and apartments, they can be desirable.)

Several manufacturers produce through-the-wall vents. Some of these devices are specially designed to have very-low resistance to airflow for their size and

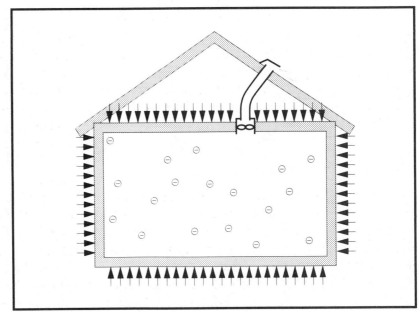

Figure 7–1. *When air is blown out of a house with an exhaust ventilation system, the house becomes depressurized.*

some have built-in compensation for excess wind. In addition, they are relatively small, so only a limited amount of air can enter through any one vent. This, and the way they disperse air into the room, reduces the chance of occupants feeling a draft. The vents are sometimes placed in closets or other out-of-the-way locations. Whenever any device is exhausting air from the house, whether it is a range hood or a fireplace, make-up air will enter the living space by way of the through-the-wall vents automatically. This approach works most effectively in tight houses that don't have natural-draft combustion appliances, and it can be especially suitable for apartments where soundproofing requirements result in relatively tight construction. In very cold climates, uncomfortable drafts can sometimes be felt near through-the-wall inlets, but this is generally not a problem in moderate climates.

There are two types of exhaust ventilation systems: local-exhaust, and central exhaust. Both can cause a house to be depressurized to some extent.

Local-exhaust ventilation

When an exhaust fan removes air from one particular room (*e.g.* a bath fan or range hood), it is referred to as *local* or *spot* ventilation because it is used primarily to remove contaminated air from one area of the house. This is the purpose of local-exhaust ventilation: to intermittently remove high concentrations of contaminants from a specific location quickly. It is always more efficient to remove a pollutant at its source than to allow it to disperse throughout the house and then try to remove it.

While they were virtually unheard of a hundred years ago, local-exhaust ventilation systems of some type are found in nearly all houses today. Bathroom exhaust fans and kitchen range hoods are typical examples. The purpose of these devices is to blow stale, polluted, or moisture-laden air from specific rooms of a house. Indirectly, bath fans and range hoods also improve the air in the rest of the house because air will tend to infiltrate into other rooms from the outdoors and eventually find its way to the room where the exhaust device is operating.

Bath fans typically have a capacity of 50–100 cfm, while range hoods can be as powerful as several hundred cfm. This will contribute to an equal volume of outdoor air entering a house, but only when the fans are operating. When they aren't running, they aren't ventilating. (Of course, there may be other pressures interacting with each other that affect the total air-exchange rate of the house.) Actually, many bath fans and kitchen

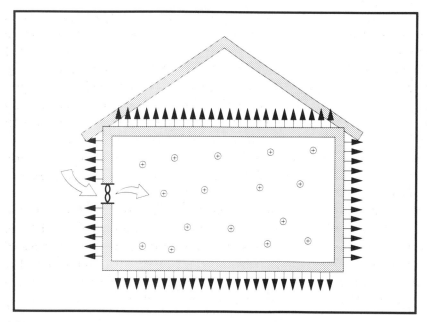

Figure 7–2. When air is blown into a house with a supply ventilation system, the house becomes pressurized.

range hoods don't have motors or bearings that are designed for long-term operation.

Central-exhaust ventilation

Exhaust ventilation can also be used to provide a continuous supply of fresh air to occupants in all rooms of a house and, at the same time, remove low concentrations of pollutants that may be generated throughout the house. This can be done using a high-quality, energy-efficient, quiet exhaust fan designed to run for extended periods.

When an exhaust fan is used to ventilate all rooms in a house, it is usually called a *central-exhaust* fan. A central-exhaust fan is an example of a general ventilation system—one designed for the whole house. Sometimes, central-exhaust systems are called *exhaust-only* systems. This isn't a very accurate term because it implies that air is only being exhausted from the house when in fact air is entering (infiltrating) as well as being exhausted. Central-exhaust fans often have a lower capacity than local-exhaust fans.

Supply ventilation systems

Like the name implies, a supply ventilation system blows air into a building. This can result in uniform pressurization (Figure 7–2). When a house is pressurized, stale indoor air will exfiltrate through random holes in the structure. If any windows are open, the stale air will leave through them; otherwise it will leak out wherever it can. Commercial buildings are often pressurized, and you can sometimes feel a rush of air escape from them as you open the door. Pressurization isn't common in residences, but it certainly has some advantages.

In a tightly built house with a central-supply ventilation system, there may not be enough random holes for air to escape easily. This will mean greater pressurization. To relieve the pressurization, small through-the-wall vents can be placed in the exterior walls to allow air to easily escape from the house. If they are located in rooms that often have high moisture or pollution levels (*e.g.* kitchens or baths), they will be more effective at expelling moisture and pollutants from the house than if they are located in less-contaminated rooms. These through-the-wall vents are the same types of vents that are often used with exhaust ventilation systems. (Some are designed only to be used as inlets.)

Because there is only a single inlet, a filter can be used with a supply ventilation system to clean incoming air. (Most filters can only be used effectively when there is a fan available to actively push or pull air through them directly, so they don't work well with passive inlets such as through-the-wall vents.)

Supply ventilation systems are occasionally called *supply-only* systems, but this is not a very accurate way to describe them, because you can't only supply air to a house—an equal volume of air must also be leaving (exfiltrating) somewhere else. In rare cases, a *local* supply ventilation system can be used to deposit fresh air in one room, but supply ventilation is almost always reserved for *central* ventilation systems.

If a supply fan blows air into a single central location in a house, the resulting draft can be very uncomfortable. Using a system of ducts to direct the incoming fresh air into several rooms will be more draft-free. Coupling a supply ventilation system with a central heating or air-conditioning system has been relatively common for a number of years.

Balanced ventilation systems

With a balanced ventilation system, fresh air is brought indoors with one fan and stale air is exhausted outdoors with a second fan. This is sometimes called *double-flow ventilation*. The fans should have the same capacity so that the amount of air coming in matches the amount of air leaving to let the house experience a neutral pressure. Actually, the *amount* of air flowing through a house with any kind of ventilation system is always balanced (the amount of air entering a house always equals the amount leaving); it is the *pressures* that can be unbalanced with some strategies.

It is possible for a balanced ventilation system to be used to ventilate a single room (local ventilation), but central systems are far more common. A balanced ventilation system can consist of two wall-mounted or duct-mounted fans (Figure 7–3), but there are also a few packaged systems available. In either case, a sys-

Understanding Ventilation

tem of ducts is generally used to distribute air evenly throughout the house (Figure 7–4).

A balanced ventilation system requires only two holes in the structure—an inlet port and an exhaust port. In contrast, exhaust and supply systems sometimes require several openings (through-the-wall vents) in various locations throughout the house. (The single inlet can be fitted with a filter to clean the incoming air.) When a balanced ventilation system is shut off, the house is more tightly sealed than with supply or exhaust strategies that have multiple openings in the exterior walls. This can be desirable when outdoor pollution levels are

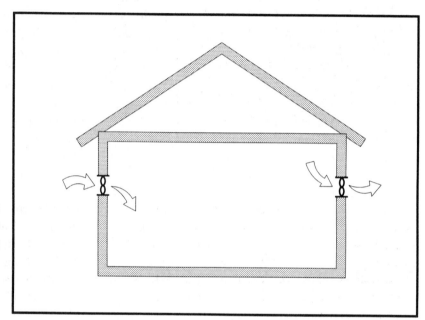

Figure 7–3. A balanced ventilation system, which leaves the house under a neutral pressure, can consist of two simple wall fans.

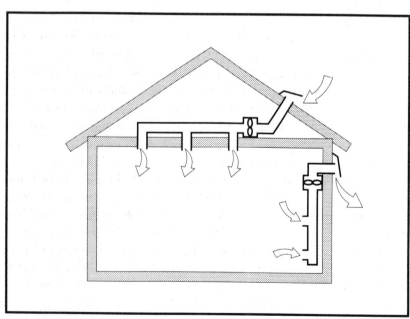

Figure 7–4. A balanced ventilation system can be used in conjunction with a series of ducts to distribute air throughout a house. All ducts should be sealed—especially those outside the conditioned part of the house.

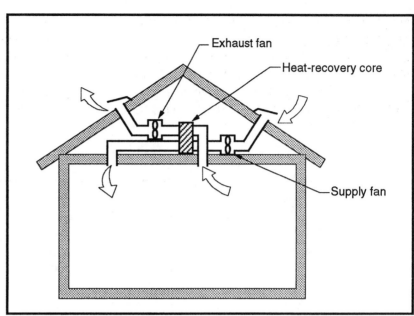

Figure 7–5. Most heat-recovery ventilators (HRVs) are simply balanced ventilation systems with the addition of a heat-recovery core. The use of such an HRV results in a neutral pressure in a house. In this illustration, the HRV is located in an attic, outside the living space, so it is important to seal all the ducts. Many installers prefer to locate the equipment and most of the ducts within the heated/cooled part of the house for easier access.

high and the occupants wish to temporarily turn the ventilation system off to prevent the contaminated outdoor air from entering the house. But balanced ventilation systems are not always installed with dampers to prevent unwanted air movement when they are shut off, so there can still be air moving through the system when it is not in operation.

A balanced ventilation system can be more costly to install and more expensive to operate than either a supply or exhaust system because an extra fan is involved. But a balanced ventilation system can minimize any problems associated with pressurization and depressurization (See Chapters 8–10).

Heat-recovery ventilation

With all of the above approaches to controlled ventilation, cold air will be brought indoors in the winter and warm air will be expelled. If the house is air-conditioned in the summer, cool air will be exhausted and warm humid air brought in. While it is definitely a good idea to ventilate a house, none of these methods offers a very energy-efficient way of doing so. In fact, they are no more energy-efficient than installing a fan in an open window. If it is 5°F outdoors, you will be bringing in air that is quite cold, and exhausting an equal volume of warm air from the living space.

To provide ventilation in a controlled manner, to bring in air that is comfortable, and to do all this in an energy-efficient manner, devices known as heat-recovery ventilators (HRVs) have been developed. Sometimes they are called air-to-air heat exchangers (AAHXs) or energy-recovery ventilators (ERVs). HRVs first became available in cold Scandinavian and Canadian climates, but they are now readily available in the U.S.

Most HRVs are balanced systems, like those described above, but they have the addition of a specially constructed core (Figure 7–5). (One type of HRV is not a balanced ventilator; it is a combination central-exhaust ventilator and water heater.) Actually, most of the balanced ventilation systems currently on the market are HRVs. During operation, both the incoming and the outgoing airstreams pass through the core. While the airstreams don't mix and contaminate each other, the design of the core allows heat to pass from the warmer airstream into the cooler airstream. In the winter, 70% or more of the heat in the warm exhaust air is transferred to the cold incoming air. "Cool" is similarly transferred in the summer in air-conditioning climates. Some HRVs are also designed to transfer water vapor from one airstream to another.

Understanding Ventilation

Some people mistakenly believe that HRVs produce heat. They don't; they simply conserve some of the heat that would otherwise be lost during the process of ventilating. Therefore, they aren't a substitute for a furnace. Their only purpose is to ventilate.

HRVs have a higher initial cost compared to other ventilation systems, but because of the energy savings feature, the operating cost can be somewhat less. They often have high-quality fans and controls that are designed and built to last for years of continuous operation. In very cold climates, an HRV might pay for itself in just a few years, but in moderate or air-conditioned climates, where there is less energy to recover, the payback period can be somewhat longer. While ventilation of some type is increasingly considered mandatory, heat recovery should be viewed as an option, depending on the climate and the cost of heating fuel. In other words, ventilation is needed for health and comfort, but heat recovery may or may not make economic sense. See Chapter 13 for a discussion of the costs associated with ventilation.

Passive ventilation

With most controlled ventilation systems, provision is made for one or more deliberate holes in a house through which air can move, as well as a fan to push the air through those holes. In effect, you are controlling two things: holes and pressure. With passive ventilation, you only control the holes and have no control over the pressure. In other words, passive ventilation is ventilation "on purpose" but without a fan. While this is an interesting idea, it is a less-than-perfect option. To operate effectively, a passive ventilation system requires a tight house with through-the-wall (or through-the-roof) vents strategically located. Then the natural and accidental pressures will push air into and out of the house

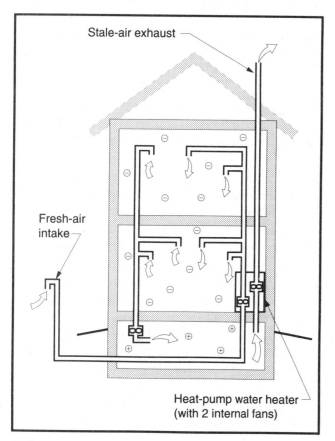

Figure 7–6. *In this unique approach, a builder purposely depressurized the living space (to prevent hidden moisture problems) and pressurized the crawl space (to keep out radon and other soil gases). This type of ventilation system needs careful thought and planning.*

through the vents. The vents must be sized carefully to prevent too much air from passing through them. Because passive ventilation depends on natural and accidental pressures—pressures that sometimes depressurize a house and sometimes pressurize—a house that is passively ventilated may not be consistently either under a positive or a negative pressure.

The big drawback to a passive ventilation system is the fact that in most situations, you can't rely on natural or accidental pressures to regularly push air through the openings. On one day, they may result in a great deal of negative pressure in one part of a house and positive pressure in another. On another day, there may be little or no pressure. At another time of the year, the pressures could be reversed.

From a technical standpoint, it is possible to use electronic monitors to sense the pressures and the airflows, then open and close the appropriate vents by the amount necessary to allow the proper quantity of air to flow through them. The vents would be controlled by the airflow, the amount of pressure, and whether the pressure is positive or negative. This is all possible, but it would be rather complicated, expensive, and most likely not very practical. Besides, it would only work when there happened to be some kind of pressure acting on the house, and the wind is usually too erratic. Still, the idea of passive ventilation is intriguing to many people. In reality, tightly constructed, small two-story houses (and apartments) can be passively ventilated effectively in cold climates, primarily due to stack effect. This strategy is often considered in places where electricity is not available, such as vacation cabins.

Partial pressurization & partial depressurization

It is possible to purposely manipulate the pressures in a house so that one part is pressurized and another depressurized. As was stated in Chapter 6, forced-air heating/air-conditioning systems often accidentally cause significant pressure changes in houses. If the wrong parts of a house are pressurized or depressurized, undesirable results can occur. But if you understand all the principles involved, you can carefully manipulate the pressures in a house for specific reasons. Though not commonly done, a partial-pressurization/partial-depressurization approach can have some definite advantages. For example, if you are in a cold climate, you may like the idea of depressurization because it can prevent hidden moisture problems (see Chapter 8), but pressurization is also appealing because it keeps out radon (see Chapter 9).

With careful construction techniques, one builder has demonstrated this approach by depressurizing the main living space of the house and pressurizing a well-sealed crawl space.[Nuess] He successfully used a ventilation system to induce opposing pressures to solve different problems in different parts of the house (Figure 7–6). Using a ventilation system to create positive and negative pressures in different parts of a house is only effective in a very tight house. This is because most ventilation systems aren't powerful enough to create significant pressures in houses of average tightness. However, forced-air heating/cooling systems are significantly more powerful and they can be designed and adjusted in this way. Thus, combining a ventilation system with a forced-air heating/cooling system can make a partial pressurization/partial depressurization strategy a more viable option.

Interactions

It is important to keep in mind that a house is a dynamic, ever-changing system. Natural, accidental, and controlled pressures are constantly fluctuating and they never act independently of one another—they continually interact. For example, on a windy day in the winter, the pressures generated by both stack effect and wind will combine as shown in Figure 7–7. The length of the arrows corresponds to the strength of the pressures. When arrows resulting from wind and stack effect point in the same direction, their lengths are added together to represent a stronger combined pressure. When the arrows point in opposite directions, the length of the shorter (weaker) arrow is subtracted from the length of the longer (stronger) arrow to represent a weaker combined pressure. In the example in Figure 7–7, most of the infiltration will be low on one side of the house, and most of the exfiltration will be high on the opposite side.

Understanding Ventilation

Of course, when it is cold outdoors, the furnace will likely be running, so there may be additional interacting pressures due to duct leakage and combustion gases going up the chimney.

Figure 7–8 shows how stack effect in the winter can be combined with the overall depressurization caused by a central vacuum that is exhausting air from the living space. Notice that the house still has a neutral-pressure plane, but it is no longer in the center of the wall height. When engineers talk about a house being depressurized in the winter, they sometimes say the neutral-pressure plane is higher. When a house experiences a positive pressure, the neutral-pressure plane is lower.

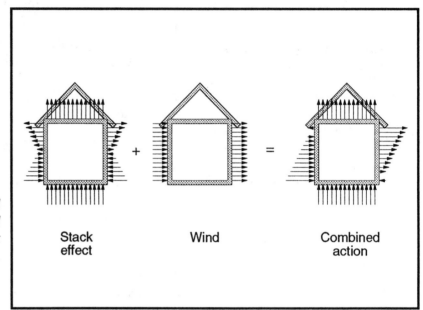

Figure 7–7. *The individual pressures affecting a house will combine into a more-complex pattern of pressures.*

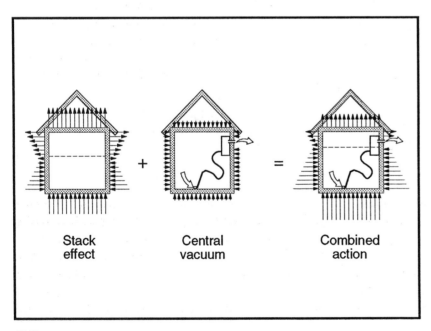

Figure 7–8. *Any combination of pressures will work together to create a specific pattern of inward and outward pressures.*

Chapter 7 Controlled ventilation

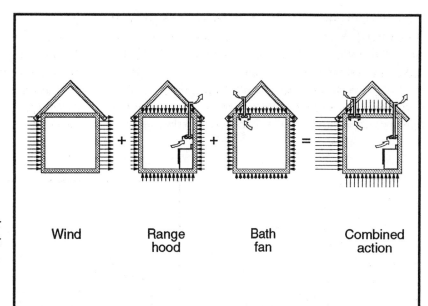

Figure 7–9. Quite a few different pressures can act on a house at any given time.

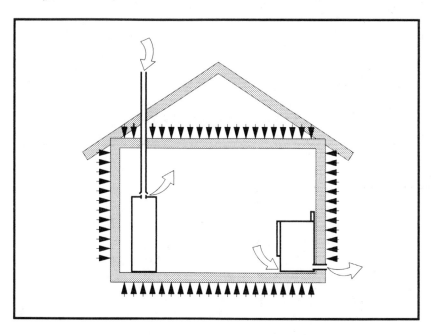

Figure 7–10. In this example, a clothes dryer depressurizes a house enough to cause make-up air to enter by way of the chimney serving a gas water heater. This can occur when a house (or the room containing both appliances) is too tight for enough make-up air to enter through the random gaps and cracks.

The effect of three different factors (wind, kitchen range hood, and bathroom exhaust fan) are shown in Figure 7–9.

If an exhaust device is running at the same time a draft is induced in a chimney, those two pressures can compete with each other. Suppose the combustion by-products from a gas water heater are flowing up a chimney in an unimpeded manner. This means that an equal amount of air must be entering the house from somewhere else. If a nearby clothes dryer is turned on, it too will blow air out of the house. If the house (or the room they are located in) is very tight, there may not be enough random holes to let in enough make-up air. Figure 7–10 shows the result if the pressure generated by the clothes

Understanding Ventilation

dryer is more powerful than the pressure of the draft in the chimney. Whichever pressure is stronger will determine which direction the air actually moves in the chimney. When the pressure generated by the clothes dryer is stronger than the draft pressure in the chimney, the combustion gases can't escape from the house as long as the dryer is running. This is called backdrafting be-

cause the direction of the flow in the chimney is backwards. Of course, if the house is leaky enough, both the chimney and the dryer can operate simultaneously (Figure 7–11) and backdrafting won't be a problem. (Some combustion devices have sealed combustion chambers that don't require a natural-draft chimney. They are immune from backdrafting.)

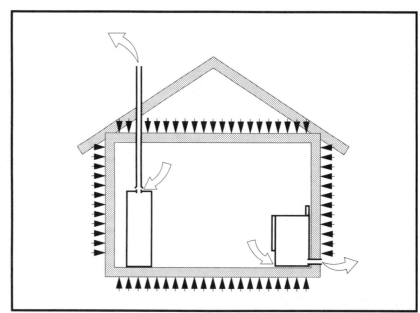

Figure 7–11. The only difference between this example and the one in Figure 7–12 is the tightness of the house. Here, the house is loose enough to allow make-up air to enter through the random gaps and cracks; therefore, the chimney functions safely.

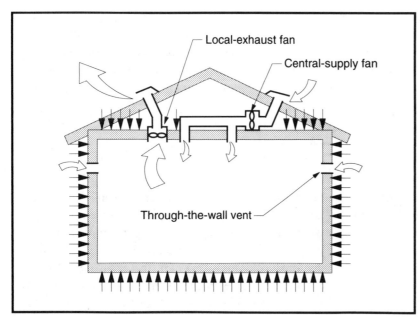

Figure 7–12. Mechanical ventilation equipment can also interact. If a local-exhaust fan and a central-supply fan are operating at the same time, the more-powerful fan will determine which direction air will flow through the through-the-wall vents.

Chapter 7 Controlled ventilation

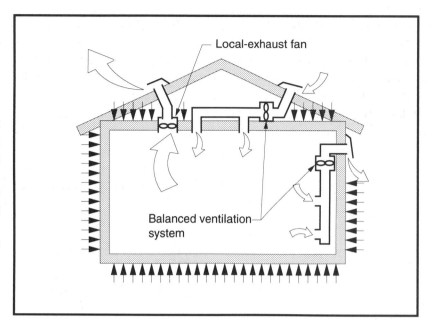

Figure 7–13. When a local-exhaust fan is used at the same time as a balanced ventilation system, a house will become depressurized. There are enough random holes in most houses for the local-exhaust fan's make-up air to enter through, so the balanced system's airflows remain in balance.

It isn't at all unusual for the pressures in a chimney to fluctuate. Even if the result isn't backdrafting, the draft may be too weak to expel the combustion by-products from the building fast enough. This means that some of the combustion gases will spill into the air inside the house. This is called spillage. In theory, backdrafting and spillage are addressed by current building codes, but in practice, combustion by-products (principally carbon monoxide) often cause serious health-related problems in houses. They will be covered in greater depth in Chapter 10.

If you factor in the wind, stack effect, clothes dryers, central vacuums, range hoods, bath fans, leaky ducts, closed doors, chimneys, etc., houses can be quite complicated when it comes to analyzing how air pressures affect infiltration and exfiltration. When mechanical ventilation and tight construction aren't a specific part of a house's design, you may have poor indoor air quality, high heating/cooling bills, uncomfortable drafts, or simply not enough air to satisfy the needs of the occupants.

When a house contains more than one type of mechanical ventilation system, they too can interact. Consider a house with a central-supply ventilation system and a powerful local-exhaust fan. If the central-supply fan is turned off and the local-exhaust fan is on, make-up air will enter passively through both the through-the-wall vents and the supply fan's inactive duct. If both fans are running, the supply fan will try to force air out of the through-the-wall vents at the same time the exhaust fan is trying to pull air in through them. If the local-exhaust fan has more capacity than the supply fan, the net result will be air entering the through-the-vents (Figure 7–12).

A house with a balanced ventilation system generally has no through-the-wall vents, so when a local-exhaust fan blows air out, make-up air will either enter through the balanced ventilation system or through the random holes in the structure. In theory, when a local-exhaust fan is running in a perfectly airtight house, the balanced system will become slightly out of balance and more air will enter the supply duct than will leave through the exhaust duct and the house will become depressurized (Figure 7–13). However, in the real world, most houses (even very tight ones) have enough random holes in the structure for the exhaust fan's make-up air to enter through, so the central system tends to remain in balance.

It is important to keep in mind that these examples show what the pressures acting on a house look like. If there are no holes where a strong pressure is located, there will be no air moving into or out of that part of the

93

Understanding Ventilation

house—you need both an air-pressure difference and a hole to have an exchange of air.

Another way of looking at interactions

The airflows resulting from wind and stack effect will interact with the airflows of a balanced ventilation system differently than they will with the airflows of a ventilation system that pressurizes or depressurizes a house. To understand this difference, we need some examples. Let's consider two houses that are identical except that one has a balanced ventilation system and the other has a central-exhaust system.

The first house only has four holes in it. (This isn't exactly realistic—most houses have many small random holes—but for explanatory purposes four holes is easy to understand.) On average, let's say air infiltrates through one hole at a rate of 50 cfm due to wind and stack effect, and air exfiltrates through another hole at a rate of 50 cfm, also due to wind and stack effect. (The average natural infiltration rate can be estimated by using a diagnostic device called a blower door that will be discussed in Chapter 11.) A balanced ventilator (it doesn't matter if it has heat recovery or not) is connected to the other two holes. When the ventilator is off, the air exchange rate is 50 cfm. Because a balanced ventilator neither pressurizes nor depressurizes a house, it will not interact with the natural pressures. Thus, when the balanced ventilator is running, the total air-exchange rate will be the sum of the natural rate plus the ventilator's rate. For example, if the ventilator has a capacity of 30 cfm, the total air-exchange rate will be 80 cfm (50 + 30). Thus, if a house has a balanced ventilator, it is easy to determine the total rate—you just add the two numbers together. But if a house has a ventilation system that is not balanced—a system that pressurizes or depressurizes the house—you can't simply add the two numbers together. The next two examples will explain why.

The second house only has three holes in it. (Again, not very realistic, but easier to understand.) Again, let's say on average, air infiltrates through one hole at a rate of 50 cfm, and air exfiltrates through another hole at a rate of 50 cfm. A central-exhaust fan is connected to the third hole. (The same principles apply if a central-supply fan is used.) When the exhaust fan is off, the air-exchange rate is 50 cfm. If the exhaust fan has a 10 cfm capacity, it will exhaust 10 cfm from the house, so 10 cfm of make-up air is needed. With only two holes for the make-up air to enter, 5 cfm will enter through each hole. One of those holes had 50 cfm already infiltrating through it, so it now has 55 cfm (50 + 5) passing through it. The other hole had 50 cfm exfiltrating through it, but now 5 cfm is trying to enter, so the net result is 45 cfm (50 − 5) exfiltrating through that hole. When we look at the whole house, we see that 55 cfm is leaving (10 cfm through the fan and 45 cfm through one of the holes), and 55 cfm is entering through the third hole. So, we started out with a house having 50 cfm of natural ventilation, we added a 10 cfm exhaust fan, but the total rate only increased by 5 cfm, not 10 cfm.

Now, let's use a more powerful central-exhaust fan in the same house, one that will blow 150 cfm outdoors. (Again, the same principles apply if a central-supply fan is used.) This house still has 50 cfm infiltrating through one hole and 50 cfm exfiltrating through another hole when the ventilator is off, so the natural air-exchange rate is 50 cfm. When the ventilator is running, it will exhaust 150 cfm from the house, so 150 cfm of make-up air is needed. With only two holes for make-up air to pass through, 75 cfm will try to enter through each one. The hole that has 50 cfm infiltrating will now have another 75 cfm coming through it for a total of 125 cfm (50 + 75). The other hole had 50 cfm leaving, but now 75 cfm is trying to enter. The net result is 25 cfm entering (75 − 50). So, in this example, 150 cfm is leaving the house through the exhaust fan and 150 cfm is entering through the two holes (125 + 25). We started out with 50 cfm of natural ventilation, we added a 150 cfm exhaust fan, and the combined rate is 150 cfm.

In the real world, air would infiltrate and exfiltrate through quite a few random holes, gaps, and cracks, not just two holes as in the above examples, but the results are the same no matter how many holes there are. Actually, the numbers in these examples are slightly inaccurate because of the specific laws of physics that apply, but without getting into complex calculations, they give a reasonably accurate picture of what occurs

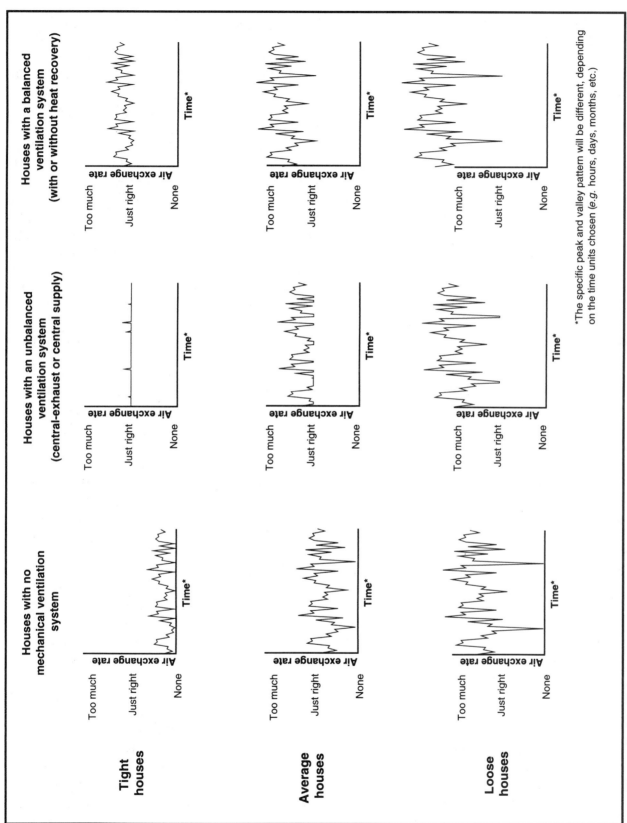

Figure 7-14. Unbalanced ventilation systems interact with the natural ventilation rate differently than balanced systems. All systems can reduce the periods of underventilation, but the only way to prevent overventilation is to tighten the house.

*The specific peak and valley pattern will be different, depending on the time units chosen (e.g. hours, days, months, etc.)

when ventilation systems interact. While these examples may seem confusing, they can be summed up in the three following rules:

> In a house with a balanced ventilator (it may or may not have heat recovery), the total ventilation rate will be the natural rate (due to wind and stack effect) plus the ventilator's rate.
>
> In a house with an unbalanced ventilator (either an exhaust fan or a supply fan), the total ventilation rate will be equal to the fan's capacity—but only when the natural rate is less than half the mechanical rate.
>
> In a house with an unbalanced ventilator (either an exhaust fan or a supply fan), the total ventilation rate will be the natural rate plus $1/2$ the fan's rate—when the natural rate is more than half the mechanical rate.

While these rules can be stated rather easily, it is often quite difficult to determine the precise ventilation rate in a house. This is because the natural pressures are constantly changing, sporadic accidental pressures occur, and the house becomes looser whenever a window or door is opened. In addition, when a central-exhaust or a central-supply ventilation system utilizes through-the-wall vents, the vents (being holes) make the house looser.

What does it all mean?

So what does all of this mean? What is the net result of all of these interactions? Let's look at a series of graphs. The three houses on the left side of Figure 7–14 show how there are periods of both too much air exchange and too little exchange in most houses. Tight houses have more episodes of no air movement and loose houses have more periods of excessive air movement. Compare this to the three houses on the right side in Figure 7–14, showing the effect of adding a balanced ventilation system. According to the first of the three rules mentioned above, you can determine the total air exchange rate in a house with a balanced ventilation system by adding the ventilator's rate to the natural rate. So, the net result with balanced ventilation is a minimum amount of air exchange at all times (the rate of the ventilator), with quite a few periods of higher flow. There are more periods of excess ventilation in a loose house, compared to a tight house.

In the center three houses in Figure 7–14, we see the net effect of using an unbalanced ventilator in a house. (It can be either an exhaust or a supply system.) When the natural rate is low (less than half the ventilator's capacity), the total rate is equal to the ventilator's rate. When the natural rate is high, the total rate is the natural rate plus $1/2$ the fan's rate. The fan provides a minimum rate continuously, and there are still periods of excess ventilation (especially in average and loose houses), but the overventilation isn't as excessive as it is in a house with a balanced ventilator.

As can be seen in Figure 7–14, the total amount of air exchange in a house with a balanced ventilator is higher than the total amount of air exchange in a similar house with a central-exhaust or a central-supply ventilator of the same capacity. As a result, it is possible to install a smaller-capacity balanced system and have a similar net result.

The bottom line

To summarize, natural and accidental pressures rarely supply the occupants of a house with the correct amount of fresh air. The quantity of air is either too small to be of much benefit, or so large that heating or cooling bills are exorbitant. In order for the amount of air infiltrating and exfiltrating to be "just right" it is often more a matter of luck or coincidence than anything else. And, even when the *amount* of air is correct, it may not be delivered to the part of the house where people are located. This second point is very important because natural ventilation often places a significant burden on the heating/cooling system and doesn't always benefit the occupants. For example, ventilating the basement does little to provide fresh air to people upstairs.

Even though the interactions can seem complex, the main point is this: A controlled ventilation system can easily eliminate the periods when there is too little air exchange in a house, but it can't reduce the periods of too much air. If you want to reduce periods of excessive air exchange, you must tighten the house. Because of the way the pressures interact, if the capacities are equal, a central-exhaust or central-supply system will result in less overventilation than a balanced system.

Chapter 7 Controlled ventilation

The best situation is to have any type of mechanical ventilation system in a tight house. That way you can have the correct amount of air exchange at all times.

The mechanical system also has the significant advantage of supplying air to and exhausting air from the right parts of the house.

Part 3

Possible adverse effects of ventilation

Chapter 8 Ventilation-induced moisture problems
Chapter 9 Pollutant entry due to depressurization
Chapter 10 Backdrafting and spillage

Chapter 8

Ventilation-induced moisture problems

To some degree, moisture problems exist in houses all over North America.[Lstiburek and Carmody] Excess moisture inside a house can cause mold growth or decay, which can damage building components and furnishings, help proliferate allergenic dust mites, increase outgassing from formaldehyde and other VOCs, and can be downright uncomfortable. The moisture levels in a house depend on a variety of different factors such as lifestyle (frequent showers), number of occupants (moisture in exhaled breath), plumbing or roof leaks, ground moisture entering basements, etc. Because excess humidity is a frequent problem in underventilated houses, one of the chief reasons to install a ventilation system is to exhaust excess moisture to the outdoors. However, if a ventilation system results in a house becoming pressurized or depressurized, there is the possibility of a *hidden* moisture problem developing. The likelihood of this happening depends on the type of ventilation strategy (supply, exhaust, or balanced), the degree of pressurization or depressurization, the tightness of the house, the materials of which the house is constructed, the indoor and outdoor temperature, and the indoor and outdoor relative humidity.

Ventilation air passes into and out of a house through two kinds of holes: deliberate holes that are specifically designed for the passage of air, and random holes that aren't specifically designed for such air movement. Hidden moisture problems can develop when moisture-laden air (all air contains some moisture) passes through the random holes in the structure. If this happens when temperature conditions inside wall cavities are just right, there can be hidden pockets of condensation. This is a very serious concern because hidden condensation can lead to hidden mold or mildew growth or rot.

As the following discussion points out, positive and negative pressures can result in hidden moisture problems in some houses. However, the pressures and

Understanding Ventilation

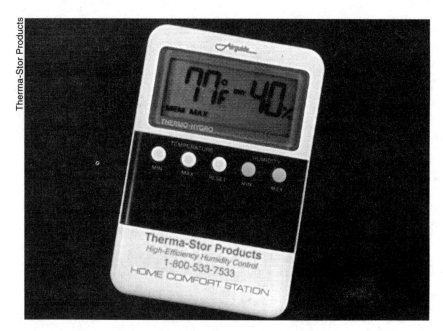

Figure 8–1. Digital relative-humidity meters offer an easy way to measure indoor humidities.

airflows due to residential ventilation systems are almost always considerably smaller in magnitude than the accidental pressures and airflows caused by forced-air heating/cooling systems due to leaky ducts and closed doors. Therefore, hidden moisture problems are less likely to be caused by ventilation-induced pressures than forced-air heating/cooling system induced pressures. Overall, pressure-induced moisture problems are more of a concern in air-conditioning climates than in heating climates, but they can occur anywhere.

Even if hidden-moisture problems aren't a concern, ventilating a house will always have an effect on the indoor humidity. This is because the act of ventilating mixes outdoor air (which contains a certain amount of humidity) with indoor air (which contains a different amount of humidity). The result is an air mixture in the living space that is either drier or more humid than it started out.

Relative humidity & temperature

We learned in Chapter 5 that warm air is less dense and more buoyant than cold air. In order to understand how and when moisture problems develop, it is important to understand another important trait of air:

Warm air can hold more moisture than cold air. This leads us to the concept of *relative humidity* (RH).

For all practical purposes, air always contains some moisture, but the precise amount of moisture it can hold depends on the temperature of the air. For example, one cubic foot of air at 70°F can contain no more than about 0.00118 pounds (about $1/10$ teaspoon) of moisture, but one cubic foot of air at 0°F can only contain a maximum of about 0.00006 pounds (about $1/180$ teaspoon) of moisture. (The exact amount of moisture that air can hold is also dependent on atmospheric pressure, so these maximums will be different at sea level compared to high in the mountains.)

When air contains as much moisture as it can hold, it is said to be saturated. If a cubic foot of 70°F air contains 0.00059 pounds of moisture, it is half-saturated. If 0°F air contains 0.00003 pounds of moisture, it is half-saturated. Another way of saying air is saturated is to say it is at 100% relative humidity (RH). If air is at 50% RH, it is half saturated. While these tiny amounts of water per cubic foot of air don't seem very large, they can add up. For example, the air in a 2,000-sq.-ft. house at 70°F and 50% RH will contain a total of about $9 1/2$ pounds of water (a little over a gallon).

Because cold air can't hold as much moisture as warm air, air at 0°F at 50% RH contains much less moisture than 70°F air at 50% RH. Even though both batches of air have the same RH, they contain different amounts of moisture. If we talk about the actual amount of moisture in the air (so many pounds of moisture per cubic foot of air, for instance) we are referring to the *absolute* humidity of the air. *Relative* humidity refers to the amount of moisture in air compared to the maximum amount of moisture air at that temperature can contain. As it turns out, in most cases, it is more useful to know the relative humidity of air than the absolute humidity.

If you have a batch of air at a certain RH and you warm it up or cool it down, the RH changes. The number of water molecules in that particular batch of air (the absolute humidity) remains the same, but because the maximum amount of moisture it can contain varies with temperature, the RH changes. As an example, lets consider a cubic foot of air at 70°F at 50% RH. It will contain 0.00059 pounds of moisture. Now, let's cool this air down to 60°F and measure the RH. The 60°F batch of air still contains 0.00059 pounds of moisture, but the RH is now 70%. If we cool the air even further, down to about 51°F, the RH will be 100%—the air will be saturated. If we warm the original batch of 70°F air, with its 50% RH, up to 80°F, we will be able to measure the RH at 35%. Because warm air has the capacity to hold more moisture, it is now only 35% saturated.

When air is saturated (100% RH) it is said to be at its dew point. If air is cooled below its dew point, condensation or dew (or sometimes fog) will form. In our example, the dew-point temperature for 70°F air at 50% RH is 51°F, because if you cool the air to 51°F it will be saturated. Cooling it further will result in condensation.

All of this is important to understand because the temperature is never uniform in every part of the house. Consider a glass of iced tea with water condensing on the outside. The air in most of the room might be 70°F @ 50% RH, but since condensation has formed on the glass, the temperature of the air close to the glass must be below the dew point (below 51°F). What is happening is that the ice has cooled not only the tea and the glass, but also the air surrounding the glass below the dew point, and moisture from the air has condensed on the cold surface. This is why moisture condenses on cold windows in the winter—the temperature of the microclimate next to the glass is below the dew point of the air in the room. In the winter, closets (especially when they are located adjacent to exterior walls) are often cooler than other parts of a house, so they often have higher relative humidities in them. This makes them more susceptible to mold growth. The temperature and RH also varies inside building cavities such as roofs, floors, and walls.

Measuring relative humidity

Several types of relatively inexpensive instruments (hygrometers) are available for measuring RH. Dial-type relative-humidity gauges often utilize a spiral-wound material that senses humidity fluctuations. They are called "strain hygrometers" and are often available in department and hardware stores. **Airguide**, **Radio Shack**, and **Therma-Stor Products** each offer digital meters (Figure 8–1). The big drawback to any of these inexpensive instruments is the fact that they aren't always very accurate. For reasonable accuracy they should be within 5% but they're sometimes off by as much as 30%.

Checking the accuracy of an RH meter

To check the accuracy of a strain hygrometer or a digital humidity meter, use the following procedure. 1) Mix one-half cup of salt in one cup of water (it isn't necessary for all of the salt to dissolve). 2) Place the cup with the saltwater mixture in a clear plastic bag. 3) Place the strain hygrometer or digital meter in the bag next to the cup. 4) Seal the bag. 5) After a couple of hours read the meter. The water and salt should create an atmosphere inside the bag of 75% RH. If the meter is within 5% of being correct, it is probably accurate enough. If a strain hygrometer is not accurate, it may be possible to adjust it, but digital meters usually can't be adjusted. However, even if an inaccurate meter can't be adjusted, it can still be used. Simply determine how much it varies from 75% while in the plastic bag, and compensate accordingly.

Predicting RH at different temperatures

In order to predict whether or not moisture will condense inside a wall, you first need to determine the temperatures inside the wall cavity. Then you can use a graph called a *psychometric chart* to find out if those temperatures are conducive to condensation.

One way of determining internal wall temperatures involves drawing a cross-section of the wall to scale—but using an R-value scale instead of a conventional scale. In other words, if a component of a wall has a thickness of 2" and an R-value of 5, then draw it so it is 5 units wide. Similarly, if 5½" of fiberglass insulation has an R-value of 19, then draw it 19 units wide. Figure 8–2 shows a wall drawn to an R-value scale that consists of ½" drywall (R = 0.45), 3½" of fiberglass insulation (R = 11), 1" of extruded polystyrene foam sheathing (R = 5), and ¾" wood siding (R = 1.05). (The air films on the inner surface of the drywall and on the outer surface of the siding also have small R-values, but they are ignored here to simplify things.)

After you have drawn a wall using an R-value scale, you can easily determine what the temperature will be at any point inside the wall. To do so, you need to know the temperatures on each side of the wall, and the difference between those two temperatures. In the example in Figure 8–2, it is 70°F indoors and 10°F outdoors, so there is a 60° temperature difference (70 – 10). Temperature difference is usually abbreviated as TD or ΔT (delta T). When you know the TD, you can divide the drawing of the wall into the same number of units as the TD. In Figure 8–2, a ruler has been laid on the drawing of the wall to divide the wall into 60 units. Each increment along the ruler represents a temperature change of one degree. In this example, the surface of the expanded polystyrene insulation that is next to the fiberglass insulation is 40°F colder than the indoor temperature, so the temperature at that point in the wall is 30°F (70 – 40).

This same technique can be used to determine the temperatures inside any wall assembly at any indoor/outdoor temperature conditions. You simply draw the wall using an R-value scale, then divide the wall into the same number of units as the TD. Once you have the temperatures of the surfaces inside the wall, you can determine if those temperatures are conducive to condensation by using a psychometric chart.

A psychometric chart is a handy graph used by engineers to predict what the RH of air will be at different temperatures. The psychometric chart in Figure 8–

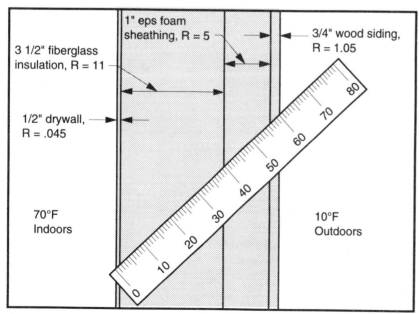

Figure 8–2. This wall section has been drawn to an R-value scale (see text for an explanation). By dividing the width of the drawing into the same number of increments as the indoor/outdoor temperature difference (in this case 60), each increment will represent one degree of temperature change at that point inside the wall. For example, in this wall, at the indoor and outdoor temperatures shown, the side of the drywall facing the insulation will be about 2°F colder than the interior of the house, or about 68°F.

Chapter 8 Ventilation-induced moisture problems

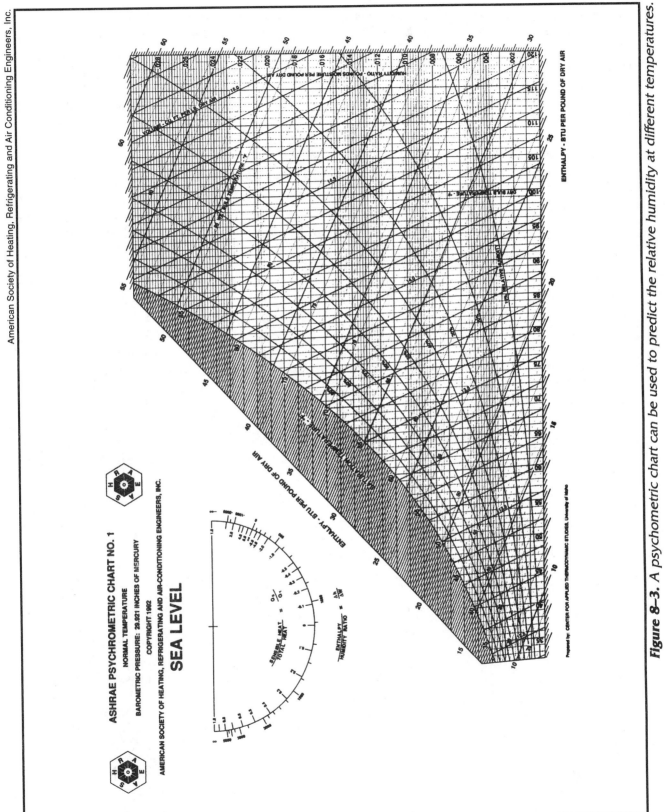

Figure 8–3. *A psychometric chart can be used to predict the relative humidity at different temperatures.*

3 can be used for many different purposes, most of which are beyond the scope of this book. A temperature scale runs across the bottom of the chart and the curved lines represent different relative humidities. The scale on the right side represents absolute humidity. The absolute humidity is often measured in pounds of moisture per pound of dry air. To show how to predict the RH of air at different temperatures, we will use a simplified version of this same chart.

To begin, let's use as, an example, 70°F air @ 50% RH. First, locate 70°F on the scale along the bottom of the chart in Figure 8–4 (point A). Next, locate the point where the vertical 70°F line intersects the curved 50% RH line (point B). All points along the horizontal line passing through point B will have the same absolute humidity. To determine what the RH will be if the air is warmed up, move from point B, directly to the right until you intersect the 80°F line (point C). This point is roughly between the curved 30% and 40% lines—so the RH of the air will be about 35% if it is warmed to 80°F. To determine what the RH will be if the air is cooled, move horizontally along the absolute humidity line back to the left. Point D is on this line at the intersection of the 60°F vertical line. By looking at the curved lines, we can see that the RH of the air will be about 70%.

Here are two more examples. If some 75°F air @ 30% RH is cooled to 55°F, the RH will be about 61%. If a batch of air at 60°F @ 80% RH is warmed up to 90°F, the RH will be 29%.

What if air at 72°F @ 60% RH is cooled down to 40°F? First, look at Figure 8–5, and find point A where the vertical 72° line crosses the curved 60% line. Now, if you move left to the vertical 40° line, you will see that you are off the chart (point B). The curved 100% RH line at the left of the chart represents the dew point, where air is saturated. In most situations, air can't have an RH that is more than 100%. So, if air is cooled beyond the dew point (seemingly off the chart), some of the moisture will condense out of the air onto cool surfaces. To see what actually happens, move horizontally from point A (72°F @ 60% RH) to the left until you run into the 100% line (point C), then move down along the curved line until you reach the 40°F line (point D). The air will still be at 100% RH, but you are now at a different horizontal absolute humidity line. The difference between this horizontal line and the one we started on (X) represents how much moisture condensed out of the air onto cool surfaces. Now the air contains less absolute humidity. If we decide to warm up the air from 40°F to 50°F, we must stay on the new absolute humidity line to determine that the new RH will be 65% (point

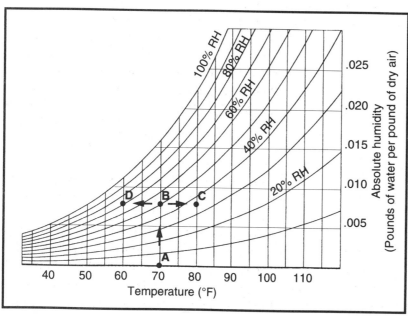

Figure 8–4. In this simplified version of a psychometric chart, it can be shown how the relative humidity changes with temperature. For example, 70°F air at 50% RH will have a 35% RH if warmed to 80°F and 70% RH if cooled to 60°F.

Chapter 8 Ventilation-induced moisture problems

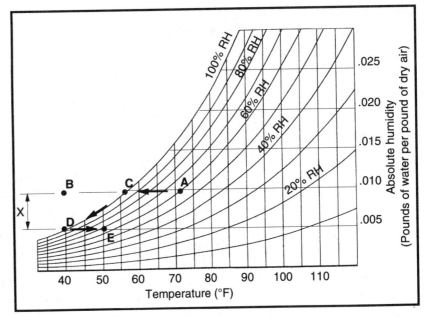

Figure 8–5. *If air is cooled beyond the "dew point" (100% RH), a certain amount of water will condense onto cold surfaces.*

E). In reality, the RH probably won't remain at 65% because some of the moisture that condensed on the cold surface may start to evaporate and be reabsorbed back into the air.

Hidden moisture problems

When moisture problems are visible, the occupants often take remedial action quickly, as in the case of a roof leak. Unfortunately, hidden moisture problems can occur that are not readily known to the occupants. Sometimes, moisture migrates into wall, roof, or floor cavities and results in mold or mildew growth, or even rot, inside the structure. On occasion, moisture exfiltrating from the living space will cause paint to peel off the siding. These problems can actually be caused by a ventilation system if the wrong system is chosen for a particular climate.

When a house becomes depressurized or pressurized for any reason (natural, accidental, or controlled ventilation), moisture-laden air (remember, all air contains some moisture) will be pushed or pulled through the random holes in the house (all houses have some random holes). If the temperature of the air changes as it travels through a building cavity, its RH will also change. For example, if the temperature goes up, the RH will go down and moisture problems will be unlikely. But if the temperature goes down, the RH will go up. If the temperature drops enough, the RH will reach the dew point (100% RH), and condensation could occur hidden within the structure of the house, not a good situation.

Diffusion can transport some moisture into building cavities, but it usually isn't a significant factor, especially if a house has a correctly installed vapor retarder. Most of the moisture that gets into wall cavities (or floor or roof cavities) gets pushed or pulled there through random holes by air-pressure differences.

In theory, if a house is constructed in a perfectly airtight manner (no holes), it won't matter how much pressure is applied to the building; there won't be any air movement, so there wouldn't be any hidden moisture problems. (There still could be problems due to something such as a hidden plumbing or roof leak.) However, a perfectly airtight house is almost impossible to build.

A loosely built house has plenty of random holes for air to pass through when the house is pressurized or depressurized. As a result, a small amount of air will pass through each of many holes. This could lead to many small moisture problems. In a tight house (fewer

Understanding Ventilation

random holes), more air may be passing through each hole, so there can be more-concentrated moisture condensation near each leakage point. If you control the air pressures, neither a tight house nor a loose house will have any hidden condensation problems.

While hidden pressure-related moisture problems won't occur in a theoretically airtight house, in reality, problems can occur in any house—but only when it is cold enough (below the dew point) inside building cavities for condensation to occur. Because the indoor temperature is fairly constant throughout the year (within a few degrees anyway), the outdoor temperature turns out to be the important determining factor in whether or not ventilation-system pressures will cause hidden moisture problems. As it turns out, hidden pressure-related moisture problems tend to occur in cold climates when a house with humid indoor air is pressurized, and in hot, humid climates when an air-conditioned house is depressurized.

Hidden winter moisture problems

In a heated house in the winter, indoor air is warmer than outdoor air. In many cases, the indoor RH is low enough (say less than 40%) so that there will be no serious moisture-related problems within the living space. However, if a house, or any part of a house, gets pressurized for whatever reason, the indoor air will be pushed through the structure into building cavities, and that is where problems can occur. Although exactly the same thing can happen inside a roof or floor system, we will consider a wall cavity. Let's say that the inside of the house is kept at 69°F and the outdoor temperature is 0°F (Figure 8–6). If you measure the temperature inside the wall cavity, it will be warm close to the interior of the house (67°F just behind the drywall), and cold near the outdoors (9°F at the sheathing). As air passes through this wall, it will become cooler and cooler as it gets closer to the outdoors. By the time it gets to

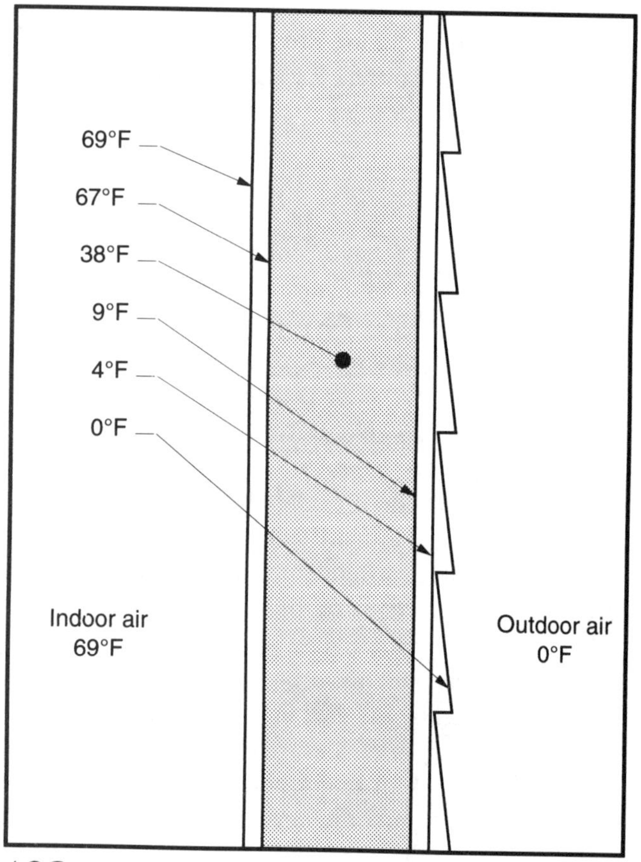

Figure 8–6. In the winter, the temperature inside a wall (or ceiling, or floor) cavity drops as you near the cold outdoors. If you know the temperatures inside a building cavity, you can predict if and when hidden condensation will occur.

Chapter 8 Ventilation-induced moisture problems

the sheathing, it will have reached the dew point and moisture will condense on the side of the cold sheathing that faces the inside of the wall cavity.

Whether or not condensation will actually form inside a wall cavity depends on how much moisture is in the air to begin with (its RH), and the actual temperature of the sheathing. This is a bigger problem in Canada than in the U.S. because Canadian winters are colder; therefore Canadian wall cavities are colder. Still, it is a potential problem in the northern U.S.

One way to minimize the chance of hidden condensation in the winter is to use well-sealed insulating sheathing, such as a foamboard. When this is done, the side of the sheathing facing the inside of the wall will be warmer than if non-insulated sheathing is used, because the sheathing's insulating ability keeps itself warm. If the sheathing is thick enough, the inside surface is never colder than the dew point of the indoor air, and there will not be any condensation on that surface. Another way to minimize condensation problems inside a wall in the winter is to construct the wall out of permeable materials, so that if it does get wet, it can easily dry (such a wall is said to be *forgiving*). For example, a permeable sheathing, such as asphalt-impregnated fiberboard, will not trap moisture inside the wall; it will allow the moisture to pass through it and be released outdoors. For more on construction techniques that will minimize hidden moisture problems, see the *Moisture Control Handbook*.(Lstiburek and Carmody)

If cold outdoor air infiltrates into a *depressurized* house in the winter, it will get warmer (and the RH will go down) as it passes toward the interior, so condensation won't occur with infiltration in the winter—only with exfiltration. Therefore, hidden condensation inside building cavities can be prevented by preventing exfiltration in the winter. Since exfiltration only occurs when a house, or a part of a house, becomes pressurized, it is possible to minimize the chance of hidden moisture problems by choosing a ventilation system that either depressurizes the house (exhaust ventilation) or operates at a neutral pressure (balanced ventilation) in the winter. In Sweden, there has been discussion of modifying the national building code to require houses to be depressurized in the winter in order to prevent hidden moisture problems. (Actually, a house can experience a variety of pressures, no matter what type of ventilation system it has—depending on various accidental and natural pressures. For example, a house with a balanced ventilation system can become depressurized when air leaves the house via a clothes dryer or natural-draft chimney.)

Because of the potential for hidden moisture problems, supply ventilation systems are not often recommended in cold climates. However, if the RH of the indoor air is low enough (or insulating sheathing is thick enough), the air that does get transported into a wall cavity will never reach the dew point. Hidden winter moisture problems are most common in pressurized houses when the indoor RH is high—especially when it is over 50%—but often when it is above 40% if a wall isn't forgiving. If the indoor RH is in the 25% range, it is doubtful if pressurization will cause hidden moisture problems in most of the U.S. Problems tend to occur when humid air exfiltrates from kitchens and bathrooms in the winter—but if the local exhaust fans in those rooms are used regularly, the humid air will be blown outdoors through deliberate outlets before it has a chance to exfiltrate through random holes.

Joseph Lstiburek, a leading moisture-control authority, has estimated that about 10% of the houses in cold U.S. climates are at risk of hidden pressure-related moisture problems. The bottom line: Pressurization has the potential to cause hidden moisture problems when it is cold outdoors and humid indoors.

Hidden summer moisture problems

In an air-conditioned house in the summer, indoor air is cooler than outdoor air. The inside of a wall (or roof, or floor) cavity will be cool just behind the drywall (or paneling, or plaster), and warm near the sheathing. If cool (air-conditioned) indoor air exfiltrates through the random holes in a pressurized house, it gets warmer and warmer as it nears the outdoors. As it gets warmer, the RH goes down because warm air can hold more moisture than cool air. So cool exfiltrating air won't cause hidden moisture problems in a warm climate. However, if warm, moist outdoor air infiltrates into an air-conditioned house, hidden condensation can occur (Figure 8–7).

Let's say it is 75°F indoors, 95°F outdoors and the outdoor air is at a fairly humid 70% RH. The back

Understanding Ventilation

of the drywall might be 76°F and the sheathing might be 92°F (Figure 8–8). If the hot humid outdoor air infiltrates through the random holes in the wall, it will get cooler and cooler as it nears the living space. The dew point of this incoming air is about 84°F (according to the psychometric chart), so when it hits the back of the cool drywall, it will condense there. This will occur unnoticed by the occupants—inside the wall cavity.

If there is a vapor retarder (*e.g.* plastic sheeting, foil-faced insulation, or asphalt-impregnated-kraft-paper-faced insulation) next to the drywall, or if the wall is decorated with a vinyl wall covering, the condensed moisture will be trapped inside the wall cavity, so it will tend to build up. This can result in hidden mold or mildew growth, or decay. A forgiving wall would have no vapor retarder (or it would be located at the opposite side of the wall, near the outdoors), to allow any moisture to pass through the wall, into the living space, and not build up inside the wall.

The potential for hidden moisture problems in depressurized, air-conditioned houses depends on the humidity of the outdoor air, the precise temperature of the back of the drywall, and how forgiving the wall is. In equatorial countries, there is more potential for problems than in Canada because the summer air at the equator is hotter and more humid. In the U.S. the greatest potential for this type of problem is in the humid coastal areas of the South, but problems can also develop along the coast in the middle latitudes of the country. Moisture-control expert Joseph Lstiburek has estimated that, primarily because of unforgiving wall assemblies, there is the potential for hidden moisture problems in half the houses in the humid southern U.S. if a house experiences a depressurization as small as 3–5 Pa. The reason so many houses are at risk is because interior moisture retarders and vinyl wall coverings (which cause no problems in cold, northern locales) commonly trap moisture inside wall cavities in hot, humid climates.

Figure 8–7. *Hidden mold growth can occur when moisture gets trapped and condenses inside a wall assembly. In this Florida house, depressurization caused hot humid outdoor air to migrate from the outdoors towards the indoors. The moisture condensed on the back side of the vinyl wall covering, resulting in severe mold growth. The wall covering was cool because the house was air conditioned.*

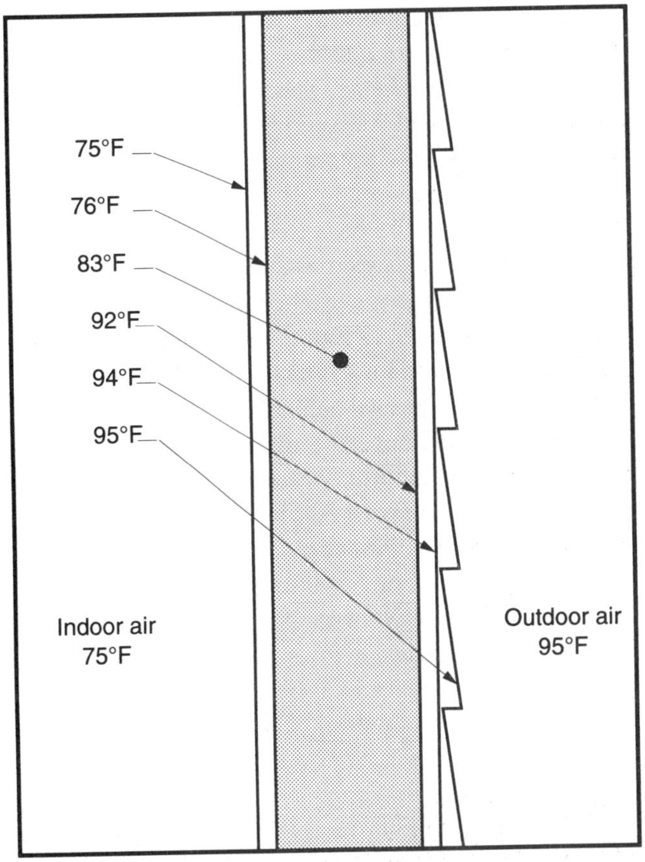

Chapter 8 Ventilation-induced moisture problems

Figure 8–8. Hidden moisture condensation can occur inside a building cavity if hot, humid outdoor air passes inward through the cavity and reaches a cool surface—such as the back side of the drywall or wall paneling.

Indoor air 75°F

75°F
76°F
83°F
92°F
94°F
95°F

Outdoor air 95°F

Because exhaust ventilation systems that depressurize houses can cause hidden moisture problems in hot humid climates, a system that pressurizes a house (supply ventilation) or operates at a neutral pressure (balanced ventilation) is a better choice in air-conditioning climates. (Of course, the use of balanced ventilation will not guarantee that pressurization will not occur due to natural or accidental pressures.) The bottom line: depressurization has the potential to cause hidden moisture problems when it is hot and humid outdoors and air conditioned indoors.

What about mixed climates?

If prevailing opinion advises against pressurizing a house in the winter in the northern U.S., and against depressurizing a house in the summer in the South, what do you do in a mixed climate? One answer is neither. If it is cold outdoors for part of the year and hot and humid for part of the year, a good choice for a ventilation system is one that has balanced airflows—one that results in the house experiencing a neutral pressure. In fact, balanced ventilation systems work well in any climate. Actually, in many mixed climates the temperature and humidities aren't extreme enough to result in hidden condensation in any season.

What about occasional pressurization or occasional depressurization in the wrong climate? What happens if moisture-laden air gets pushed or pulled into building cavities for only a few months at the wrong time of the year? In many cases, not much.

Wetting potential and drying potential

The reason occasional pressurization or occasional depressurization in the wrong climate doesn't always result in hidden moisture problems has to do with two related concepts called *wetting potential* and *drying potential*. Wetting potential refers to the amount of water that will condense inside a building assembly and be absorbed by the construction materials. Drying po-

Understanding Ventilation

tential refers to the ability of that same building assembly to dry out after it gets wet. If a building assembly has greater drying potential than wetting potential, there may not be any problems. In other words, even if a building assembly gets wet, no problems will develop if it dries out quickly enough. In reality, many building materials can get wet occasionally without moisture-related problems occurring. For example, wood absorbs moisture slowly, and it can take a considerable amount of time for dry wood to absorb enough moisture to be in danger of decay. However, chronic, continuous wetting can result in serious hidden moisture problems.

Consider a house in a mixed climate with a central-exhaust ventilation system. The house will be depressurized all year, even during the air-conditioning season. During the summer, perhaps for a period of 2 or 3 months, there is some wetting potential inside leaky building cavities. However, the building cavities will have the rest of the year to dry out. During mild weather, there should be no problems because there are no hidden cold surfaces, and during the winter the incoming cold air will dry the cavities.

As it turns out, in many climates, building cavities routinely absorb moisture during an entire season, then dry out during another season. Even though the building cavities are absorbing moisture for several months, they tend to dry before hidden moisture problems occur. So, in the real world, a house may not have a serious hidden moisture problem even though it theoretically could. Most of the time, hidden moisture problems only occur when there is a great deal of very moist air pushed through building cavities on a regular basis. Smaller amounts of less humid air passing occasionally through building cavities are often not serious considerations. With a knowledge of how a house dries out (see the *Moisture Control Handbook*(Lstiburek and Carmody)), it is possible to predict if a particular house will exhibit problems. But, to be on the safe side in a questionable situation, you might choose a balanced ventilation system.

Pressurization and depressurization that only last for a few hours at a time are even less of a problem. For example, in a climate where it is hot and humid outdoors most of the year, houses should typically have either a central-supply ventilation system or a balanced ventilation system to avoid hidden pressure-related moisture problems. But most houses also have local-exhaust fans such as bathroom fans or a kitchen range hood. In addition, natural and accidental pressures may cause the house to become temporarily depressurized. Again, if the building cavities in such a house have more drying potential than wetting potential, hidden pressure-related moisture problems will be unlikely. Therefore, if the local-exhaust devices are not operated for extended periods—and in most cases they aren't—hidden moisture problems probably won't occur. After all, the combined operating time of range hoods, bath fans, clothes dryers, etc. is typically only a small percentage of the number of hours in a year. Anyway, because one of the main functions of bath fans and range hoods is to ventilate moisture-laden rooms quickly, they usually expel more moisture from the house through deliberate outlets than they cause to be sucked in through random holes, so they tend to reduce moisture levels indoors.

Solutions for mixed climates

Central-exhaust ventilation will never cause a hidden moisture problem in Minnesota, but it might in southern Florida. On the other hand, central-supply systems are very risky to use in Minneapolis but safe in Miami. But what about mixed climates? The good news is that if you have an understanding of the ways in which a house can dry out (as discussed in the *Moisture Control Handbook*.(Lstiburek and Carmody)) you can design a house for any climate with any type of ventilation system. However, because construction practices vary from region to region, it can be difficult to predict what will happen in a "typical" house in mixed climates.

But there is more good news. As it turns out, depressurization systems are safer to use in mixed climates than pressurization systems. In fact, depressurization systems can generally be used in the northern two-thirds of the United States. But they have also been used further south. This can be done safely, for example, by building a very tight house with through-the-wall vents for inlets. In a tight house, there aren't many random holes for the humid outdoor air to pass through, so most of the incoming air will enter through the vents. Because very little air passes through hidden pathways, there is little potential for condensation.

To be extra cautious, pressurization systems are probably best reserved for the humid areas of southern

Florida and Texas. However, pressurization systems have been used successfully in Canada. One way this is possible is to keep the indoor RH low and use enough well-sealed insulating sheathing so there aren't any surfaces inside the building cavities that are below the dew point temperature.

For the entire country, balanced ventilation is probably the most conservative recommendation. However, in the real world, houses get pressurized and depressurized every day in mixed climates without resulting in any hidden moisture problems. Of course, if a problem is hidden, it may be there and you just don't see it. But, when the walls of houses are opened up during remodeling, most of the time no one finds any hidden moisture problems—most likely because of a combination of factors such as low RH and the presence of more drying potential than wetting potential.

When does decay occur?

Mold and mildew belong to the large botanical classification known as fungi. They often grow on surfaces when the RH near the surface is above 50%. The RH of the air can rise quickly, for example, in a bathroom after a shower, but it can also drop just as quickly if you use a local exhaust fan. If a local exhaust fan is not used, and the RH remains high for several days, mold or mildew can quickly start growing.

Decay organisms are another type of fungi, but they typically takes longer to get started than mold or mildew. Decay fungi can attack wood and other materials when the moisture content (m.c.) of the material itself (not the RH of the air) is over 20%. It is important to note that % RH is different from % m.c. (In most houses, the m.c. of wood is almost always considerably less that the RH of the air surrounding the wood.) The m.c. of wood is expressed as a percent of the wood's oven-dried weight.

When a tree is growing, it contains a certain amount of water. For example, the m.c. of a growing birch tree might be 75%. By the time the boards and lumber reach builders and consumers, the wood has usually been dried to about 19% m.c. The wood in a house continues to lose moisture over time until it reaches equilibrium with the climate it is in; it is typically less than 10% m.c. in most houses. There is generally a difference between the m.c. of the lumber of a house in the winter than there is in the summer, but because wood takes up and loses moisture rather slowly, the difference usually isn't significant, and the m.c. is almost always less than the 20% needed for decay organisms to attack the wood.

Decay problems occur when wood remains wet for an extended period. For example, if there is a plumbing leak under a bathtub, the floor may be wet for many months. When there is enough moisture present for a long enough time, the m.c. of the wood can rise above 20%, and decay is a strong possibility. If a leak is temporary, the wood won't be wet long enough for decay organisms to attack it. (It has more drying potential than wetting potential.) So, decay is generally only a concern with a long-term moisture problem, not a temporary moisture problem. But plumbing leaks aren't the only cause for worry; excessive pressurization in cold climates, and excessive depressurization in hot, humid climates can both transport enough moisture in building cavities for hidden decay to be a occur.

How ventilation affects the indoor relative humidity

Because indoor and outdoor temperatures and relative humidities are rarely the same, ventilating a house will always have some effect on the indoor RH. This effect is most pronounced in the winter in cold climates and in the summer in hot, humid climates when the indoor/outdoor temperature and humidity differences are greatest.

When cold, dry outdoor air enters a house and replaces warm indoor air, it will always lower the RH indoors. This often occurs in winter. When too much cold, dry outdoor air is brought indoors by whatever mechanism (natural, accidental, or controlled ventilation), the indoor air can become too dry. If this happens, the occupants may complain of skin dryness or irritation of mucous membranes, and they may have an increased susceptibility to respiratory illness.

When hot, humid outdoor air is added to indoor air at room temperature, the RH indoors will always

Understanding Ventilation

rise. This often happens in the summer and can generate complaints of mugginess. If the indoor RH rises too much, you could end up with a proliferation of dust mites or mold growth. In some cases, when excessive amounts of hot, humid air are brought indoors, there can be so much moisture in a house that the air conditioner can't remove it fast enough, or can't keep the temperature in the house comfortable. Under normal circumstances the RH of air can be no more than 100%. This means the air can hold no more moisture; it is saturated. But fog can actually contain much more moisture because it is made up of tiny water droplets as well as 100% RH. So when warm foggy air, which is common in some coastal climates, is brought indoors through a ventilation system, it can add a tremendous amount of moisture to the air inside a house.

The best solution to indoor winter dryness and summer humidity is not to overventilate. If a ventilation system is sized to match only the needs of the occupants (people pollutants), comfortable humidity levels are much easier to maintain. On the other hand, if a ventilation system is sized to dilute large quantities of pollutants outgassed from building materials, or to dilute large quantities of pollutants sucked indoors by pressure differences, the air in a house may need to be exchanged so rapidly that a comfortable indoor humidity is difficult and expensive to maintain. The bottom line: If source control and separation are used to reduce indoor pollution levels, a small-capacity ventilation system will suffice, and it will be easier to keep indoor humidities at a comfortable and healthy level.

When adding a ventilation system to an existing house, it is important to make sure the heating/cooling system has enough capacity to handle the added load imposed by the ventilation air. This is especially important in a humid climate, where the air conditioner must have the ability to both cool and remove the moisture from the incoming air. If you are adding a ventilation system to an existing house, and the existing air conditioner won't be able to remove the additional moisture brought indoors by the new ventilation system, you may need to consider a new air conditioner with more capacity. Or, perhaps you could use an HRV that is capable of removing some of the humidity from the air before it enters the living space (See the next section, *Humidity control with HRVs*).

Humidity control with HRVs

Sometimes a ventilation system will cause a house to become too dry in the winter or too humid in the summer. This can occur when a ventilation system is very powerful, or when the climate is extreme.(Barringer)

In the winter, a ventilation system will bring in cold air that contains very little moisture. As this air enters the living space, it warms up, its RH goes down, and it contributes to drying the indoor air. At the same time, moisture-laden indoor air is exhausted to the outdoors. This also contributes to drying the indoor air. In moderate climates, a moderate amount of ventilation will result in a comfortable and healthy indoor RH being maintained. However, in an extremely cold climate, when the cold outdoor air contains very little moisture, it isn't unusual for a ventilation system to overdry the air. This happens when the combination of bringing in dry outdoor air and exhausting humid house air dries the air in the house faster than moisture is generated indoors. Overdrying tends to occur in the winter when a house is ventilated at a high rate, especially at rates above 15 cfm per person.

In the summer, a ventilation system can bring in air that contains a great deal of moisture. When this incoming moisture combines with the moisture generated indoors, the RH in a house can go up significantly. Some of this excess moisture will be exhausted by the ventilation system, and some will be removed by the air conditioner. If the combination of moisture entering through the ventilation system and moisture generated indoors exceeds the amount of moisture removed by the ventilation system and air conditioner, the result will be an indoor RH that is too high.

One type of heat-recovery ventilator (HRV) can minimize the overdrying effect in the winter, or excessive indoor humidity in the summer. Most HRVs bring in fresh air complete with all the moisture it contained outdoors and exhaust stale air complete with all of the moisture it contained when it was indoors. But some HRVs allow a certain amount of moisture to pass between the two airstreams.

In the winter, these HRVs will allow a percentage of the moisture to pass from the house's exhaust air, through the core, into the dry incoming air; thus, some of the humidity returns to the indoors that would otherwise have been blown outdoors. In hot, humid climates, these HRVs will transfer some of the incoming moisture from the fresh air, through the core, into the drier exhaust air, rather than letting all the moisture enter the house; thus, the indoors remains drier, and the air conditioner doesn't need to work as hard.

When an HRV is capable of transferring moisture *and* heat between the airstreams, you can have higher ventilation rates in harsh climates without as much effect on indoor humidity. The HRVs that are capable of transferring moisture between airstreams have either rotary cores or crossflow cores made of treated paper. They will be discussed in Chapter 23.

What is the correct amount of humidity to have indoors?

An understanding of RH is important because microorganisms such as mold, mildew, dust mites, etc. all thrive at higher relative humidities. They don't sense the absolute humidity of air (the number of molecules per cubic foot), but they know if air is more saturated, and when it is, they reproduce faster. Dust mites and mold don't need moisture in a liquid form (water) to start growing, but just a high RH. Microbial contamination is most likely when the relative humidity is above 70%. A high RH can also mean prolonged survival of bacteria and viruses, and VOCs such as formaldehyde outgas at faster rates when the RH is high.

The RH in a house will vary from place to place. For example, it will be higher in a bathroom after a shower than in an empty bedroom. But at night it may be higher in the bedroom due to the occupants' exhaled breath, than in the empty living room. (The RH can also be higher in the bedroom if it is kept cooler at night.) In addition, the RH is rarely uniform throughout a room. Because RH changes with temperature, the RH will be lower in warm parts of a room (near a baseboard heater, for example) and higher in cool parts of a room, such as near a window, or the base of an exterior wall on a concrete floor slab.

Moisture problems in unventilated or underventilated houses tend to occur first in the cool locations where the RH is high. For example, the air in the center of a room may be 72°F @ 50% RH, but the temperature might be 55°F in a corner during the winter if the walls aren't insulated very well. Because the water vapor molecules are usually fairly evenly distributed throughout the room, the 55°F air will have an RH of about 92%. When the RH is that high, mold or mildew growth are likely. If the air gets below 50°F, say near a cold window, the RH will be at 100% and condensation will form on the cold surfaces. This often happens near windows when outdoor temperatures are low, and it can easily lead to mold growth or rot. Ice or condensation on the inside of windows in the winter is often a sign that the humidity in a house is too high. Condensation on windows in the winter can be reduced by either lowering the overall RH indoors or warming the window. When you warm the window, you also lower the RH near it. A window can be warmed by increasing its insulating ability (*e.g.* adding a storm sash) or by blowing warm air on it from a heating register.

Most houses have quite a few different microclimates of localized low temperature and high RH in the winter. If water condenses on a window pane, it is obvious that the RH is 100% near the window. But if the RH is 90% inside a cool closet or near a concrete floor slab, it won't be so obvious, unless you notice a colony of mold growing.

To lower the RH near a cool surface, you can either remove some of the moisture molecules from the air or warm up the surface. Carpeting on a cool concrete floor slab can be home to millions of dust mites and mold spores because of a high RH near the floor. To minimize microbial contamination on a cool concrete floor slab, you can lower the overall RH in the entire house either by ventilating (in the winter) or by using a dehumidifier. Warming up an existing concrete floor slab can be done by raising the overall air temperature in the entire house. In new construction, concrete floor slabs can be kept from getting too cold by placing insulation around the perimeter, underneath them, or by embedding radiant heating pipes in them. But it is still possible to have mold or dust mites in carpet on a warm floor if the overall RH in the house is too high.

So, what is the correct relative humidity to strive for in a house? Human beings can be comfortable be-

Understanding Ventilation

tween relative humidities of 15% and 85% depending on the temperature of the air, the amount of air movement, how active they are, and the amount of clothing they have on.

At typical indoor temperatures, an RH in the 40–60% range is generally associated with health and comfort. But when it is 60% RH in most parts of the living space, there could easily be localized areas of 90% RH in the winter. Therefore, many experts are now recommending that the indoor RH be kept below 40% during cold weather. When outdoor temperatures are extremely low in a poorly insulated house, the RH may need to be somewhat lower than 40%, perhaps as low as 20%, in order to prevent condensation.

But you don't want the relative humidity to be too low. If the RH is too low, mold might not cause any microbial growth, but occupants could start complaining of dryness. For example, when RH levels are below 20%, there may be an increased incidence of respiratory infection, skin dryness, allergies, and asthma, because of more sensitive mucous membranes. In addition, there can be considerable shrinkage in wooden furniture or musical instruments resulting in loose joints and damage. Many house plants also suffer when the RH is very low, and house dust will tend to remain airborne longer. The bottom line: In most houses, it is best to keep the RH between 25 and 40% in the winter and below 70% in the summer.

Chapter 9

Pollutant entry due to depressurization

A variety of pollutants originate outside the living space (*e.g.* radon, termiticides, insulation, etc.) that can get sucked into a house when it becomes depressurized. For this to occur, three conditions must be present: 1) there must be contaminants outside the living space waiting to be sucked indoors, 2) there must be some holes in the structure for the contaminants to pass through, and 3) there must be a negative pressure to pull the contaminants through the holes. In most cases there are at least some contaminants outside the living space, the majority of houses have at least a few holes, and all houses are subject to a variety of positive and negative pressures.

Inasmuch as ventilation systems directly affect the pressures in a house, they can either contribute to this type of pollution problem (if they depressurize a house) or prevent such a problem (if they pressurize a house). However, because forced-air heating/cooling systems induce significantly higher pressures (both positive and negative) in houses due to leaky ducts and closed doors than ventilation systems, they are usually a bigger contributor to pressure-induced pollution problems than ventilation systems. Still, a ventilation system can cause pollutants to be sucked indoors. But a ventilation system often improves the air quality in a house because the overall quality of the incoming air (which contains fresh air along with the pollutants) is better than the air in the house that it is replacing.

It would certainly be desirable if all the incoming air were pollutant-free. However, this is only possible when a house is very tightly constructed, and all the air is filtered as it passes through deliberate inlets. In the real world, most houses have some uncontrolled air movement through random holes in the structure (infiltration), which brings some pollutants along with it. Loosely built houses have a great deal of air entering randomly and none entering through deliberate openings. Tighter houses have very little air entering ran-

Understanding Ventilation

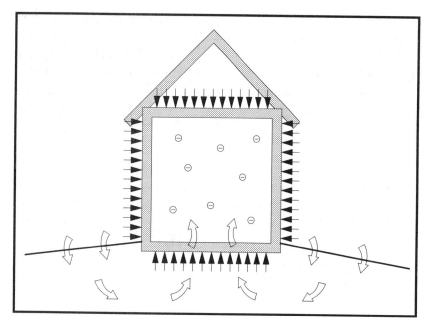

Figure 9–1. *If a house gets depressurized for any reason, air can be pulled through the soil through the foundation and into the living space. This occurs more readily when there are numerous holes in the foundation, the soil is sandy or porous, and with stronger depressurizations.*

domly and most of it entering through deliberate openings. To ensure that the air entering is as clean as possible, you should minimize the random infiltration and maximize the amount of air entering planned openings. But in any case, it is generally better to bring in slightly polluted air than to bring in none.

The contaminants we are talking about can be divided into two categories: those found below ground in the soil, and those found above ground either outside the living space or within the structure of the house.

Below-ground pollution sources

A number of pollutants are present in the soil around houses that are not good to breathe: radon, pesticides, herbicides, sewer gas, etc. They often get sucked into the living space through random cracks, holes, and gaps in the foundation when a house becomes depressurized. Excessive amounts of moisture can also be pulled from the soil into the living space.

Actually, only the a lower portion of a house—the part closest to the soil—needs to be depressurized for below-ground pollutants to be sucked indoors. If the upstairs is depressurized and the basement is pressurized, the soil pollutants will not enter the living space.

Also, a house must be in contact with the ground for significant amounts of soil pollutants to be sucked indoors. Houses built up in the air on piers, or second-floor apartments are at much less risk than houses with basements.

Depressurization of the lower level of a house can occur for any reason (*i.e.* natural, accidental, or controlled ventilation). For example, stack effect in the winter causes the lower portion of a house to become depressurized, and a powerful kitchen-range hood can cause the entire house to experience a negative pressure. In new construction, the most important way of preventing below-ground pollutants from entering the living space is to build in a healthy manner in the first place: Install a radon-removal system, don't use toxic chemicals in the soil, use proper moisture-control techniques, etc. In existing construction, it will be necessary to analyze each situation separately. However, when below-ground pollutants are entering the living space, it usually better to deal with them first, then consider a ventilation strategy.

We don't normally think of dirt or soil as being porous but it often is, and sandy soils are more porous than clay. Being porous, soil contains a certain amount of air, and that air can become contaminated by whatever is in the soil. It has been shown that negative pres-

Chapter 9 Pollutant entry due to depressurization

sures in basements can pull air through the soil and into a house from as far away as 40–50 feet from the foundation.(Garbesi) For this to happen, there must be some holes in the foundation for the air to travel through, and there must be a negative pressure in the house to pull the air through the soil (Figure 9–1). The author of one study(Lstiburek 1994) has reported that soil gases can be pulled into the living space within *seconds* of a house becoming depressurized.

Soil gases can also diffuse through the solid materials that make up a foundation (even if a house is experiencing a neutral pressure), but this is much less important than depressurization. However, because of diffusion, a basement might contain higher levels of soil

RADON RISK FOR NONSMOKERS		
Radon level	If 1,000 people who never smoked were exposed to this level over a lifetime...	The risk if cancer from radon exposure compares to...
20 pC/l	About 8 people could get lung cancer	The risk of being killed in a violent crime
10 pC/l	About 4 people could get lung cancer	
8 pC/l	About 3 people could get lung cancer	10 times the risk of dying in an airplane crash
4 pC/l	About 2 people could get lung cancer	The risk of drowning
2 pC/l	About 1 person could get lung cancer	The risk of dying in a home fire
1 pC/l	Less than 1 person could get lung cancer	(Average indoor level)
0.4 pC/l	Less than 1 person could get lung cancer	(Average outdoor level)

Figure 9–2. The only known negative health effect of radon is lung cancer, but the degree of risk varies, depending of the amount of radon to which you are exposed.

RADON RISK FOR SMOKERS		
Radon level	If 1,000 people who smoked were exposed to this level over a lifetime...	The risk if cancer from radon exposure compares to...
20 pC/l	About 135 people could get lung cancer	100 times the risk of drowning
10 pC/l	About 71 people could get lung cancer	100 times the risk of dying in a home fire
8 pC/l	About 57 people could get lung cancer	
4 pC/l	About 29 people could get lung cancer	100 times the risk of dying in an airplane crash
2 pC/l	About 15 people could get lung cancer	2 times the risk of dying in a car crash
1 pC/l	About 9 people could get lung cancer	(Average indoor level)
0.4 pC/l	About 3 people could get lung cancer	(Average outdoor level)

Figure 9–3. Smokers have a substantially higher risk of contracting lung cancer due to radon than nonsmokers.

pollutants than a single-story house built on a concrete slab simply because a basement has more surface area in contact with the ground for soil gases to diffuse through. Diffusion can be minimized by utilizing a diffusion retarder such as polyethylene sheeting during construction and by using a high-strength (low-water) concrete mix. Any soil gas that does diffuse through a retarder or dense concrete will be minimal and can be diluted by a general ventilation system. Depressurization has little effect on diffusion rates.

Radon

Radon is a radioactive gas that is a natural decay product of the element radium, which is itself a natural decay product of uranium. Because uranium and radium are found in tiny quantities in the soil virtually everywhere on the planet, radon can also be found in the soil everywhere. Radon is invisible, you can't smell it, and you can't taste it, but breathing it can result in lung cancer. Radon often contaminates the air in houses when it is pulled from the soil, through the random cracks and holes in the foundation, and into the living space by negative house pressures.

Actually, radon gas itself doesn't cause lung cancer—if you inhale radon and then exhale it quickly, nothing bad will happen. But radon decays naturally in a very short time, and it is radon's decay products, microscopic respirable particles sometimes called radon daughters or radon progeny, that cause lung cancer. Radon progeny carry a small static charge, so they attach themselves to dust or water vapor molecules in the air. Then, when they are inhaled, the tiny particles lodge inside our lungs. Once in the lungs, the progeny continue to decay. As they decay, they give off small bursts of alpha, beta, and gamma energy. It is this energy that can damage our tissues and result in lung cancer.

Because it is naturally occurring, there is always some radon in the air we breathe. Most of the time it dissipates into the atmosphere outdoors, so we don't breathe very much of it. But if radon gets into an enclosed space, and it can't easily escape, the concentration can build up to dangerous levels. This is an especially serious problem in underground mines. There are reports that radon killed about half the miners in Europe in the late 1800s. Radon can also build up in caves, but its buildup in houses potentially affects the health of millions of people.

The amount of radon in the air is often measured in metric units of picoCuries per liter (pC/l). The average outdoor level is less than 0.5 pC/l. The average indoor level in the U.S. is 1.3 pC/l, but indoor levels over 2,000 pC/l have been measured in some houses. The USEPA suggests that if the level in a house is above 4.0 pC/l, you should consider remedial action. While there is some risk posed by a radon level of 4.0 pC/l (if you have never smoked, it is about equal to your lifetime risk of drowning), the USEPA feels that most Americans are willing to accept some degree of risk. If you don't smoke, a radon level indoors of 20 pC/l carries about the same risk of being killed in a violent crime—about eight out of a thousand people who are exposed to this level over their lifetime are predicted to die of lung cancer (Figure 9–2). Perhaps as many as 10–20% of our houses have radon levels in them above 4.0 pC/l. Inexpensive radon test kits can be found in many building-supply stores, or your local Board of Health can help you locate one.

If you are a smoker, your risk of getting lung cancer from radon is substantially higher (Figure 9–3). This is because radon daughters cling to particles in tobacco smoke, so if you inhale smoke, you inhale more particles and, thus, you inhale more radon. Because of this, a smoker's risk of getting lung cancer from radon is about 15 times higher than that of a nonsmoker.

On rare occasions, radon can be released from building materials such as concrete because the ingredients (*e.g.* Portland cement, sand, gravel) occasionally contain tiny amounts of uranium or radium. Radon can also be found in granite countertops, well water, and in heating fuels that come out of the ground such as oil, natural gas, or coal. However, in the vast majority of houses, radon originates in the soil, and it gets pulled into the living space through random cracks and holes in the foundation when the part of the house near the soil becomes depressurized.

As it turns out, it is best to address a radon problem separately from the house's ventilation system. For example, radon mitigation is often handled by using a high-pressure, low-volume fan to depressurize the soil beneath the house. (Ventilating the living space, on the other hand, typically uses a lower-pressure, higher-volume fan.) During operation, the radon fan pulls air from beneath the house and blows it outdoors into the atmosphere. A certain amount of air from the living space flows through the random holes in the foundation, into the soil, then the fan blows it outdoors. Even if there is a central-exhaust ventilation system in the house, there is rarely any danger of radon being pulled indoors.(Exhaust) The bottom line: You should consider radon and ventilation to be two separate issues. If radon is mitigated effectively, typical house depressurization levels should not affect radon concentrations.

Pesticides and herbicides

Pesticides and herbicides are poisons. They are formulated to kill living creatures and plants. Some are nerve poisons, some cause birth defects, and some cause genetic damage. Yet, we purchase hundreds of million pounds every year and apply them directly inside the living space of our houses to control fleas, ants, roaches, and other pests. We also apply them to our lawns so they will look green and picture-perfect, then we track them indoors on our shoes and they build up in carpeting. We are now learning the dangers of indiscriminate use of poisonous chemicals inside our houses(Whitmore)(National) and are beginning to switch to less toxic methods of controlling pests.(Olkowski) However, even if you personally quit using these chemicals, you may still be at risk because of long-past applications. And, you may be exposed when your neighbors apply these chemicals to their house or lawn.

The vast majority of houses in the U.S. have been treated with termite-killing chemicals. Though many cases go unreported, there are thousands of people whose health has been damaged by these products. The most widely used termiticide, chlordane, belongs to the chlorinated hydrocarbon class. It is a known carcinogen and is no longer on the market, but over the years it has been applied to millions of houses. Chlordane was typically injected into the soil around the outside of a house where it killed any termites with which it came in direct contact. Chlordane is an extremely long-lasting chemical, so it poisons the soil permanently, forming a highly toxic barrier that termites won't cross. Even though chlordane is supposed to remain in the soil, studies have shown that it can easily contaminate the living space of a house.

A 1987 study of 5,038 chlordane-treated houses on a U.S. Air Force base found that chlordane was detectable in the indoor air of 44% of them.(Lillie) Various other studies have found similar contamination. Though not all of these houses have significantly high levels of

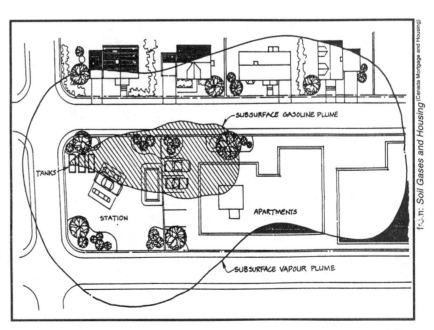

Figure 9–4. Soil gases can often travel underground for hundreds of feet from a source.

termiticides in them, the potential exists in too many houses to ignore the issue. Because it is so toxic, chlordane has received the greatest amount of negative press coverage, but other termiticides, notably those in the organophosphate class such as Dursban (chlorpyrifos) are now also coming under fire. Reported symptoms of organophosphate poisoning include drooling, sweating, nausea, diarrhea, abdominal pain, weakness, fatigue, and anxiety.(Hodgson)

In search of the perfect lawn, many people have their yard saturated with chemicals several times a year to kill weeds and pests.(Abrams) Pets and children are especially vulnerable to these products when they play on the grass. Though lawn-care chemicals must be registered with the USEPA, the registration does *not* mean the chemicals are safe; it only means that they are on a list of chemicals currently in use. Many negative health effects have been recorded from the use of these products, yet they continue to be marketed without proper disclosure. Central nervous system or neurotoxic effects (convulsions, bizarre behavior, paralysis, memory loss, fuzzy thinking, slowed reflexes, etc.) are some of the more serious health problems associated with pesticides. Many are actual nerve poisons, yet they are rarely tested for neurotoxicity.

Poisonous lawn-care chemicals tend to soak down into the soil and poison it deeper than we realize. Termites are then forced to live deeper in the ground because they cannot tolerate the lawn poisons any more than other pests. Sometimes, termites must take refuge underneath a basement floor slab where they are more difficult for pest control operators to eradicate.

Just because lawn and termite chemicals are applied to the soil *outside* a house doesn't mean that they won't contaminate the interior—they can. These chemicals enter living spaces in exactly the same way as radon. They are pulled indoors through random holes in the foundation by negative pressures in the living space. The good news is that they can be kept out of the living space by depressurizing the soil under a house with a radon-mitigation fan. While this can prevent additional chemicals from entering the living space, house depressurization in the past might have caused various building materials to have become permanently contaminated. In such a case, removal of the contaminated materials and furnishings may be necessary.

If you suspect that pesticides or herbicides are being pulled through the soil into the living space, you can have the indoor air tested by a local analytical laboratory. Information about the potential health effects of different levels of specific products can be obtained by calling the National Pesticides Telecommunications Network at (800) 858–7378. Funded by the USEPA, this network can also tell you what type of testing is appropriate for your situation.

Ground moisture

Most soils contain a certain amount of dampness. Excessive dampness can be caused by a high water table in the ground, downspouts that dump water near the foundation, broken or plugged foundation drainage tiles, underground plumbing leaks, etc. This ground moisture can be picked up by air passing through the soil and be brought indoors through random holes in the foundation. Once indoors, this moisture will affect the indoor relative humidity, possibly resulting in a proliferation of microorganisms or increased outgassing of formaldehyde and other VOCs from building materials or furnishings. Ground moisture can also enter the living space by diffusion—wicking through a concrete slab or foundation wall.

Moisture entering a house from the ground is one of the most significant ways water vapor gets into houses. In fact it often dwarfs the amount of moisture generated indoors by the occupants. If moisture entering the living space from the ground is a significant problem, it should always be dealt with separately from the house's ventilation system. Moisture-control strategies are covered in *The Moisture Control Handbook*(Lstiburek and Carmody) listed in Appendix C.

Other soil gases

The soil can also contain several other pollutants that can get be pulled through random holes in the structure into the living space when a house becomes depressurized. For example, soil gases include the VOCs given off by mold as a part of its metabolism, and sewer gases escaping from cracked or improperly sealed drain lines. The ground around a house located near a landfill—hazardous or conventional—may be contaminated with methane or a variety of other potentially harmful chemicals. And leaking underground storage tanks (*e.g.* fuel oil, gasoline, etc.) can easily release contaminants into the soil (Figure 9–4).

Above-ground pollution sources

Above-ground outdoor pollutants can also be pulled into the living space when a house is depressurized. Again, for this to occur three conditions are necessary: 1) there must be contaminants outside the living space waiting to be sucked indoors, 2) there must be some holes in the structure for the contaminants to travel through, and 3) there must be a negative pressure to pull the contaminants through the holes. All three factors are present to some degree in most houses.

It is important to keep in mind that even if above-ground pollutants are sucked indoors by a ventilation system, the indoor air quality with ventilation will generally be better than if no ventilation system were used. Even though the air being brought in is not perfect, it is considerably cleaner than the air already in the house. In most cases, leaky forced-air heating/cooling ducts cause significantly higher pressures in houses, and significantly more pollutant transport, than modestly sized ventilation systems.

For extremely sensitive people, it may be necessary to have air in the house that is as clean as possible. For them, it may be important to reduce pollutant levels to a bare minimum and also to filter any incoming air—but for average healthy people, even though some pollutants could be brought indoors by ventilation-induced depressurization, doing something is often better than doing nothing.

Pollution sources in the outdoor air

Pollutants found in outdoor air (*e.g.* automobile exhaust or smoke from a neighbor's chimney) can easily contaminate a house when they infiltrate indoors. Other outdoor contaminants include smoke from barbecues, herbicides and pest-control chemicals sprayed into the air (*e.g.* on residential ornamental or fruit trees, or agricultural spraying), contaminants escaping from the vents of underground gasoline or fuel oil tanks when the tanks are filled, industrial smokestack emissions, odors from a neighbor's clothes-dryer exhaust or kitchen-range hood, etc.

Unlike underground pollutants, above-ground pollution sources are often sporadic and not continuous over a 24-hour day. For example, orchard spraying may only take place at a certain time of the year, underground gasoline tanks may only be filled once a week, a neighbor's range hood may only be used for the evening meal, etc. However, in a major city or in a heavily industrialized area, outdoor pollution can be a significant concern year-round.

Because these pollutants are floating around in the air outdoors, they can enter a house through the deliberate openings of a controlled ventilation system, as well as through the random holes in the structure. The only way to prevent above-ground pollutants from entering the house via a ventilation system is to filter the incoming air (see Chapter 15) before it reaches the living space (this is a good option if an outdoor pollution problem is continuous) or by temporarily shutting off the ventilation system (this often works well for occasional outdoor pollution problems).

Pollution sources in building cavities

Insulation has the potential to yield the most significant pollutants found inside building cavities, and there are negative health concerns associated with the two commonest insulating materials: cellulose and fiberglass.[Bower 1989b, 1991b] Fiberglass is a possible carcinogen and fiberglass batt insulation contains a formaldehyde resin. Cellulose contains boric acid and printing-ink residues. Dry cellulose insulation is very powdery and easily inhaled.

If air infiltrates through a wall, floor, or ceiling cavity that has been insulated, it is possible for particles or gases from the insulation to pass into the living space. However, in most houses, the pressures aren't powerful enough to pull particles of insulation into the living space, so it usually stays inside the building cavities where it was placed. Of course, if a very strong negative pressure is applied to a house, for instance by a powerful kitchen exhaust fan in a tight house, particles of insulation can be pulled indoors. Some of the largest pressures experienced by a house are due to leaky

Understanding Ventilation

forced-air heating/cooling ducts, and it isn't unusual for attic insulation to be sucked directly into the leaky return-air ducts of a forced-air furnace or air-conditioner. This is generally only a problem with loose, blow-in insulations. Batt-type insulations contain a formaldehyde-resin binder that holds the fibers together.

Though the pressures generated by most ventilation systems generally aren't enough to cause particles of insulation to be sucked into the living space, the odor from the formaldehyde resin on fiberglass batts can be pulled into a house. Because fiberglass insulation is not a significant formaldehyde emitter, this isn't often a significant concern, except for very sensitive people. Some sensitive people report being able to smell formaldehyde leaking in around loose-fitting window frames when an exhaust fan causes house depressurization.

If a house is constructed with chemically treated lumber to prevent termite or microorganism attack, it is possible for air infiltrating through random holes in the structure to become contaminated with the treatment chemicals and bring them into the living space. This would be more of a problem with pentachlorophenol or creosote preservatives than with water-borne arsenic salts. In one study, the occupants of treated log houses had seven times as much pentachlorophenol in their systems as occupants of untreated homes.(Centers 1980)

Other pollutants found inside wall cavities could include construction debris, mold, mildew, and the conglomeration of dust that can accumulate over the years in an old leaky house. Inasmuch as animals such as rodents and birds often nest inside walls, attics, and crawl spaces, these cavities can also be contaminated with feces, urine, or decaying matter from dead animals. Insects (*e.g.* termites, ants, cockroaches, etc.) can also be found inside building cavities. Particles and gases from any of these materials or creatures can infiltrate indoors through the random holes in the structure whenever a house gets depressurized.

Preventing pollutants from being pulled into the living space

Pollutants will only be pulled into the living space if there are holes in the structure and an air pressure to push the pollutants through the holes. Therefore, there are two methods that can be used to prevent this from occurring: You can either tighten the house (plug up the holes) or manipulate the pressures. A great deal has been written about preventing radon from entering the living space of a house,(USEPA 1986, July 1991) and both of these approaches are widely practiced by radon-mitigation contractors. Tightening a house and manipulating pressures are also effective at preventing other soil gases and above-ground pollutants from being drawn indoors through random holes.

Tightening the house

Radon contractors often seal the random holes in a foundation to prevent the gas from being pulled from the soil into a basement. For example, they use caulking, grout, or expanding-foam insulation to seal cracks in floors and walls, the joint between the floor and the wall, holes around electric wires or plumbing lines, etc. They also plug open drains, put lids on open sump-pump pits, and seal any other holes between the living space and the ground.

In a house with a concrete slab-on-grade foundation, many of the cracks in the floor slab are relatively easy to caulk, but there can be some holes in the slab—under showers, cabinets, bathtubs, or partition walls—that can be difficult to close up. When heating ducts are located under a floor slab, it is virtually impossible to seal the leaky joints between each section of duct.

Crawl spaces offer two places to stop soil gases from passing from the soil to the living space. First, the gases can be prevented from actually entering the crawl space by sealing any openings in the foundation walls and by covering the floor of the crawl space with well-sealed plastic sheeting (or a concrete slab). Or, if the soil gases do get into the crawl space, they can be kept out of the living space by sealing any openings in the floor system such as plumbing, electrical, or duct penetrations. It is also important to seal all the joints in any ducts (especially return-air ducts) located in the crawl space. In older houses having individual boards for a subfloor (instead of sheet products such as plywood) there can be a potential air leak between each board—something that can be difficult to seal.

Above-ground leaks exist around window and door frames, the joints between the floor and the wall, around electrical outlets, as well as the many hidden holes in the walls, ceiling, and floor that were cut during construction of the house to run plumbing lines, wires, and ducts. These openings can be sealed in new construction by utilizing airtight construction practices, or in existing houses by using standard weatherization techniques. See also Chapter 11.

Manipulating pressures

There are several ways radon contractors manipulate pressures in a house. One of the most common involves using a fan to pull air through a perforated pipe that is located in the ground under a house and blowing that air outdoors. As air moves through the soil, it picks up the radon and any other soil gases. The contaminated air is then pulled into the pipe and exhausted into the atmosphere before it can seep into the living space. This approach only works if the negative pressure generated by the fan in the pipe is greater than any negative pressure generated (by natural, accidental, or controlled ventilation) in the living space. If the radon fan is sized properly, this technique generally works quite well. Because this approach depressurizes the area beneath a house, it will not only pull air from the soil, but also from the house (through the random holes in the foundation) into the perforated pipe. This can contribute to depressurizing a house that might otherwise be under a neutral pressure, so it could lead to other depressurization-related problems in the living space.

When a house is pressurized (*e.g.* with a central-supply ventilation system), air will continually be exfiltrating through the random holes in the structure, so soil gases and contaminants from within the building cavities will be prevented from entering the living space. If the incoming air is adequately filtered before it passes through the ventilation system's inlet, outdoor airborne pollutants will be prevented from entering.

If a house has a balanced ventilation system, the goal is to create a neutral pressure indoors—but the house will not experience a neutral pressure at all times. Sometimes, because of natural or accidental pressures, a house with a balanced ventilation system will be pressurized, sometimes it will be depressurized. So, a house with balanced ventilation system isn't necessarily immune from having pollutants pulled in through the random holes in the structure (unless it is a tightly constructed house that doesn't have many random holes). As a rule, local-exhaust fans (bath fans, range hoods, etc.), though they may be powerful, operate for relatively short periods compared to central ventilation systems, so their contribution to pulling pollutants into the living space is usually not significant.

AirPro, Inc. produces an automatic ventilation-fan controller that senses the pressure in a house and compares it to the pressure in the soil under the house. This controller then varies the speed of a ventilating fan to maintain a slight positive pressure indoors to keep radon and other soil gases out of the living space. AirPro also has a manually operated system. The company's approach won a USEPA Grand Prize Award in 1992 for innovation in controlling radon. While this approach works well at keeping radon out, it may or may not bring in enough fresh air to satisfy the needs of occupants. The control for AirPro's system is a "differential pressure transducer." These devices are discussed in Chapter 14. In general, radon control and house ventilation are two separate issues that should be addressed independently of each other.

What works the best?

In most cases, a combination of tightening a house and reducing negative pressures works well in preventing pollutants from being sucked indoors. Tightening should generally be the first step because it has other advantages (*e.g.* increased comfort, better energy efficiency, etc.). With a tight house, there are fewer random holes for pollutants to be sucked through; therefore, air will be more likely to enter through deliberate openings. In new construction, it is often possible to use special building techniques to ensure that the house will be extremely tight. It is usually easier to tighten a house during construction than after it has been built. See Chapter 11 for a discussion of tightening houses.

Does exhaust ventilation always cause pollutants to be sucked indoors? Not always. After all, a well-designed central-exhaust ventilation system in a tight house will have inlets, such as through-the-wall vents, for make-up air to enter the living space. Even though air will enter both the through-the-wall vents and through random holes in the structure, in a well-designed system the total area of the through-the-wall vents should be greater than the total area of the random holes. Therefore, more air will enter the through-the-wall vents. (This is discussed in more depth in Chapter 16.)

Because there are quite a few possible pollutants that can be sucked indoors by negative pressure, you might decide to tighten the house as much as possible, assume you have done a reasonable job of keeping out pollutants, and not worry about whether a ventilation system causes any depressurization. Or, you might decide to combine a central-exhaust ventilation strategy with a radon-removal system installed in the ground under the house. Or, you might decide to minimize house depressurization by installing a central-supply ventilation system.

If a house is expected to be pressurized most of the time, tightening is less important for control of soil pollutants. (Tightening is still important for comfort and energy-efficiency.) When a house is pressurized, air will leak out of the random holes in the structure and pollutants won't be allowed to enter. This is a definite advantage of supply ventilation systems. However, this approach has its limitations because it won't be possible to significantly pressurize a loose house, and supply ventilation has the potential to cause hidden moisture problems in the winter in many parts in the U.S., especially when the indoor air has a high RH and the outdoor temperature is low.

With a balanced ventilation system, natural and accidental pressures can occasionally cause pressurization or depressurization, so a tight house is still a good idea. This is because air can also enter and leave through random holes because of natural and accidental pressures. Though still vulnerable to occasional episodes of depressurization, a balanced-ventilation system itself won't pull pollutants into the living space

What about a house that has a central ventilation system that either pressurizes the house or results in a neutral pressure, and also has some local ventilation fans that depressurize the house whenever they are running? During episodes of depressurization, make-up air will try to enter through the inlets or outlets of the general ventilation system. But some will no doubt also try to enter through the random holes in the structure. How much air enters through the structure depends on how many holes there are to pass through (the house tightness) and how much negative pressure is created. With a powerful 600 cfm kitchen exhaust fan, there will be 600 cfm of infiltration—and the possibility of significant depressurization. On the other hand, a 50 cfm bathroom fan will only have 50 cfm of infiltration and considerably less depressurization. To minimize excessive depressurization with a powerful local-exhaust fan, see Chapter 20.

The bottom line: Tighten up the house first, then whatever ventilation strategy you choose will be less likely to cause pollutants to be sucked indoors. If you are building a new house that is extremely airtight, exhaust ventilation will probably cause no problems. If you have an existing house that is of average tightness, an exhaust ventilation system, even though it will pull in some pollutants, will generally result in better air quality than existed in the unventilated house. A very loose house may have plenty of accidental and natural ventilation, so a mechanical system is less important, but such a house will likely be uncomfortable and it will consume a considerable amount of energy for heating or cooling. In a situation where there are high levels of pollutants in the soil or in the outdoor air waiting to be sucked indoors, either a balanced ventilation system or a supply ventilation system will be less likely to result in pollutants being pulled indoors. But it is usually more effective to deal with pollutants such as radon or excess ground moisture as separate issues from ventilation.

Already-contaminated houses

If a pollution problem exists in a house and it is caused by depressurization, it is often possible to manipulate the pressures and/or tighten the house to reduce the occupants' exposure. For example, when less radon is allowed to enter, the radon remaining indoors will quickly decay, leaving the house uncontaminated. If excess water vapor was being pulled indoors by a negative pressure, and the problem is corrected, the excess indoor moisture will be removed quickly with a general ventilation system.

Unfortunately, some highly toxic contaminants that are pulled indoors by pressure differences can create long-lasting problems—even after random holes are sealed and pressures are modified. For example, chlordane or pentachlorophenol can permanently contaminate a house. This is because, once they get indoors, they are absorbed by building materials and furnishings, where they remain toxic for decades. In one study, pentachlorophenol was found to have permanently contaminated dust, curtains, furniture, carpets, clothing, and food in a pentachlorophenol-treated house.(Gebefugi) In such a situation, you can prevent additional pollutants from entering the house by sealing random holes or manipulating pressures, but you may also need to get rid of contaminated furnishings and do some serious cleaning of floors, walls and ceilings.

Chapter 10

Backdrafting and spillage

A healthy human being can tolerate small exposures to many indoor pollutants, so our primary concern should be about exposures over an extended period of time. For example, a short-term exposure (say, a month) to a moderate level of radon isn't nearly as serious as several years of exposure to the same level. There is one particular pollutant, carbon monoxide, that we should be very concerned about, even at low levels for short periods of time—because it can kill swiftly. Carbon monoxide (CO) is one of several combustion by-products that should go up the chimney—but sometimes finds its way into the living space.

Backdrafting and *spillage* are two related ways that combustion by-products get into the living space of houses. They are very closely tied to two things: 1) the amount of depressurization to which a house is subjected (this depends on the tightness of the house and the exhausting capacity), and 2) the design and condition of the chimney. The type of ventilation system in a house can have a direct bearing on backdrafting and spillage if it causes the house to become depressurized.

In most houses, combustion by-products are expelled through a chimney having a *natural draft*. The draft is called "natural" because there is no fan—the only pressure involved is naturally occurring. (The warm combustion gases rise because of stack effect.) Sometimes this is called an atmospheric-draft chimney and it has been the traditional approach to getting combustion by-products out of buildings for centuries. In today's houses a natural draft often doesn't work very well, especially when a house is depressurized. It takes surprisingly little negative pressure in a house (3–5 Pa.) to cause backdrafting and spillage.

Combustion by-products are warm and buoyant so they have a tendency to rise up into cooler air. As the warm gases rise inside a chimney, they create enough upward pressure that they pull additional air with them. Thus, a natural draft in a chimney will contain both com-

Understanding Ventilation

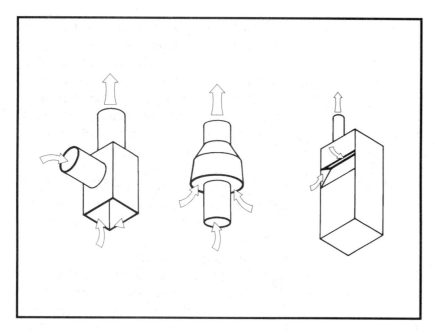

Figure 10–1. Dilution air flows into a chimney through a draft hood, which can take several different shapes. The one at the far left might attach to the side of a furnace, the one in the center to the top of a water heater. The furnace at the right has a built-in draft hood.

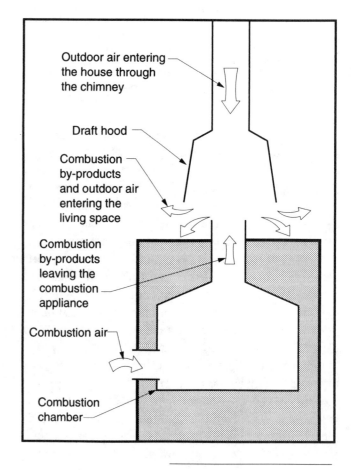

Figure 10–2. When a chimney is backdrafting, there is so much air coming down the chimney that none of the combustion by-products can escape. Instead, the combustion by-products flow around the draft hood and into the furnace room.

Chapter 10 Backdrafting and spillage

bustion by-products as well as some extra dilution air from the furnace room. With gas furnaces and gas water heaters, this extra air enters the chimney around a device called a draft hood. Draft hoods come in a variety of shapes such as those shown in Figure 10–1.

When a house becomes depressurized for any reason (natural, accidental, or controlled pressures), make-up air can enter the house from the outdoors by coming down the chimney. After all, a chimney is just a hole in a house that air can move through in either direction. *Backdrafting* occurs when so much air flows down a chimney that none of the combustion by-products can be expelled from the house (Figure 10–2). *Spillage* refers to a situation where some of the combustion by-products are expelled and some spill into the room around the base of the chimney or the bottom of the draft hood (Figure 10–3). After a while, a spillage problem often corrects itself because, as the inside of a chimney gets warmer, the draft becomes more powerful.

If a house has any type of fuel-burning appliance connected to a chimney, a water heater, furnace, wood stove, etc.—it is important to make sure that the chimney is operating correctly. A water heater connected to a chimney with a small-diameter pipe that runs horizontally for several feet can be more susceptible to depressurization-induced backdrafting or spillage than a furnace or wood stove having a vertical connection to a chimney—but any natural-draft chimney can be at risk. Proper chimney operation must be checked before and after installing a ventilation system because anything that changes the pressures in a house can cause backdrafting or spillage.

Fortunately, there are solutions to this problem. There are fuel-burning appliances that are immune from backdrafting and spillage. Some don't rely on a natural draft because they expel their combustion by-products outdoors without using a conventional chimney. Others are installed in such a way that they are not affected

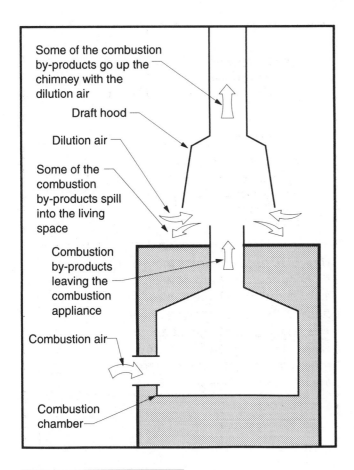

Figure 10–3. Spillage occurs when the upward pressure due to the natural draft in the chimney is too weak. With spillage, some of the combustion by-products flow up the chimney and some spill around the draft hood into the furnace room.

Understanding Ventilation

by negative pressures in the house. Safe installations will be discussed later in this chapter.

Combustion gases

One of the by-products of the combustion of common heating fuels is water vapor. Almost a pound of water vapor is released by a pound of burning fuel. If this moisture isn't expelled from the house—preferably up the chimney—it can easily contribute to a high indoor humidity problem.

Other combustion by-products include carbon monoxide, carbon dioxide, formaldehyde, nitrogen oxides, particulates, sulfur oxide, and various hydrocarbons. In general, burning wood and oil is more polluting than burning natural gas or liquefied petroleum gas (propane), but no combustion appliance is problem-free. For example, the combustion gases released by burning natural gas are more noxious when an appliance is out-of-tune. A gas appliance should have a blue flame—a yellow flame indicates incomplete combustion and high CO production

Backdrafting and spillage are very serious problems because combustion gases can kill if their concentration is high enough. While there are negative health effects associated with many different combustion gases, carbon monoxide (CO) is by far the most deadly. Being colorless, odorless, and tasteless, it is virtually impossible to detect with your senses.

Small amounts of CO in the air are deadly because the gas is so easily absorbed by our bodies. When you inhale air containing CO into your lungs, your blood cells are always attracted to the CO molecules more readily than to oxygen molecules. As more CO is taken up by the blood, less oxygen gets absorbed. Fortunately, not many houses have CO levels high enough to kill the occupants; however, excessive CO in houses does

Figure 10–4. Sometimes there can be enough downward flow of air in a chimney to blow the flames outside the combustion chamber. This is called flame rollout and it can create a very serious fire hazard. It is more common in water heaters than in furnaces

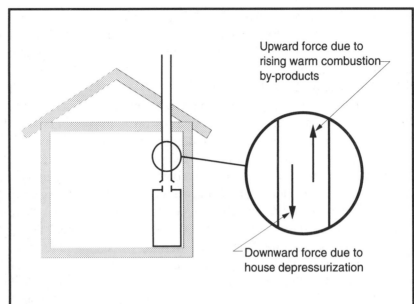

Figure 10–5. Chimneys often experience a "battle of pressures"—an upward natural draft pressure vs. a downward pressure due to house depressurization. Whichever pressure is stronger determines if combustion by-products will enter the living space.

kill several hundred people each year in the U.S. The following CO levels are considered dangerous enough to require an immediate remedy:

>9 ppm for an 8-hour exposure
>35 ppm for a 1-hour exposure
>200 ppm for a single exposure

At levels that aren't considered deadly, CO can cause flu-like symptoms such as headache, dizziness, nausea, vomiting, diarrhea, weakness, general malaise, or shortness of breath. People with heart disease are more susceptible to these symptoms. One study reported that 23.6%—nearly a quarter—of patients reporting flu-like symptoms were actually suffering from low-level carbon monoxide poisoning.(Dolan) Many more people are probably affected but, no doubt, assume that when they get sick each heating season, they have the flu. A national survey conducted in 1974 estimated that as many as 700,000 houses in the U.S. have elevated CO levels during the winter months.(Schaplowsky)

A classic example of how CO can enter the living space involves a house with a wood-burning fireplace and a forced-air natural-gas furnace, each connected to a separate natural-draft chimney. After supper, on a cold winter evening, the family builds a romantic fire in the fireplace. As the evening wears on, the fire dies down, and the family retires to bed. The draft in the chimney becomes weaker a few seconds after the flame dies out. Only glowing embers are left and they produce a great deal of CO due to incomplete combustion. As the house cools down, the gas furnace, which is connected to a different chimney, starts up. As the furnace's combustion gases move up the chimney and out of the house, the house becomes depressurized. (Closing bedroom doors at night often contributes to depressurization.) Make-up air enters the house by the path of least resistance—down the fireplace chimney. The make-up air picks up the CO from the dying embers and distributes it throughout the house. By morning, family members complain of headache and nausea. This scenario is most common when the fireplace has a masonry chimney on an exterior wall because such a chimney can cool off and lose its draft quickly.

Sometimes enough make-up air rushes down a chimney to blow the flames of a fuel-burning appliance outside the combustion chamber. This is called *flame rollout* and it can represent a very real fire hazard (Figure 10–4). At other times the air flowing down a chimney causes a pilot light to inconveniently blow out. This is more common in water heaters than in furnaces.

What if you never use your fireplace? Even if a chimney is no longer being used to expel combustion gases from the house, it isn't a good pathway for air to

Understanding Ventilation

enter the living space because the incoming air can become contaminated with soot, creosote, or other pollutants. If a fireplace is unused, a tight-fitting damper or an inflatable pillow (see Chapter 6) can prevent unwanted infiltration down the chimney.

If a combustion appliance (furnace, water heater, boiler, kitchen range, etc.) that burns gas or oil is in tune and working properly, there should not be a significant amount of CO being produced. CO production is only high when there is incomplete combustion of the fuel. Even if backdrafting or spillage aren't a problem, it is important to keep CO production to a minimum because it means more efficient fuel consumption—and a lower heating bill. Burning wood always produces CO, especially when there are glowing embers and no flame is present.

How house tightness affects backdrafting and spillage

It can sometimes be difficult for a natural draft to form in the chimney of a tight house. This is because when air leaves the house through the chimney, an equal volume of air needs to enter the house from somewhere else. If the house is tightly constructed, the make-up air will have a difficult time entering because there aren't enough random holes in the structure. If not much air can enter, then not much air can leave. Therefore, in a very tight house, a natural draft can have a difficult time getting started. In an extremely tight house it can be impossible for a draft to form. Therefore, fuel-burning appliances that rely on natural-draft chimneys may not function correctly in very tight houses. Building codes generally require a combustion-air duct so outdoor air can reach a combustion appliance (see below), but sometimes there can be enough depressurization to cause backdrafting or spillage in a house that meets the requirements of the code.

Backdrafting and spillage occur in loosely built houses if the negative pressure generated in the house is stronger than the upward pressure created by the natural draft in the chimney. The negative house pressure can be the result of natural pressures (wind or stack effect), accidental pressures (clothes dryers, leaky ducts, etc.), or controlled pressures (exhaust fans). Many houses have a continuous "battle of pressures" going on in a chimney—house pressures vs. the natural-draft pressure. The stronger pressure wins the battle and the right to determine the direction the air moves in the chimney (Figure 10–5). If the house pressure is stron-

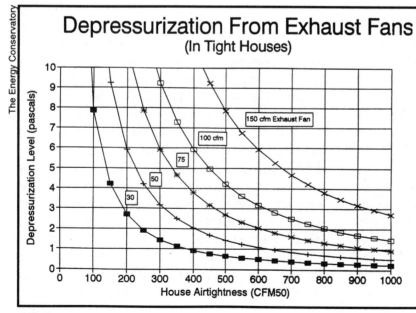

Figure 10–6. This chart shows how much an exhaust fan will depressurize a tight house. (House tightness can be expressed as an airflow at a certain pressure, (CFM_{50}), something that will be covered in Chapter 11.) Based on this chart, a house that has been measured to have a tightness of 500 CFM_{50} will experience about 4.2 Pa. of depressurization when a 100 cfm exhaust fan is used.

ger than the natural draft in the chimney, the occupants are the real losers because they will be exposed to the combustion by-products that can't escape from the house. While it can be difficult for house pressures to depressurize a large loosely built house enough to cause a backdrafting or spillage problem, it often occurs in houses of average tightness when strong negative pressures are present.

Backdrafting and spillage are more serious concerns today than in our grandparents' era for several reasons. First of all, when houses were uninsulated, they required very large heating systems which operated with very powerful natural drafts—drafts that were strong enough to overcome any non-mechanical negative pressure applied to the house. (Of course, back then, mechanical devices such as clothes dryers and exhaust fans weren't available to apply negative pressure to a house.) Second, fuel was burned inefficiently and more heat went up the chimney. (A hotter chimney has a more powerful draft than a cooler chimney.) Third, many of the chimneys were centrally located (not on exterior walls), so stack-effect pressures experienced by the house itself provided a boost to the natural draft in the chimney. Fourth, houses used to be fairly leaky, so there were rarely any strong negative pressures in a house—air easily blew right through the walls. And fifth, today's houses often have very powerful exhaust fans in them (*e.g.* kitchen-range hoods and barbecue fans) that have the capacity to create a great deal of depressurization.

The graph in Figure 10–6 shows the amount of depressurization an exhaust device can induce in a house of given tightness. In this chart, the house tightness is expressed as CFM_{50}, something that can actually be measured. (100 CFM_{50} is extremely tight and 1000 CFM_{50} is moderately tight.) CFM_{50} will be discussed in depth in Chapter 11. What this chart shows is that tighter houses are more easily depressurized than loose houses, and tight houses can be depressurized with smaller-capacity exhaust fans.

During the normal start-up of a natural-draft chimney, there will often be a small amount of spillage. Then, as the chimney warms up, the draft pulls extra dilution air around the draft hood and into the chimney along with the combustion by-products. When it does so, it pulls most of the spilled gases from the air around the base of the chimney and sends them outdoors. This small amount of spillage has not been considered a problem in the past; in fact, it has been considered normal. However, any amount of spillage can be a serious concern for the growing number of people who are hypersensitive to even tiny amounts of combustion by-products. For them, it is very important to get all of the combustion by-products out of the house—not just most of them.

Other causes of backdrafting and spillage

Not all backdrafting and spillage occurrences are directly related to house depressurization or house tightness. Sometimes a problem is due to poor chimney design or weather conditions. For example, with today's energy-efficient heating systems, less heat is wasted. This means that the combustion gases being exhausted aren't as hot; therefore, the draft isn't as strong as it is with an inefficient heating system. A draft will be even weaker if a chimney is incorrectly sized. For example, a chimney that is too small may not have enough capacity to do its job properly. On the other hand, it can be difficult for an adequate draft to be produced by a small furnace connected to a very large diameter chimney, or a chimney that has angled jogs in it rather than a straight vertical run. It isn't uncommon for an older house to be made more energy-efficient by adding insulation and weather-stripping. If the house needs a new furnace, it will probably need less capacity than the original furnace, so it may require a chimney liner to reduce the size of the original chimney.

A strong wind blowing across the top of a chimney can affect the pressures that cause a draft to form. If a tall chimney reaches above the peak of the house, the wind can create a suction in the chimney, helping a draft to form. But if the chimney is shorter, and it terminates below the peak of the house, it is possible for wind blowing over the roof to produce eddy currents that will inhibit a draft.

Chimneys can become blocked by a buildup of creosote (a common problem with burning wood), an animal nest, or even a dead animal such as a raccoon. And cracked, corroded, incomplete, misaligned, or miss-

Understanding Ventilation

ing liners can adversely affect chimney function. Even if a chimney is functioning correctly, some combustion by-products can still enter the living space on occasion. For example, opening the door of a wood stove to stoke the fire or to add wood will result in some combustion by-products entering the house.

All the factors that affect chimney operation will be more serious whenever an exhaust device is used in a house. If a draft is weak for whatever reason, it becomes more likely that any negative pressure in the house will be greater than the draft—even in a loosely built house.

Chimney function in the winter

As was mentioned earlier, in the winter a chimney located in the center of a house will work better than one attached to a cold exterior wall. In fact, chimneys on exterior walls only work marginally well under the best of circumstances in cold climates. This is because the outward pressure acting on the entire house due to stack effect is strongest at the top of a house, and a chimney that is primarily within the insulated part of the structure will exit through the top of the house. Thus, the outward stack-effect pressures acting on the top of the house will cause air to leave the chimney whenever it is cold outdoors. This enhances the chimney's ability to sustain a draft. Actually, the colder it is outside, the greater the indoor/outdoor temperature difference, and the better such a chimney will function. On the other hand, if a chimney penetrates an outside wall (instead of the ceiling), the house's outward stack-effect pressures at that point will be less (in winter) than they would be if the chimney penetrated the ceiling (Figure 10–7).

In the winter, the worst place for a chimney to penetrate an outside wall is below the neutral-pressure plane. This is because the house's stack-effect pressures are inward, and air will try to enter the living space through the chimney. Fireplaces are often hard to start or smoky in the winter when they have a chimney outside an exterior wall—a basement fireplace with a chimney outside the insulated wall may never function correctly during the winter. A chimney that penetrates the insulated wall below the neutral-pressure plane will function worse and worse the colder it gets outdoors, because the inward pressures are stronger during the coldest weather. And this is when space-heating combustion appliances are more likely to be in use, so a chimney that penetrates an outside wall below the neutral-pressure plane functions at its worst when it is needed the most. (See Chapter 5 for a discussion of how stack effect in the winter causes air to enter a house

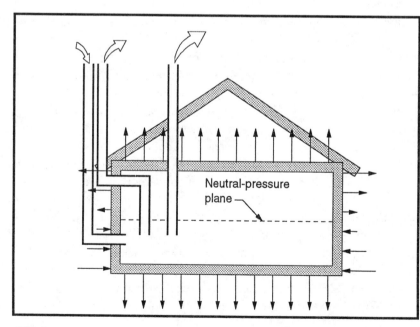

Figure 10–7. The point where a chimney penetrates the shell of a house determines how the house's stack-effect pressures will interact with the chimney's operation. In the winter, a chimney that penetrates the ceiling will have its draft enhanced by the outward stack-effect pressure acting on the house itself. The lower a chimney penetrates the shell, the less the outward pressure. When a chimney penetrates the shell of the house below the neutral pressure plane, the house's stack-effect pressures will cause air to flow down the chimney, making a draft difficult, sometimes impossible, to form.

Figure 10-8. Unvented combustion appliances, such as portable kerosene space heaters, have no chimney, so their combustion by-products flow directly into the living space. To avoid exposure to combustion by-products, you should avoid unvented combustion appliances.

through low openings and leave a house through high openings.)

For a natural draft to be sustained, the inside a chimney must be warm. If enough hot combustion gases are trying to leave a house through a chimney that is outside the insulated part of the house, they can eventually cause the chimney to become warm enough so that it will begin to function correctly. Once such a chimney is warm, its own internal outward pressure will overpower the inward stack-effect pressure experienced by the house itself. But this may only occur after an extended period of backdrafting or spillage.

Chimney function in the summer

In an air-conditioned house in a warm climate, a chimney that penetrates the ceiling will function worse than one that penetrates low on a wall, because the stack-effect pressures push inward at the top of the house (working against the draft) and outward at the bottom of the house (enhancing the draft). So, a chimney in the center of a house that penetrates the ceiling will function more poorly in a southern air-conditioning climate than in a northern heating climate. Even though it is unlikely that a furnace or fireplace will be operating during the air-conditioning season, chimneys serving gas water heaters are in use year-round.

Solutions to backdrafting and spillage

There are several ways to prevent combustion by-products from entering the living space of a house. The easy answer is to avoid the use of fuel-burning appliances altogether. In some instances, this is the best thing to do, but there are alternatives. For example, in new construction, a good choice is to select fuel-burning appliances that are immune from backdrafting and spillage. In an existing house where it isn't economically feasible to replace the furnace or water heater, there are other solutions.

Avoid unvented fuel-burning appliances

Unvented fuel-burning appliances are not connected to the outdoors with a chimney. They deposit combustion by-products directly into the living space. Vent-free gas fireplaces and portable kerosene space heaters are examples (Figure 10–8). The only way for the combustion by-products from these devices to reach the outdoors is through dilution by natural, accidental, or controlled ventilation. Manufacturers often recommend "adequate ventilation" or "opening a window," but this rarely provides enough dilution. Because there are negative health effects to most combustion by-products—CO has the potential to kill the occupants—they should always be expelled from a house as soon as possible. Because this is difficult to do with an unvented space heater, they have been banned in several locales. Most indoor air quality experts highly recommend that unvented fuel-fired space heaters simply not be used. Many government-funded weatherization programs will not do any work on a house without physically removing an unvented combustion appliance.

There is one unvented fuel-burning appliance that is much more common than a kerosene space heater—a natural-gas (or propane) kitchen range. It can release pollutants into the kitchen whenever there is a flame present. If a range hood is vented to the outdoors and it is operating, then the burned gases will be expelled from the house. But range hoods aren't always operating because they are too noisy, some aren't very efficient, and some range hoods are not vented to the outdoors. Thus, combustion by-products often enter the living space whenever a gas kitchen range is used. A typical gas range can actually release more combustion by-products than a typical gas water heater when the oven and three burners are being used. If a range has a pilot light, it will emit small amounts of various combustion gases whenever it is lit—which may be 24 hours per day. A range with electronic ignition will at least eliminate the continuously burning pilot light.

The bottom line: Don't use unvented combustion appliances; always ensure that combustion by-products are vented outdoors.

Controlling negative house pressures

If a house is pressurized (*e.g.* with a central-supply ventilation system), air will continually leave any random holes in the structure. Most of the air will leave through the largest holes in the house. Because the chimney is basically just a hole in the house, air will often escape from the house by flowing up it. This will enhance the natural draft and minimize the possibility of backdrafting or spillage.

If a house is depressurized (*e.g.* with an exhaust ventilation system), backdrafting and spillage are definite possibilities, but only if the depressurization is stronger than the draft in the chimney. Therefore, it isn't just depressurization that causes the problem—it is the *amount* of depressurization. As it turns out, the natural draft in most chimneys, in most situations, has enough upward pressure to overcome 3–5 Pa. of house depressurization. This means that as long as a house isn't depressurized more than 3–5 Pa., backdrafting and spillage are unlikely. However, the amount of depressurization that a chimney can tolerate and still function is dependent on a number of variables. For example, chimneys can backdraft at lower house pressures in the summer. Safe operation is also dependent on the design, size, and condition of the chimney, and where the chimney penetrates the insulated shell of the house (*e.g.* low on a wall or through the ceiling). The precise pressure necessary to cause a problem also depends on the chimney temperature and its height. The bad news is the fact that many houses are tight enough for depressurization-induced backdrafting or spillage to be a serious long-term problem. For example, in a very tight house, it wouldn't be unusual for a powerful bath fan to generate 5 Pa. of negative pressure, or a range hood to depressurize the house by 10 Pa.

When a house has a balanced ventilation system (neutral pressure), there can still be negative pressures due to natural or accidental pressures, so balanced ventilation isn't a guarantee that backdrafting or spillage won't occur. But a neutral-pressure ventilation system will reduce the likelihood of a continuous depressurization. Of course, occasional depressurization, due to local-exhaust devices or natural or accidental pressures, can cause backdrafting or spillage in any house, no matter what kind of general ventilation system it has. For example, when a strong local-exhaust fan causes more than 5 Pa. of depressurization in a house, you could have a problem. Because the strength of house depressurization is related to the house's tightness, this is more likely to occur in a tight house than in a loose house. With CO poisoning being such a serious issue, even occasional episodes of backdrafting and spillage should be of concern.

Appliances that are immune from backdrafting and spillage

Various studies have found that the majority of houses containing natural-draft fuel-burning appliances have the potential for backdrafting or spillage. This is because most houses are depressurized to some extent at least occasionally. Also, many houses are being tightened up for energy conservation reasons. A house that only occasionally becomes depressurized can, after tightening, be much more easily depressurized. As a result of the potential seriousness of CO poisoning, many manufacturers are now producing fuel-burning appliances that either tolerate a greater degree of house depressurization, or are immune from pressure changes in a house altogether. This is all part of an evolutionary process going on in the housing industry: We expect more comfort, energy efficiency, and safety—and products are evolving to meet that demand.

There are also electric appliances—heat pumps, electric furnaces and water heaters—that don't even require a chimney, but some of these are significantly more expensive to operate than their combustion-fueled counterparts.

Fuel-burning appliances

Any time a house with a fuel-burning appliance connected to a natural-draft chimney gets depressurized, backdrafting or spillage are possibilities. Figure 10–9 shows a typical natural-draft furnace installation. Actually, even though these low-efficiency (62%) natural-draft gas furnaces still exist in many houses, they are no longer being manufactured. Today, gas furnace manufacturers are offering medium-efficiency (80%) models with *induced-draft* or *forced-draft* fans that push or pull air through the combustion chamber (Figure 10–

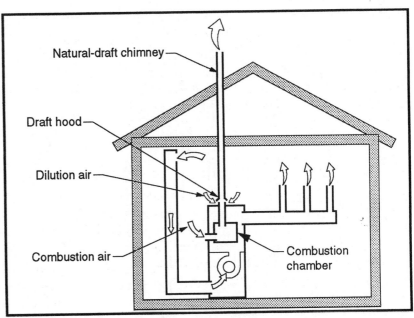

Figure 10–9. With a natural-draft appliance, combustion air enters the combustion chamber where it mixes with the fuel and is burned. The combustion by-products leave the combustion chamber, mix with dilution air, which enters through the draft hood, then they flow up the chimney as a result of being warm (warm air rises).

Understanding Ventilation

10). The terms power-vent and fan-assist are sometimes used to describe these appliances. These approaches (which added about $200 to the cost of a new furnace) sometimes reduce the likelihood of backdrafting and spillage, but not always; it depends on the specific approach taken. Some manufacturers also offer induced-draft water heaters.

Induced-draft and forced-draft appliances pull air from inside the house into the combustion chamber where it mixes with the fuel and burns, then the combustion by-products are expelled outdoors (sometimes they can be vented through a sidewall). With air leaving the house, the house becomes depressurized to a certain extent. Thus, make-up air will enter wherever it

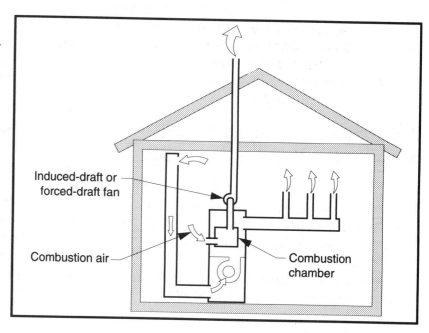

Figure 10–10. Medium-efficiency combustion appliances use an induced-draft or a forced-draft fan to push or pull combustion by-products through the combustion chamber. Some of these devices require a well-sealed, pressurized exhaust that reduces the likelihood of backdrafting or spillage. But some utilize a natural-draft chimney, so backdrafting and spillage are possible. Because combustion air is drawn from the living space, house depressurization occurs with all these devices.

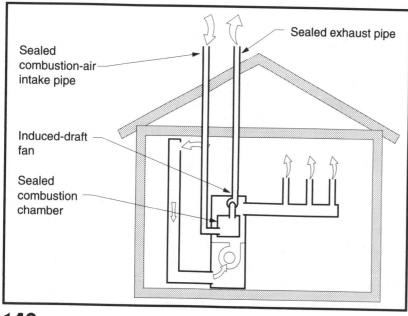

Figure 10–11. The highest-efficiency combustion appliances often have totally sealed combustion chambers. They are characterized by both a sealed combustion-air intake pipe and a sealed exhaust pipe.

Chapter 10 Backdrafting and spillage

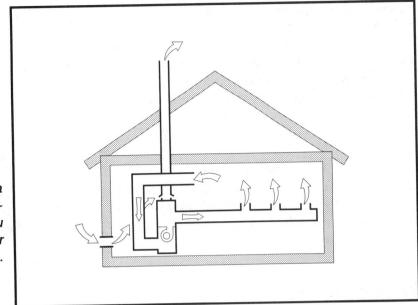

Figure 10-12. By adding a combustion-air duct near a natural-draft combustion appliance, you create a path for the combustion air to enter the house.

can—perhaps through the old, no-longer-used chimney. This depressurization can also lead to other problems, such as pollutants being sucked into the living space (see Chapter 9).

Some high-efficiency appliances have *sealed combustion chambers*. Thes term *direct-vent* is often used to describe them, and they draw air from the outdoors rather than from the house. Sealed-combustion furnaces, boilers, and water heaters will not affect the pressures in the house, nor will their operation be compromised by negative house pressures. During operation, air is pulled from the outdoors through a sealed ABS or PVC plastic pipe into a sealed combustion chamber. (Other types of plastic are being used, but they aren't as durable.) Inside the chamber, the air blends with the fuel and the mixture is burned. A great deal of heat is then extracted from the combustion by-products through a series of sealed heat exchangers, then the combustion by-products are cool enough to blow outdoors through a second sealed plastic pipe (Figure 10–11). The entire process is totally sealed.

Appliances with sealed combustion chambers are the most energy-efficient (90% or more) as well as the healthiest on the market. One of these furnaces is about $600–800 more than a medium-efficiency furnace. Most forced-air gas furnace manufacturers produce sealed-combustion models. Sealed-combustion gas boilers, water heaters, and fireplaces are also available.

Sealed-combustion forced-air oil furnaces are less common, but they are available from **Dornback Furnace and Foundry**. The **Riello Corporation of America** produces *high-pressure oil burners* that are used in boilers and forced-air furnaces. These burners generate more draft pressure in a chimney than a conventional oil burner, so they are much less susceptible to backdrafting or spillage.

Because of serious concerns about backdrafting and spillage, energy-efficient houses that qualify under guidelines such as Canada's R-2000 program and the Energy Crafted Homes program in the northeastern United States are not allowed to have natural-draft fuel-burning appliances. Sealed combustion, though not required elsewhere, is a very good idea.

Non-fuel-burning appliances

Another way of ensuring that backdrafting and spillage won't occur is to use appliances that do not burn fuel. Electric furnaces, space heaters, boilers, and water heaters have no internal combustion, so they have no need to be connected to a chimney. Neither backdrafting nor spillage will be a problem with electrical resistance forced-air heating, self-contained baseboard heaters, radiant ceiling panels, radiant heating cables in the floor, and heat pumps. To further avoid combustion by-products, electric kitchen ranges are better choices than conventional gas ranges. Unfortunately, electric heating can be expensive, and most electricity is generated by burning coal, a process that releases large amounts of combustion by-products into the atmosphere that contribute to outdoor air pollution and the greenhouse effect.

There is no possibility of backdrafting or spillage with solar heating either. This might take the form of either active solar heating with fluid-filled panels on the roof, or passive solar heating with strategically located south-facing windows.

Remedial measures for natural-draft appliances

The first strategy to eliminating a backdrafting or spillage problem in a natural-draft chimney should be to relieve the depressurization in the vicinity of the base of the chimney. This is often done by fixing leaky return-air ducts of a forced-air furnace/air conditioner or providing make-up air for a powerful downdraft kitchen-range exhaust.

Another solution is a combustion-air duct—something that is currently required by most building codes. This is simply a connection between the outdoors and the mechanical room (Figure 10–12). A combustion air duct allows make-up air to enter the house easily in the vicinity of the base of the chimney so a natural draft won't have difficulty forming. With this type of installation, outdoor air will enter the duct whenever the mechanical room is depressurized.

A combustion-air duct is a deliberate opening in the building, and air is free to flow through it. In the winter, the stack-effect pressures experienced by the house itself can cause air to enter a combustion-air duct continuously if it penetrates the insulated shell of the house below the neutral-pressure plane. This can lead to a cold mechanical room, which can lead to a cold fuel-burning appliance and perhaps frozen water pipes. Too much cold air entering a natural-draft appliance can result in condensation and rusting inside the unit. If the air is cold enough, it can actually crack a hot heat exchanger. Fortunately, the air reaching a heat exchanger is rarely cold enough to cause damage—but it can happen. To be on the safe side, air entering a combustion chamber should be warmer than 12°C (53°F), but it is always a good idea to consult the equipment manufacturer to determine how cold the combustion air can be

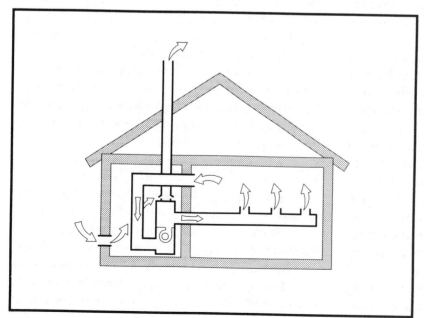

Figure 10–13. *When a combustion appliance is placed in a sealed room having a combustion air duct, the appliance is immune from pressure changes in the rest of the house. The combustion-air duct supplies air to both the combustion chamber and the draft hood. The combustion by-products and the dilution air then escape up the chimney. Of course, any leaks in the return-air ducts within the sealed room will cause the room to become depressurized.*

Chapter 10 Backdrafting and spillage

Figure 10–14. This fan can be used to blow enough air into a room containing a combustion appliance to avoid the problem of air flowing down the chimney.

before damage is possible. Appliances that have sealed combustion chambers are specially designed to handle cold incoming air.

If a house becomes significantly depressurized (perhaps due to leaky ducts or closed doors between rooms), a large amount of air can enter a combustion-air duct—far more than is needed for combustion. In effect, the duct becomes both a combustion-air supply and a fresh-air supply, and the amount of air entering is being controlled by accidental pressures. This isn't a good idea.

Supplying air to a combustion appliance to prevent backdrafting and spillage is not the same thing as supplying air to a house for the occupants. *Combustion air* and *ventilation* air should always be considered two separate issues, but they are often confused with each other. They should always be handled separately—never with the same duct—and in a way that they do not compete with each other. Because combustion air ducts can result in a cold mechanical room, occupants sometimes block them off, thinking they are saving energy. Therefore, it is always a good idea to label any ducts that run to the outdoors so that the occupants will know they have an important purpose.

A variation of this approach involves combining a combustion air duct with an airtight mechanical room.

First, the room containing the furnace, boiler, or water heater is made airtight by using caulking, gaskets, and weather-stripping, then a combustion-air duct is run between the sealed room and the outdoors. All the ducts, pipes, and wires that pass through the walls, floor, and ceiling of the mechanical room must be sealed in an airtight manner (Figure 10–13).

A sealed mechanical room is only connected to the outdoors with the combustion-air duct and the chimney. Therefore, it won't be able to sense the fact that a house is depressurized, so the natural draft in the chimney won't be affected. In other words, the furnace will be in a location that is immune from house depressurization. For this approach to work correctly, the combustion-air duct must be sized appropriately for the requirements of the combustion appliance. From a practical standpoint, sealing a mechanical room is easier to do with a boiler or water heater than with a forced-air furnace because it is easier to seal around small pipes than large ducts, but it can be done with any type of combustion appliance.

Combustion appliances are often placed in attached garages. This can be a good location, but it isn't immune from problems. Because of loose-fitting overhead doors, garages are usually not subjected to the same degree of depressurization as houses, but they can be, particularly as a result of leaky ducts. Because a garage is not considered living space, a certain amount of backdrafting or spillage is assumed to be of minimal concern. However, it isn't unusual for polluted air from an attached garage to be pulled into the living space because of negative pressures in the house. It is also quite common for garage air to be sucked into leaky return-air ducts that are located in the garage, then distributed throughout the living space.

If you plan to install a furnace in an attached garage, the best way to keep combustion gases out of the living space is to build the garage so air cannot move between it and the house. This means an airtight ceiling, airtight walls, and an airtight service door. (Actually, this is a good idea even if there isn't a combustion appliance in the garage, because automobile exhaust gases can be equally deadly and should be prevented from entering the living space as well.) It is also very important that any ducts (especially return-air ducts) located in the garage be well-sealed.

Products are available to reduce backdrafting and spillage problems. **Exhausto, Inc.** has several sizes of chimney fans that mount on top of either a new or an existing chimney to suck combustion by-products out of the house. These are high-quality devices capable of tolerating temperatures up to 650°F and they can be used in a variety of applications, with any type of fuel. By pulling air up the chimney, they enhance the draft to ensure that all the combustion by-products are removed from the house efficiently. In doing so, they cause the house to become depressurized; thus, they act like a central-exhaust fan whenever they are operating. Periodic cleaning is necessary to remove any accumulation of soot, but the biggest drawback is cost—they retail for over $1,000.

Tjernlund Products, Inc. has a Combustion Air In-Forcer fan system and a Power Combustion Air Intake that blow air indoors to supply combustion air whenever a furnace is operating (Figure 10–14). They are available in different sizes for combustion appliances of different capacities. Unfortunately, they may not deliver enough combustion air if house depressurization is excessive.

Outdoor furnaces and boilers

Another way to prevent combustion by-products from entering the living space is to place the fuel-burning appliances outdoors. While this may seem like an unusual approach, it has been done successfully. It involves constructing a small mechanical-equipment building (perhaps something like a metal lawn-equip-

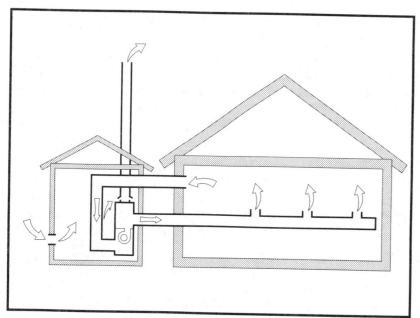

Figure 10–15. A combustion appliance can actually be housed in a small building separate from the house. The two buildings would only be connected by sealed ducts (in the case of a forced air furnace) or sealed piping (in the case of a water heater or boiler).

Chapter 10 Backdrafting and spillage

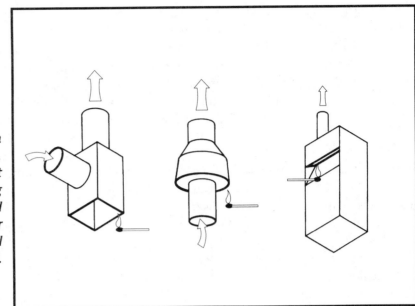

Figure 10–16. To determine if a chimney is functioning correctly, place a lighted match near the draft hood. When the draft is functioning correctly, the flame will be pulled into the hood. If backdrafting or spillage are occurring, the flame will be blown away from the hood.

ment storage building) next to the house and placing the fuel-burning furnace, boiler, or water heater inside. With a forced-air system, there will need to be well-insulated, well-sealed ducts carrying air from the house to the furnace and back again. With a boiler or water heater, the pipes running between the house and the mechanical building should be insulated (Figure 10–15). This approach is more practical with a boiler, but it can be done with a ducted system as well. In fact, some forced-air furnaces are designed to mount on a rooftop, outside the living space.

Gas-Fired Products, Inc. offers a Seahorse gas water heater that is designed to mount outside a house, and **Spacemaker Co.** manufacturers an uninsulated outdoor enclosure that can be used to house a gas water heater in a mild climate.

The advantage of having a separate mechanical building is that it can contain natural-draft appliances, and their operation won't be affected by pressure changes in the house. However, because leaky ducts can cause significant accidental pressures, it is very important that all ducts containing air going to or coming from the house have all joints sealed well with high-quality aluminum-foil tape or mastic (mastic is preferred because it is more durable and longer lasting). It is also important that the chimney not be located where combustion by-products can reenter the house through an open window or a ventilation inlet.

Evaluating a house for backdrafting and spillage

In order to determine if a particular house has, or has the potential to have, excessive combustion by-products in the living space, a number of tests can be performed. For example, it is possible to measure the worst-case depressurization in the room containing the fuel-burning appliance, to check a furnace for a cracked heat exchanger, and to evaluate a chimney for safety, blockage, and sufficient draft. These tests should generally be performed by a technician familiar with how house pressures affect chimney operation.

Measuring the worst-case depressurization will tell you whether the operation of the house's exhaust devices will result in enough depressurization to cause backdrafting or spillage. Basically, this test involves doing everything at once that can cause depressurization, then seeing if it results in backdrafting or spillage in the chimney being tested. For example, to check a furnace chimney, all the exhaust fans, the clothes dryer,

Understanding Ventilation

and central vacuum, etc. are turned on at the same time. A fire should be built in any fireplace that is connected to a separate chimney. Windows and exterior doors should be closed, as well as doors between rooms. Then, with the furnace operating, the chimney serving the furnace can be checked for backdrafting or spillage by holding a lighted match near the draft hood. If the match's flame is drawn toward the draft hood, this indicates that the draft is working correctly. If the match's flame is pushed away from the draft hood, it indicates air is coming down the chimney (Figure 10–16). If the match goes out, it could indicate insufficient oxygen and excessive combustion by-products.

While a match test is useful, there is a much more accurate way to evaluate a natural-draft chimney. You can actually measure the negative pressure inside the chimney. This test should also be performed when the house is experiencing a worst-case depressurization. First, you drill a small hole in the chimney vent pipe (to be plugged after the test). Then you insert a probe that is connected to a pressure-measuring device such as a Magnahelic gauge. After the combustion appliance has been operating for about five minutes, there should be a negative pressure inside the chimney with respect to the room around the chimney. Because the negative pressure in a chimney is dependent on a temperature difference, the outdoor temperature has a bearing on the amount of negative pressure necessary for safe chimney operation.

Following are minimum draft pressures that are considered safe during a worst-case depressurization test at different outdoor temperatures:(The Energy)

Outdoor temperature	Minimum draft pressure
below 20°F	−5 Pa. (0.020" w.g.)
20°F to 40°F	−4 Pa. (0.016" w.g.)
40°F to 60°F	−3 Pa. (0.012" w.g.)
60°F to 80°F	−2 Pa. (0.008" w.g.)
above 80°F	−1 Pa. (0.004" w.g.)

Gary Nelson of **The Energy Conservatory** has estimated that 50–80% of houses containing natural-draft combustion appliances will experience backdrafting or spillage under a worst-case depressurization test. The percentage is so high because many exhaust devices (*e.g.* clothes dryers, range hoods, etc.) are capable of causing more than 5 Pa. of negative pressure in a tight house. In a loose house, an individual exhaust device may not be a problem, but several operating at the same time might be.

A Canadian study examined how much air would need to be exhausted from a house to cause a depressurization of 5 Pa.(Hamlin) Not surprisingly, the results depended on the tightness of the house. The tightest house in this study registered a 5 Pa. depressurization when only 30.5 l/s (64.6 cfm) was blown out of it; the loosest house didn't experience −5 Pa. until 338.5 l/s (717.2 cfm) was exhausted. The study suggests that about half of all houses will experience at least a 5 Pa. depressurization if 110 l/s (233 cfm) is blown out of them. That is the amount of air expelled by a typical clothes dryer, bath fan, and kitchen-range hood combined, or by a fire burning in a fireplace.

By turning different devices off during a worst-case test, it can be possible to determine if one specific device is a major contributor to house depressurization. If, for example, backdrafting and spillage only occur when a powerful kitchen range hood is running, the homeowners can avoid a potential problem by opening a window whenever using the range hood. An open window can also be a solution when a roaring fire in a fireplace is a cause of excessive depressurization.

Checking a heat exchanger for cracks is often a routine part of furnace maintenance. With cracks in a heat exchanger, combustion gases can get into the air that is distributed into the house. Though not related to pressure changes in the house, this is a good test to perform when evaluating a furnace for backdrafting or spillage. To ensure an adequate draft will form, a chimney also needs to be checked for damage (corroded pipes, cracked brick or tile), obstacles (dead animals, nests), correct size (too big or too small), soot or creosote buildup, etc. Simplified instructions are available for checking the draft of a chimney and providing combustion air from several sources,(Minnegasco)(Combustion Air) as well as the National Fuel Gas Code.(American National) In-depth information is available from the *Chimney Safety Tests Users' Manual*(Scanada) and the *Combustion Venting Training Course Manual*.(Geddes and Scanada) All these references are listed in Appendix C.

If anything is done to a house that can affect the house tightness (*e.g.* weatherization or insulation) or the amount of pressurization/depressurization (*e.g.* a ventilation device), chimney operation should be evaluated. Testing for CO is highly recommended. In fact, some utilities are proposing that all houses with natural-draft fuel-burning appliances contain a CO detector to warn occupants if CO levels ever rise too high in the living space. These devices are similar to smoke detectors but sense CO rather than smoke. There are several manufacturers, including **First Alert** and **American Sensors, Inc.**, and they are often available through local hardware stores, department stores, and building-supply stores.

Even if combustion by-products aren't entering the living space, combustion appliances should always be kept in tune and in proper working order. Not only will CO production be minimized, but the fuel-utilization efficiency will be higher, and the utility bills will be lower.

Part 4

Design considerations

Chapter 11 House tightness
Chapter 12 Capacity of a ventilation system
Chapter 13 The cost of ventilation
Chapter 14 Controlling a ventilation system
Chapter 15 Air-filtration equipment
Chapter 16 Fresh-air inlets and stale-air outlets
Chapter 17 Distribution: moving air around a house
Chapter 18 Miscellaneous considerations

Chapter 11

House tightness

Today's houses can't help but be tighter than houses built a hundred years ago. In fact, many of our houses are tighter than we realize. One of the reasons houses are tighter today is because we now build with sheet goods such as drywall and plywood rather than individual boards, so there are fewer random holes. Also, most modern windows and doors are inherently tighter than those of the past. Of course, workmanship is also an important factor; certainly there are sloppily built houses of all ages that are quite loose.

Is tighter better? In many ways, yes. There are some definite advantages to tight construction. For example, tight houses are less costly to heat and air condition. Tight houses have fewer drafts, making them more comfortable. (Canada's R-2000 energy-efficiency program emphasizes both the words *draft-free* and *comfort*.) Tight houses also often have fewer hidden moisture problems. So, tight houses can be energy-efficient, draft-free, comfortable, and free from mold and rot.

The term *airtight* is often used to describe tight houses. Unfortunately, the word can be used in two ways: 1) it can mean there is no air at all flowing through the house by any means (not a good idea), or 2) it can refer to a lack of infiltration and exfiltration as a result of a tight structure (a good idea). By the second definition, you can have an airtight house with plenty of air flowing through it—if it has a ventilation system. Because of these two conflicting meanings, *airtight* really isn't the best word to use. But there doesn't seem to be a better word available. In this book airtight applies only to *the tightness of the building structure itself*, not whether air is being exchanged between the indoors and the outdoors.

When considering the installation of a ventilation system, it can be very helpful to know precisely how tight a house is. A very leaky house will have a great deal of infiltration and exfiltration due to the various natural and accidental pressures acting on it, so a central ventilation system isn't as important in a loose house as it is in a tight house having only a few random holes in it. But infiltrating air may be bringing pollutants in with it, or causing moisture problems. So, it is often a good idea to tighten a loose house, then add a mechanical ventilation system.

How to measure tightness

Even though houses, in general, are tighter than they used to be, there is still a wide variation from house to house. Some new houses are extremely tight, some aren't quite as tight, and some are surprisingly loose. Because airtightness is a function of how many random holes there are, and because random holes are hidden, you simply can't tell how tight a house is by its looks. Rather than guessing, researchers and contractors use some proven techniques to measure house tightness. Once you know how tight a house is, you can estimate how much infiltration and exfiltration there will be on an average basis.

One of the problems with estimating average infiltration and exfiltration rates is that those rates vary tremendously from day to day and season to season. This is because the pressures a house experiences are inconsistent, the weather constantly changes, and houses are all built differently. Even when you know how much air is being exchanged in a house, it can be difficult to determine if it is enough, because occupants all have different metabolisms, life-styles, furnishings, and hobbies. Still, an estimate of the average air exchange rate can help determine if a ventilation system is extremely critical or somewhat less important.

Tracer-gas testing

One method of estimating the tightness of a house involves using an inert *tracer gas* such as helium, perfluorocarbon, or sulfur hexafluoride. The goal in tracer-gas testing is to determine how fast the polluted stale air in a house is flushed out by the incoming fresh air. Because it would be impractical to measure the concentration of all indoor pollutants, the dilution of an inert gas is used as a substitute, or surrogate, for the other contaminants. Sometimes carbon dioxide is used as an indicator, but because it is released by people, occupancy can affect the test results. Tracer-gas testing is used by researchers more often than by contractors, so it can be very difficult to locate someone who can do tracer-gas testing.

Tracer-gas testing is used to determine how quickly the outdoor air replaces the air within a house under normal conditions. The rate of air exchange will depend on whatever natural, accidental, or controlled pressures are acting on the house during the test. Tracer-gas testing is done by 1) injecting a specific amount of the inert gas into the air of a house and letting it diffuse throughout the living space, 2) measuring the indoor concentration of the gas, 3) waiting awhile, then perhaps mixing the air in the house by turning on the central furnace/air-conditioner fan (remember that simply running a furnace or air-conditioner fan can increase the infiltration/exfiltration rate), then 4) measuring the gas concentration a second time. If any outdoor air entered the house or if any indoor air has been expelled during the test, the second measurement will show the concentration of tracer gas to be lower due to dilution with the incoming outdoor air. Usually, several measurements are taken over a period of time and plotted on a graph. By knowing the beginning concentration, the ending concentration, and the time interval, it is possible to calculate the actual indoor/outdoor air-exchange rate.

The air-exchange rate measured with tracer-gas testing is only accurate for the time period the test was performed. The rate may be entirely different on a different day or at a different time of the year because of changing weather conditions or different accidental pressures. Therefore, to obtain the most accurate results, this type of testing should be done when weather conditions and mechanical pressures are as close to typical as possible. Because tracer-gas testing takes longer to perform than blower-door testing (see below), it is generally only used as a research tool. In spite of the drawbacks, tracer-gas testing can be useful in estimating the tightness of a house.

Blower-door testing

A *blower door* (often called a fan door) is a popular diagnostic device that can be used to manipulate air-

Chapter 11 House tightness

Figure 11-1. A blower door can be used to temporarily pressurize or depressurize a house, under controlled conditions, in order to determine house tightness.

flows and pressures in a house.(Rosenbaum, 1994) With a blower door, you can estimate average infiltration rates based on the tightness of the house, then gauge the importance of a controlled ventilation system. Blower doors are available from **Eder Energy**, **Infiltec**, **Retrotec, Inc.**, and **The Energy Conservatory** (Minneapolis Blower Doors).

A blower door contains a large fan, a fan speed controller, and one or more pressure gauges mounted in an adjustable framework that can be temporarily placed in an exterior door opening of a house (Figure 11-1). These devices are used regularly by weatherization contractors (especially those working on low-income housing for state agencies), and builders specializing in super-insulated houses. However, they are not yet commonly found in the rest of the construction industry.

To use a blower door, the entry door is opened, its frame is installed in the opening, then the frame is expanded to form a seal around all four sides of the opening. When the fan is turned on, it will blow air out of the house, causing the house to be depressurized. (Blower doors can also be installed to blow air into a house to pressurize it.) A pressure gauge, with both an indoor and an outdoor sensor, will tell you how much negative pressure the interior of the house is experiencing in relation to the outdoors. By knowing the quantity of air that is leaving the house (as read on an airflow gauge) when the entire house is subjected to a certain pressure (as read on a pressure gauge), you will know how much air is moving through the house at a given pressure. This is a measure of how tight the house is.

For the sake of uniformity, most contractors use a blower door to measure house tightness at the same negative pressure. If a house is especially leaky, there will be a great deal of airflow at that pressure. If it is a tight house, there will be much less airflow when the same pressure is reached. Pressures induced in houses by blower doors are measured in Pascals (Pa.) and the standard pressure that blower door operators use to de-

Understanding Ventilation

pressurize houses is 50 Pa. This pressure was chosen because it is typically higher than any pressures normally acting on a house. While a house could occasionally experience localized pressures as high as 50 Pa., when an entire house is depressurized to 50 Pa., every square foot of every exterior wall, floor, and ceiling will feel that amount of pressure.

One of the advantages of blower-door testing over tracer-gas testing, is that blower doors enable you to compare several different houses at the same 50 Pa. pressure. A tight house may have only a few hundred cfm of air infiltrating when it is depressurized to 50 Pa. while a loose house could have 8,000 cfm or more blowing through it. A house that is purposefully constructed extremely tight may have less than a hundred cfm moving through it at 50 Pa.

Understanding blower-door data

There are several ways to use the data derived from a blower-door test. Because 50 Pa. is more pressure than a house is normally exposed to, researchers have devised a way to estimate the amount of infiltration in a house under average conditions. They simply divide the cfm measured at 50 Pa. by 20. (Actually, the number you divide by can range from 14 to 26, depending on local climate conditions.$^{(Meier, Alan)}$ Blower-door operators will use a number appropriate to their location, but for the purposes of this book, we will use 20.) For example, if 2,400 cfm are measured at a 50 Pa. depressurization (often abbreviated as 2,400 CFM_{50}), the rate of infiltration under average conditions over an entire year would be estimated at 120 cfm (2400 ÷ 20).

The number you get when you divide the CFM_{50} by 20 is often referred to as the amount of *natural* infiltration (it is usually expressed as $CFM_{natural}$), because it only accounts for the amount of air entering and leaving a house because of natural pressures—wind and stack effect. It should be noted that because of so many variables this method of estimating natural infiltration under average conditions is not extremely accurate. It could easily vary as much as 50% in either direction.

Sometimes it is helpful to convert the CFM_{50} data to air changes per hour (ACH). To do so, multiply the CFM_{50} measurement by 60 minutes, then divide by the volume of the house. For example, if 2,000 ACH_{50} is measured in a house containing 20,000 cu. ft., then 2,000 x 60 ÷ 20,000 = 6 ACH_{50}. In other words, there would be 6 air changes per hour at 50 Pa. Dividing this number by 20, just as we did above, will give you a rough idea of how many ACH there would be under average conditions over the course of a year. In this case 6 ACH_{50} ÷ 20 = 0.3 $ACH_{natural}$. So, in this example, the house would have just under the *American Society of Heating, Refrigerating, and Air-Conditioning Engineers' (ASHRAE)* recommendation of 0.35 ACH. (See Chapter 12 for *ASHRAE*'s ventilation-rate recommendations.) But, you should keep in mind the fact that this is a very rough way of estimating the average amount of infiltration in a house. Because it could be off as much as 50%, the real average annual infiltration rate might be anywhere from 0.15 ACH to 0.45 ACH.

It is also possible to convert blower-door data into an estimate of how many square inches of leakage area a house has. This is called the *effective leakage area* or ELA. The ELA is the size of opening a house would have if all of the random holes were combined into a single round-edged hole. (A hole with square edges will have slightly different flow characteristics than an equally-sized hole with rounded edges.) *Lawrence Berkeley Laboratories* in Berkeley, California has developed a procedure to estimate the ELA in a house.

Without going into the mathematics involved, to determine the ELA of a house, you first use a blower door to determine the airflow through the house at 4 Pa., then multiply that number by 0.2825. This will give you the ELA in square inches. For example, if a house has 550 cfm flowing through it when depressurized to −4 Pa. (this can be abbreviated 550 CFM_4), it will have an ELA of 155.9 sq. in. (550 x 0.2835). There being 144 sq. in. in 1 sq. ft., if you added up all of the random holes, cracks and gaps in this particular house and combined them into a single hole, that hole would be a little larger than 1 sq. ft. in size. (One square foot of ELA is somewhat typical for many houses.) The smaller the ELA, the tighter the house.

In Canada, ELA stands for *equivalent leakage area,* and although it is an estimate of the same thing—the leakage area of a house—it is determined somewhat differently. A Canadian ELA is based on a square-edged hole and is determined by multiplying 0.2939 times the airflow at −10 Pa. It is important to keep in mind that an ELA is an *estimate* of the total area of the random holes, and that estimate varies depending on how it was calculated. For example, a Canadian ELA is typically twice as large as an ELA estimated using the Lawrence-Berkeley technique.

Limitations of natural-infiltration estimates

Using a blower door to calculate $CFM_{natural}$ (the infiltration rate driven by wind and stack effect) has some basic drawbacks. First of all, it represents the annual average natural infiltration rate, based on average wind conditions and average temperatures, so there will be many times of the year when the rate is significantly higher and many times when it is zero. When a house is being ventilated naturally, the *average* rate over the entire year may be adequate, but there will no doubt be periods of several days (or more) when there is no natural ventilation.

The CFM_{50} of a house, as determined with a blower door, is a very accurate and reproducible way to gauge the tightness of a house. However, $CFM_{natural}$ is only an *estimate* of the annual air exchange rate—and it isn't a very accurate one. Let's say a house is tested with a blower door at 1,400 CFM_{50}. That would translate to 70 $CFM_{natural}$ (1,400 ÷ 20), but in reality the annual average rate might be anywhere from 35 $CFM_{natural}$ to 105 $CFM_{natural}$ (70 ± 50%). If this house only has 35 $CFM_{natural}$, based on a need of 15 cfm per person (See Chapter 12), there would have been enough air exchange (annual average) for two people. If it has 105 $CFM_{natural}$, there would be enough air exchange for 7 people. Does this house have enough natural air exchange? It depends on how you choose to interpret the data—and how many people are in the house.

A house with forced-air heating or air conditioning will experience stronger pressures whenever the heating/cooling system is operating (due to leaky ducts and closed doors between rooms) than a house without forced-air heating or air conditioning. This means that a house with a forced-air heating/cooling system will have more accidental ventilation whenever the system is running, and $CFM_{natural}$ doesn't take this into consideration. So the actual ventilation rate (natural plus accidental) will be higher than the $CFM_{natural}$ estimate. Of course, a forced-air heating/cooling system will usually only be operating when it is either cold or hot outside—when stack effect is most pronounced. But the extra accidental ventilation is needed more in the spring and fall when there is very little stack effect.

When air moves into and out of a house through random holes, it will always affect the heating bill and cooling bill because outdoor air is exchanged with in-

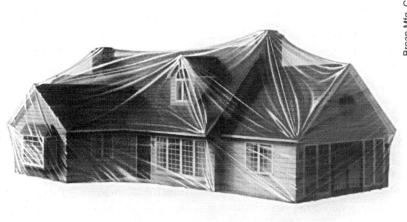

Figure 11–2. You can't tell by looking whether a house is tight or loose. In general, the houses being built today are tighter than those built a few decades ago, but there are still many exceptions.

door air. However, this air exchange doesn't always benefit the occupants. For example, what if most of the air exchange takes place in the basement, or inside walls and other building cavities? Air is moving between the indoors and the outdoors, but it isn't reaching occupants. A blower door can estimate the average amount of air exchange in a house, but it can't measure how much that air will benefit the occupants.

In short, tightness estimates can give you a better idea of how much air is moving into and out of a house than simply guessing, but there are limitations. There is no doubt that a mechanical system is a more effective way of ventilating a house than relying on natural and accidental ventilation. However, all houses do have some natural and accidental ventilation, and a tightness estimate will help you evaluate whether a mechanical ventilation system is needed "a lot," "somewhat," or "not much."

Evaluating a house without a blower door or tracer gas

Many states have weatherization programs that hire private contractors to weatherize houses. In order to determine if their work is paying off, these contractors often use a blower-door to test a house before and after weatherizing. In order to locate someone capable of doing blower-door testing, contact your local weatherization or community-action program.

What if you don't have access to anyone who can test a house for tightness? How can you tell if you have a tight or loose house? You can look for clues. A loose house will often have higher heating or cooling bills than a similarly sized, similarly insulated tight house. Tight houses feel stuffy. Loose houses feel drafty. A tight house often has condensation forming on the windows in the winter because of excessive moisture indoors.

Loose houses often feel very dry in the winter because too much moisture is escaping. Cooking odors tend to linger in a tight house. Tight houses sometimes smell like wood smoke if the chimney hasn't been able to develop enough natural draft. Tight houses are often quieter because they block out traffic and other outdoor noises. A houses with a simple geometrical design (the basic box) is often tighter than one with many niches, angles, and wings.

How tight is too tight?

When you consider the benefits of tight construction (*e.g.* comfort, energy efficiency, quietness, fewer hidden moisture problems, etc.), it becomes apparent that a tight house is really a good idea. On the other hand, a tight house prevents very much air from infiltrating and exfiltrating. So if a tight house has no controlled ventilation system, it will not have very much fresh air in it at all. Even though there are drawbacks to infiltration and exfiltration, there is no doubt that some air is better than no air. Therefore, a good question to ask is, How tight can a house be before it doesn't get enough air naturally or accidentally? In other words, how tight can you make a house before a controlled ventilation system becomes critical?

Even though you can't tell by looking how tight a house is, many new houses are fairly tight. (Of course, not all new houses are tightly constructed. It depends of the size of the house, construction techniques, workmanship, whether detached or multifamily, etc.) Using computer analysis, a 1989 study of 200 new houses in Canada found that 70% were too tight to provide 0.3 air changes per hour under "normal" conditions (*i.e.* without controlled mechanical ventilation) during the entire heating season.(Hayson) (An air exchange rate of 0.3 is often considered an acceptable minimum. For a discussion, see Chapter 12.) Almost 90% of the houses were too tight for at least a full month. And 99% were too tight for at least one 24-hour period to be able to provide the occupants with a minimum amount of ventilation. When the results of this study were compared to a 1982 survey, it was found that over the intervening seven years new houses had become 30% tighter.(Hamlin)

The bottom line: While some houses are loose enough to provide some fresh air some of the time, many houses are too tight to allow sufficient fresh air to infiltrate and exfiltrate through them on a regular basis to meet the basic needs of occupants. If a house is very tightly constructed, it needs a mechanical ventilation system—period.

But how tight is too tight? Many weatherization contractors feel that if a house is tighter than 1,500 CFM_{50}, it needs mechanical ventilation. (A good evaluation is based on local wind conditions, occupancy, etc., so the maximum degree of tightness a weatherization contractor feels is appropriate for a specific house may be higher or lower than 1,500 CFM_{50}. However, we will use that figure in our discussion below because it is an average figure that is often used.) This degree of tightness translates to an estimated 75 cfm under average conditions (1,500 ÷ 20). Keep in mind that due to the roughness of the estimate (±50%), the actual average rate of infiltration could range between 37 and 112 cfm. And, of course, there will definitely be times when the rate is zero and others when it is quite high—all depending on which natural and accidental pressures are affecting the airflow into and out of the house. In new construction, builders specializing in energy-efficient building techniques have found it possible to build houses that can be blower-door tested at less than 100 CFM_{50}. With houses this tight, mechanical ventilation is considered mandatory.

The goal of many weatherization contractors is to insulate, caulk, and seal up houses to make them more comfortable and energy-efficient—but only up to a point. They are careful to leave a house loose enough (*i.e.* no tighter than 1,500 CFM_{50}) so that all the pressures—both natural and accidental—will supply the occupants with a sufficient amount of fresh air most of the time.

Thousands of houses have been weatherized in this country that have been tightened up "only so much" (looser than 1,500 CFM_{50}), based on blower-door testing. In most cases this has not resulted in any serious problems as far as poor air quality is concerned. In other words, if you don't tighten a house too much, you probably won't make the occupants sick. But this is certainly no guarantee—especially if the occupants are particularly sensitive, or if furnishings, building materials, or cleaning products are highly polluting.

Until recently, there has been a certain amount of agreement within the weatherization community that a house needs to be looser than 1,500 CFM_{50} to ensure good indoor air quality. But as more is learned about indoor pollution, many professionals are coming to the conclusion that house tightness, ventilation, and air quality should be viewed as three separate issues. This thinking was a hot topic at a 1994 conference sponsored by Affordable Comfort, Inc. of Philadelphia, PA, attended by weatherization contractors from across the country. A draft policy statement[Affordable] was circulated among attendees that contained, among other things, the following points:

- Pollutant source strength, not an arbitrary amount of fresh-air exchange, is the most powerful determinant of indoor air quality.
- Natural infiltration/exfiltration alone cannot be relied upon to consistently maintain recommended levels of ventilation, even in leaky homes.
- If possible, houses should be as tight as economically feasible and be equipped with whole-house mechanical ventilation systems.
- Relying on any minimum tested ventilation rate is not a guarantee of health and safety.

So, how tight is too tight? Is 1,500 CFM_{50} too tight? The consensus seems to be in a state of flux. A decade ago tight houses were thought to be the cause of poor indoor air. Today, with a better understanding of how air is exchanged in houses, as well as the consequences of that air movement, it is becoming apparent that tight houses have many advantages over loose ones. It is also obvious that the tighter a house is, the more mechanical ventilation becomes mandatory.

Ventilation for loose houses

In a loosely built house a mechanical ventilation system may not be extremely critical, but it may still be a good idea because it will cause air to enter and leave in a more reliable manner. What if a house is looser than 1,500 CFM_{50}? Can it get by without a controlled ventilation system? In many cases, it probably can, but there are still benefits to installing controlled ventilation. For example, if a house contains new building materials, such as carpeting or kitchen cabinets that are capable of high outgassing rates, then the 1,500 CFM_{50} guideline probably isn't high enough, so controlled mechanical ventilation, or a looser house, is probably necessary. But what if your house doesn't have any new materials in it? It may not contain anything new today but might tomorrow. People occasionally remodel their houses and periodically purchase new furnishings, and some of the raw materials used in furniture and remodeling (*e.g.* particleboard) can outgas for several years.

Understanding Ventilation

Controlled ventilation may also be required if the people living in a loose house are especially sensitive to indoor pollutants and need a continuous supply of fresh air to maintain their health. The number of people living in a house is another important factor to consider. The 1,500 CFM_{50} guideline is based on 4–5 occupants.

If a house is leakier than 1,500 CFM_{50}, it will, no doubt, have a reasonable amount of air moving into and out of it over the course of a year. But it probably won't provide the occupants with sufficient fresh air every day of every week because the pressures involved are so erratic. Also, because the random holes are rarely in optimum locations, a certain percentage of the air may never reach the occupants. But it is definitely better than nothing.

In an ideal world, mechanical ventilation is a good idea in all houses because it can reliably supply a consistent amount of air to the occupants at all times. In many cases the 1,500 CFM_{50} guideline will result in enough air moving through a house under average conditions but in many other cases it won't. The bottom line is this: The 1,500 CFM_{50} airtightness guideline is a practical rule of thumb—but nearly everyone agrees that it isn't perfect.

Tightening a house

If a ventilation system is installed in a loosely built house, there will be some air movement through the inlets and outlets of the ventilation system, as well as some air movement through the random holes in the structure. Thus, air will enter and leave the house both through the structure and through the ventilation system. Some of the air movement will be driven by the ventilation system, some will be driven by natural and accidental forces. The tighter the house, the less air will infiltrate and exfiltrate randomly, and the greater the percentage of air that will enter and leave through the inlets and outlets.

If you have a loose house and want to minimize the random infiltration and exfiltration, you should consider tightening it up. By doing so, you will be more likely to know where air is entering and leaving. In other words, it is easier to predict where air movement will take place in a tight house.

Tightening a house is definitely a good idea. A tight house is more energy-efficient, it has fewer hidden moisture problems, and it has fewer pollutants being sucked into the living space. Tight houses are also more comfortable, less drafty, and they are often healthier. However, tight houses always need a controlled ventilation system. While it costs money to tighten up a house and to pay for the ventilation system, the occupants generally benefit because of lower heating/air-conditioning bills and better health and comfort, so it is money well spent.

Any house can be made tighter, but because all houses are different, there can be variable results. For example, because there are more possibilities for leaks in a large house, it is more difficult to tighten a large house to the same CFM_{50} than a small one. Therefore, a large house having 1,000 CFM_{50} is considered tighter than a small house at 1,000 CFM_{50}. So, besides knowing the CFM_{50}, it is also helpful to know the size of the house. In any case, a 30% improvement in tightness isn't difficult to achieve in most existing houses using standard weatherization techniques.

Where to tighten

Typical house-tightening measures include weatherstripping doors and windows, caulking, and adding insulation. Though they are more difficult to visualize, there are much more significant air leaks around the band joists and sill plates that sit on top of the foundation. Other important leakage points include penetrations where the plumbing, electrical, cable TV, and telephone lines enter the house. And as was mentioned in Chapter 6, leaky ducts are a significant source of infiltration and exfiltration.

Several different types of materials that can be used to seal up a house: Caulking can be used for small gaps, and caulking applied over a foam backer rod works for wider gaps (always use long-lasting products). For medium-sized holes, special gaskets, expanding aerosol foam, or duct-sealing mastics can work well. For large holes, patching plaster, sheet metal, rigid insulation, or drywall are often needed.

Many houses have what are called *bypasses* hidden within them that are significant air leaks. For example, there may be an opening around a chimney that connects the basement (or crawl space) with the attic. When you are in the attic, you can sometimes shine a flashlight through such a bypass and see into the basement. This represents a significant leak that air can pass through. Other large bypasses exist around recessed ceiling light fixtures, near heating or cooling ducts, or around plumbing chases. There are often large leaks connecting the rear side of a bathtub to an attic or crawl space. Significant hidden leaks often exist wherever there is a change in ceiling height (*e.g.* near dropped ceilings, or where a cathedral ceiling meets a flat ceiling) and under knee walls in a 1 1/2-story house. In older balloon-framed houses, there are no upper plates on the walls, so air can rise up through the walls into the attic. Large gaps also often exist above pocket doors, around attic hatches, and through soffits over bathtubs or kitchen cabinets. Any building cavity that opens into an unconditioned space such as an attic, crawl space, or basement is a potential air leakage site that, when sealed, will help to tighten the house.

Sealing attic and basement bypasses is often the most significant way of tightening a house. Air moving through bypasses can significantly affect heating and cooling bills, but it often doesn't provide the occupants with fresh air because the air is moving within hidden cavities and not within the occupied space where the people are located.

Another important way of tightening a house is to seal up large holes that are visible in the living space. This might include obvious holes such as missing or damaged plaster on a wall. Large holes in an old ceiling aren't readily visible if a dropped ceiling has been installed. The list of possible leakage areas could also include laundry chutes that connect different floors, or closets without ceilings. A blower door can be an invaluable tool to use in locating all these leaks.

If fiberglass insulation is loosely placed inside a wall cavity and there are holes in the inner and outer wall surfaces, air can move right through the insulation because of the wind or some other pressure. Air can also move through loose fiberglass insulation in an attic in the same way if there are holes in a ceiling. If cellulose insulation is packed into a wall tightly enough, the insulation can, in effect, plug up the holes. So an easy way to tighten an existing house is to blow cellulose insulation into the wall cavities at a high density.[Fitzgerald] The technique for packing cellulose into building cavities at a higher-than-standard density is not widely used by insulation installers, but it is gaining popularity in the low-income weatherization community. In new construction, a "wet-spray" cellulose installation technique can be used to minimize infiltration.[Reiss] With this process a tiny amount of water is mixed with the cellulose to hold it in place when it is sprayed into open wall cavities. The insulation is then allowed to dry somewhat before drywall is installed.

For more information

Many techniques have been developed over the years for tightening houses. Some are applicable to existing houses, some lend themselves primarily to new construction, and some techniques work in any house. To seal hidden air leaks in an existing house, there are a number of informative articles.[Blandy][Sullivan January 1993] In addition, *Contractor's Guide to Finding and Sealing Hidden Air Leaks*[Massachusetts] is a good illustrated reference. While they are geared primarily to new construction, *Advanced Air Sealing*[Iris], the *Builder's Field Guide*[Bonneville 1991], and *The Airtight House*[Lischkoff] are also good sources for information about tightening houses. All are listed in Appendix C.

Chapter 12

Capacity of a ventilation system

One of the first questions that must be answered when designing a ventilation system and selecting equipment involves capacity. How much air should a ventilation system be able to move into and out of a house to do its job effectively?

It is fairly easy to determine the capacity of a central ventilation system for a relatively uncontaminated house in which the occupants are the principal pollution sources. This is because the amount of pollution caused by by-products of metabolism is well understood. It is also fairly easy to select the capacity of a kitchen or bathroom fan because the amount of contamination found in those rooms due to odors and moisture is also well-understood. However, it is more difficult to determine how much ventilation is needed to adequately dilute the hundreds of other pollutants in the indoor air. While any amount of ventilation will certainly help, there are so many possible pollutants that it is impossible to study them all (or in combination with each other) in order to determine precisely how much ventilation is enough for all situations. Based on our present lack of scientific knowledge about many of these pollutants, it is virtually impossible to accurately select an appropriate ventilation rate—but there are some guidelines that can help.

Calculating cfm and ACH

There are two basic ways to size a ventilation system—air changes per hour (ACH) or cubic feet per minute (cfm). Both refer to moving a certain volume of air in a certain period of time.

Cubic feet per minute refers to the amount of air a fan moves each minute. A 150 cfm fan will move 150 cubic feet of air every minute it is operating—or 9,000 cubic feet of air per hour (150 x 60). A 300 cfm fan will move twice as much air. While most of the fans in the

Understanding Ventilation

U.S. are rated in units of cfm, countries on the metric system often use liters per second (l/s). (Cubic meters per hour or cubic meters per second are also used.) One l/s is equal to approximately 2 cfm. A fan's capacity will either be marked on the housing or listed in the manufacturer's product literature.

Air changes per hour (ACH) refers to how many times per hour a fan moves enough air to change all the air in a house. A ventilation system capable of 2 ACH will have a fan with enough capacity to change the air in the entire house twice every hour. A fan capable of 4 ACH will move twice that amount, changing the air in the house four times an hour.

Actually, because of mixing and dilution, a fan can have a capacity of 1 ACH, but it will not be able to replace all the air in a house in an hour. This is because the incoming air always mixes with some of the stale air already in the house and then some of this mixture leaves the house. Some stale air is exhausted and some remains indoors.

To convert a fan's capacity to ACH, you need to know two things: the cfm rating of the fan and the volume of the house. To calculate the volume of a house, multiply the interior square footage of the floor plan by the ceiling height. As an example, suppose a house has a rectangular floor plan 36' x 52' and 8' high ceilings. It will have a volume of 14,976 cubic feet (36 x 52 x 8). For a ventilation system to be capable of 1 ACH in this house, it will need to move 14,976 cubic feet of air per hour. To convert that amount of air to cfm, divide it by 60 minutes (14,976 ÷ 60 = 249.6 cfm). So, in this example, a 250 cfm ventilation system will provide approximately 1 ACH. A 500 cfm fan will move twice as much air, so it will provide about 2 ACH.

To work backwards, let's determine how many ACH will be provided by a 300 cfm fan in a 24' x 36' two-story house that has 9' high ceilings on the first floor and 8' high ceilings on the second floor. The volume of the first floor is 24 x 36 x 9 or 7,776 cubic feet, the volume of the second floor is 24 x 36 x 8 or 6,912 cubic feet, so the total volume of the house is 7,776 + 6,912 or 14,688 cubic feet. The 300 cfm fan will move 300 cubic feet of air every minute, so if it runs for an hour (60 minutes) it will move 18,000 cubic feet of air in a hour (300 x 60). To determine the ACH, divide the hourly capacity of the fan by the total volume of the house. In this case, the fan will provide 1.23 ACH (18,000 ÷ 14,688), so it will change the air in the house about $1\frac{1}{4}$ times every hour.

How much capacity is enough?

Human beings need only about $\frac{1}{2}$ cfm of clean air to adequately supply their red blood cells with sufficient oxygen. This is about how much air scuba divers consume when they are swimming at the surface (they use up more oxygen as they go deeper). When people are indoors they need considerably more air than $\frac{1}{2}$ cfm for two reasons. First, it is impossible to deliver fresh air to a person indoors as effectively as it is through a scuba diver's mask. In a house, air may be delivered to one room while people are in another room. With a scuba diver, air is delivered directly to the mouth and nose. Second, the fresh air entering a house always gets diluted with house air which—even in a very unpolluted house—has been contaminated by the people themselves. We exhale by-products of metabolism directly into the same air we are breathing. Air from a scuba diver's tank is always fresh.

If you only consider the amount of air we need to replenish the oxygen in our blood (about $\frac{1}{2}$ cfm), we require 720 cubic feet of air per day. This is more important to us than food or water. We can go without food for weeks, without water for days, but we can only go without air for a few minutes. We typically need approximately 1 pound of food and 5 pounds of water per day, but 720 cu. ft. of air weighs about 54 pounds. One half cfm may supply a person with enough oxygen, but it isn't nearly enough air to dilute the pollutants found indoors. A house ventilated at that rate will feel very stuffy very quickly.

The *American Society of Heating, Refrigerating and Air-Conditioning Engineers* (*ASHRAE*) is a well-respected trade organization for professional engineers who specialize in ventilation and related issues. *ASHRAE* has been grappling with the question of, How much fresh air is enough?, for a number of years. Between 1973 and 1989 they recommended 5 cfm per person of continuous ventilation. Many buildings were constructed based on this recommendation, but as it turned out, it wasn't enough fresh air to maintain good air quality in buildings. Even respected professional organizations sometimes make mistakes, and *ASHRAE* has since revised their recommendations.

ASHRAE's current guideline for residences calls for either 15 cfm per person or .35 ACH (whichever is greater). Because *ASHRAE* is not a regulatory agency, this is only a recommendation—it is not a requirement. This amount of fresh air has been determined to be enough to adequately dilute the pollutants generated indoors by people. It has been found that 15 cfm per person is enough to keep carbon dioxide levels (primarily from exhaled breath) to below 1,000 parts per million (ppm) in a house—if people are the only source of carbon dioxide. Fifteen cfm per person will also reduce the likelihood of moisture condensation problems within the living space—if people and their normal activities are the principle moisture sources. If a loosely-built house becomes pressurized or depressurized, it can still experience *hidden* moisture problems in some climates, even when it has a 15 cfm-per-person ventilation rate. For a discussion, see Chapter 8.

Fifteen cfm may not sound like a great deal of air, but it is. When you realize that that amount of air must be provided every single minute of the day, it can add up quickly. Over a 24-hour period, 15 cfm means a total of 21,600 cubic feet of air (15 cfm x 60 minutes x 24 hours). That amount of air—one day's worth for one person—will fill a cube measuring over 27' on each side.

Most experts agree that 15 cfm per person seems to be enough ventilation air to flush "people pollutants" out of a house. But some people are more sensitive than others. When you read *ASHRAE*'s recommendation in its entirety, they say that 15 cfm per person will only satisfy the comfort needs of 80% of the population. Twenty percent of us need more fresh air than 15 cfm to feel comfortable. Also, according to *ASHRAE*, the 15 cfm-per-person figure should be considered a minimum—many situations require more. For example, if a house contains a number of pollutant sources besides people (*e.g.* new carpeting or new kitchen cabinets), then more ventilation air may be needed. Determining how much more can be difficult because every house is different. (*ASHRAE* has an "Indoor Air-Quality Procedure" that can be used to determine a ventilation rate that will control a variety of indoor pollutants. This is a daunting task because, as *ASHRAE* says, "Uniform governmental policies regarding limits on exposure to environmental carcinogens have not yet emerged.")

There are other recommendations besides those of *ASHRAE*. For example, the **Home Ventilating Institute (HVI)**, a ventilation-equipment trade organization, suggests .5 ACH minimum. The **Bonneville Power Administration**, a electric power utility in the Pacific Northwest, recommends 10 cfm for each bedroom plus 10 cfm for one other room (*e.g.* 30 cfm for a two bedroom house) or .35 ACH, plus a 100 cfm kitchen-range hood, plus 50 cfm in each bathroom. The base rate should be continuous but the range hood and bath fans are assumed to be intermittent.

The Canadian National Building code requires a capacity of .5 ACH if the fresh air is not distributed or .3 ACH if it is distributed throughout the house. In addition, Canada has an R-2000 program that certifies extremely tight, energy-efficient houses. An R-2000 house is required to have at least 10 cfm per room, plus 20 cfm for a master bedroom, plus 20 cfm for an unfinished basement (all continuous capacity), plus an additional 20 cfm continuous (or 50 cfm intermittent) for a bathroom, plus 60 cfm continuous (or 100 intermittent) for the kitchen.

Sweden has been grappling with ventilation issues longer than either the U.S. or Canada because of tighter construction practices warranted in its cold climate. Sweden's 1988 building code specifies at least 0.35 l/s of ventilation per square meter of floor area. For a 1,500 square foot house, this translates to 103 cfm or about 0.5 ACH.

It is important to bear in mind that ventilation codes often cite ventilation capacity, but they don't always require that the ventilation system be running continuously. For example, a building code may require that your house have a 50 cfm exhaust fan in each bathroom, but it doesn't say you have to use them!

Understanding ASHRAE's recommendations

If you adhere to *ASHRAE*'s recommendations, you will certainly improve the air quality in a house, but there is no guarantee you will have healthy indoor

Understanding Ventilation

air. In fact, the *ASHRAE* guidelines were established for odor control—not for health—and not for thermal comfort. Thermal comfort depends on 1) the air temperature, 2) the average radiant temperature of walls, floor, ceiling, etc., 3) the RH, and 4) the amount of air motion felt by people. It is very possible to feel comfortable in air that is odorous and unhealthy.

Because *ASHRAE*'s 15 cfm-per-person guideline will dilute the stale indoor air with enough fresh outdoor air to keep the concentration of CO_2 below 1,000 ppm, many people assume that a CO_2 level above 1,000 ppm is unhealthy. That is not the case. For example, sailors function well in submarines with CO_2 levels as concentrated as 10,000 ppm. But submarines have very elaborate filtering systems to remove all the other unhealthy and uncomfortable air pollutants. Houses rarely have that kind of filtration, so *ASHRAE* uses 1,000 ppm of CO_2 as a surrogate, in other words an indicator, of satisfaction, not health.

Water vapor and CO_2 are two major components in our exhaled breath. On average, a person produces 0.011 cubic feet of CO_2 per minute. Therefore, if one person is in a small unventilated bedroom (*e.g.* 75 sq. ft.) with the door closed, after an hour the CO_2 level will build up to about 1,000 ppm (based on normal activity). If the bedroom is larger, say 225 sq. ft., it will take a little longer for the CO_2 level to reach 1,000 ppm, perhaps 3 or 4 hours. As the CO_2 level rises, the relative humidity in the room will also rise. If the bedroom contains any pollutants besides people—new carpeting, cigarettes, etc.—those pollutants will build up as well.

Many people begin feeling uncomfortable as a room starts to get "stuffy." Stuffiness is often associated with a higher relative humidity (RH) at a higher temperature, so it may take a several hours of being in a closed room (depending on the size of the room) before it starts feeling uncomfortable. In the meantime, the person has been exposed to whatever pollutants are released into the room from the building materials, furnishings, or cleaning products.

It can be difficult and expensive (if not impossible) to measure the concentration of all the pollutants in a house. What *ASHRAE*'s recommendation of 15 cfm per person does is use the CO_2 level as an indicator of occupant satisfaction. If CO_2 is kept below 1,000 ppm, then the RH will also be kept below the level where people feel stuffy (in a cold climate) and body odors will not be offensive—assuming that people are the primary source of CO_2 and water vapor. So, 15 cfm reduces CO_2 and other by-products of metabolism to a level that 80% of people find satisfactory, and it also reduces the concentration of whatever other pollutants happen to be in the room—but it is difficult to say whether they are reduced to a healthy level. To put this another way, if there is 15 cfm of ventilation air per person, only 20% of the population will be able to sense too much body odor.

An interesting point that must be made is the fact that people are very adaptable. If you walk into a room and it seems stuffy, you will get used to the air quality and, after a while, no longer notice the stuffiness. Even though stuffiness isn't necessarily harmful to your health, you can also adapt to more dangerous air pollutants. This explains why someone can walk into a room and say, "This room stinks," while a person who has been in the room for some time might respond, "I don't smell anything." In other words, a person entering an enclosed space (a visitor) will notice poor air quality more readily than an adapted occupant. Researchers took this adaptability into consideration when they determined that 80% of the population would be satisfied with 15 cfm per person. What they did was have test subjects spend some time in clean air for a while, then they brought them into an occupied room with a certain ventilation rate. Soon after entering the room, the subjects were asked how satisfied they were with the air quality. The primary pollutants in the room were by-products of metabolism. The researchers found that 80% of test subjects (as visitors, entering a room) are satisfied with the level of bad breath, body odor, and other metabolic by-products when a room is ventilated at a rate 15 cfm per person.

If a house contains no pollution sources other than people, 15 cfm per person will probably be sufficient in most cases—in fact 80% of us will be satisfied with this rate. However, because each of us has different degrees of tolerability, a few of us are satisfied when 10 cfm of fresh air is provided, but a few may require 25 cfm, or more.

Because only a handful of the thousands of possible indoor pollutants have been studied to determine what levels are healthy, neither *ASHRAE*, nor anyone else, can say that 15 cfm per person is enough fresh air if there are pollutants other than by-products of metabolism in the house. Yes, that rate of ventilation will prevent a house from feeling stuffy for most people, but there is no way that anyone can guarantee it will be healthy. What it boils down to is the fact that 15 cfm per person is a *body-odor* standard, not a *health* standard. This is an important distinction, because 15 cfm of ventilation per person is no guarantee that you will have healthy air. The best way to protect your health is to build, furnish, and maintain a house using healthy materials, and make sure pollutants do not get sucked indoors from outside the living space.

It should be pointed out that recommended ventilation rates for non-residential spaces often differ from the rates above. For example, *ASHRAE* recommends 20 cfm per person in offices and lobbies, 25 in beauty shops and discos, 30 in cocktail lounges, 50 in public restrooms, and 60 in smoking lounges.

Ventilation for formaldehyde removal

Because the above recommendations are based on occupant satisfaction, not health, what ventilation rate is required to reduce unhealthy pollutants? While many indoor pollutants have not been studied very well, because it is very commonly found in houses, often at high concentrations, a number of studies have looked at formaldehyde. Typical formaldehyde concentrations in houses are often in the 0.03–0.15 parts per million (ppm) range; in new mobile homes, peak levels can be 0.20–0.50 ppm. Though it is impossible to avoid formaldehyde (it is possible to measure a concentration of 0.01 ppm in clean outdoor air) concentrations below 0.03 ppm are often tolerable for sensitive people. Workplace exposures of as much as 0.75 ppm, averaged over 8 hours, are allowed by the Occupational Safety and Health Organization (OSHA), but the World Health Organization (WHO) recommends that residential exposures be below 0.05 ppm. Other organizations recommend levels below 0.10 ppm. So, what ventilation rate does it take to reduce high formaldehyde concentrations to an acceptable level?

A number of researchers have measured formaldehyde concentrations in houses and mobile homes before and after the use of a ventilation system. Because there are so many variables (indoor and outdoor temperature, indoor and outdoor relative humidity, ventilation rate, distribution patterns, effectiveness, outgassing rate, etc.) the results have been mixed.(Godish, 1986, 1989) However, it seems that a continuous ventilation rate of about 1.4 ACH is needed to achieve about a 70% reduction in formaldehyde levels. This is 4 times **ASHRAE**'s recommended minimum rate of 0.35 ACH! Based on these findings, if you have a formaldehyde concentration in an unventilated house of 0.15, and you ventilate that house at a rate of 1.4 ACH, you would end up with a concentration of about 0.045 ppm, or just under the WHO recommended maximum. If you want to reduce the level further, you must ventilate at an even higher rate. Of course, these high ventilation rates could easily result in indoor air that is to dry in the winter and too humid in the summer. High-capacity systems are also noisier, more expensive to install, and costlier to operate. Source control (don't build with formaldehyde-containing products) is a more effective and cheaper way of reducing formaldehyde concentrations.

Should a ventilation system be based on cfm per person or ACH?

There are drawbacks to basing a ventilation-system's capacity on either cfm or ACH. In general, it is good to base a ventilation rate on 15 cfm per person if people are the primary pollutants in a house. Because ACH is related to the size of the house, it is a better guideline if the building materials and furnishings are more polluting than the people.

If you decide to design a house based on 15 cfm per person, you must realize that occupancy of a house is rarely constant. Consider a three-bedroom house with a single parent and two children. Based on 15 cfm per person, this house should have a continuous ventilation rate of 45 cfm. But more than that will be needed when guests are visiting. And, of course, not all three bedroom houses have only three permanent occupants. The three-bedroom house next door could have two parents,

Understanding Ventilation

4 children, and a grandparent. Seven people @ 15 cfm per person would equal 105 cfm. If a contractor is building several three-bedroom houses, he may find it is very difficult to predict how many people will be living in each of them, so how much capacity should he specify?

When the occupancy isn't known, .35 ACH can often yield a reasonable average amount of ventilation—but it works best for average-sized houses. ACH may not provide enough ventilation in a small house, and it may provide too much in a large house. For example, a small 1,000 square-foot house (with 8' ceilings) will contain 8,000 cubic feet. Using *ASHRAE*'s guideline of .35 ACH, this house would need a ventilation-system capacity of 46.6 cfm (8,000 x .35 ÷ 60). This is enough air to provide 15 cfm for each of three people. If there is a family of six living in the house, and new carpeting had just been installed, then .35 ACH won't provide enough fresh air. On the other hand, in a very large house, say one containing 45,000 cubic feet, .35 ACH would translate into a capacity of 262.5 cfm (45,000 x .35 ÷ 60). This is enough air to provide 15 cfm for each member of a family of 17. A fan with that much capacity is more expensive to purchase, and if it is left running for 24 hours a day, it will result in high heating or cooling bills and very poor humidity control (the RH will be too low in the winter and too high in the summer). Besides, a large house usually has more natural ventilation simply because there is more surface area exposed to the pressures induced by wind and stack effect. (Actually, the concept of ACH was originally developed for engineers to predict heating and cooling costs due to random air leakage, something for which it can be used quite accurately. As a general concept, ACH is less valid for specifying ventilation rates—even though it is used for that quite often.)

The cfm-per-room approach was developed to counter some of the drawbacks to specifying either ACH or cfm per person. Ventilation engineer Marc Rosenbaum of Energysmiths in Meriden, NH, often uses a rule-of-thumb approach to sizing a ventilation system. He prefers to select a slightly oversized system based on the number of bedrooms in a house.(Rosenbaum) He does this because the number of bedrooms is a reasonable way to predict how many people will live in the house. For a two-bedroom house, he suggests a central ventilation system with 100 cfm capacity. For a three-bedroom house he goes with 150 cfm, and for a four- or five-bedroom house he recommends 200 cfm. With Rosenbaum's oversizing strategy, there is enough extra capacity to accommodate occasions when guests are visiting, when there is a temporary pollution problem such as spilled paint or burned toast, or to counteract some of the outgassing of building materials. An oversized ventilation system need not be operated continuously—it can be shut off for several hours a day and the 24-hour average will still be above 15 cfm per person (with average occupancy). Or, it can be operated at a slower speed when the full capacity isn't required. For example, a ventilator with 150 cfm of capacity might only be operated at 75 cfm by a family of four during normal occupancy.

Continuous operation vs. intermittent operation

There is a difference between the *capacity* of a ventilation system and the *rate* at which air is delivered to the house. For general ventilation, the most efficient way to deliver fresh air to the house and to remove stale air from it is to have a system that runs continuously. For example, if a house has 4 occupants, and you have determined that 15 cfm per person is sufficient, a continuously running 60-cfm ventilation system will suffice. This system has a 60-cfm *capacity* and it delivers air at a *rate* of 60 cfm. If you use a 120-cfm fan and run it at half-speed, you will still be ventilating at a 60-cfm rate, but the system has a 120-cfm capacity (when operated at high speed). An advantage to this is that a slower-running ventilator will be quieter than one operating at full speed. The ventilation *capacity* is the maximum amount of air that can be delivered to and from the house by the system. The ventilation *rate* has to do with the *average* amount of air actually exchanged.

Rather than run a ventilator at half-speed, you could use a timer to turn it on and off at regular intervals, but operate it continuously whenever extra air is needed. For example, the 120-cfm system described in the last paragraph could operate on a 15-minute cycle, controlled by an automatic timer—on for 15 minutes, off for 15 minutes, on for 15 minutes, etc. Because the

Chapter 12 Capacity of a ventilation system

Figure 12–1. A room-to-room fan can be used to distribute air in a house when there is no central ventilator or forced-air heating/cooling fan operating.

cycle is so short (15 minutes), this is almost as effective at moving air as a continuously running system. And, you have the option of bypassing the automatic control (with a simple switch) and allowing the system to run continuously (at twice the normally required rate) when there is extra pollution indoors. This might be necessary when several guests are visiting, while cleaning (many household cleaners should not be inhaled), or if you have done some recent remodeling and the new building materials are not completely outgassed. While a system that cycles on and off has advantages, it also has disadvantages. For example, intermittently-running systems have higher capacities than continuously-running systems, so they will be noisier and they will result in larger pressures indoors. And the continuous cycling on and off can be annoying.

It is also possible to operate a ventilation system at double the required rate for 24 hours, then shut it off the next 24 hours. But a 24-hour on/off cycle isn't nearly as effective as a 15-minute on/off cycle. During a long off cycle, pollutants will build up, and the air quality will decline. However, relatively long off cycles may be necessary for people who are hypersensitive to periodic outdoor pollutants. For example, wood smoke from a neighbor's fireplace may only be a problem at night in the winter when the smoke hangs close to the ground.

In such a situation, a ventilation system could pull wood smoke indoors at night, so some people sensitive to wood smoke may opt to run their ventilation system only during the daytime. People bothered by daytime traffic fumes may prefer to run their ventilation system only at night when there are fewer cars on the road.

For hypersensitive people who are not planning to operate a ventilation system either continuously or with short off/on cycles, a useful rule of thumb is to design the *capacity* 3 to 4 times larger than the *ASHRAE* recommendations. That way, the fan will only need to run for $1/3$ to $1/4$ of the day (6–8 hours out of 24) in order to provide an adequate 24-hour average ventilation *rate*. Even though "people pollutants" will build up during that part of the day when the ventilation system is shut off, there is usually enough total volume of air in even a small house to prevent their concentration from rising excessively. (After 24 hours, the CO_2 level in a hermetically sealed, unventilated, 1,200 sq. ft. house due to two people performing normal activities will be just over 3,000 ppm.) Actually, even in an extremely tight house, it can take two or three days of being closed up before it starts getting uncomfortably stuffy indoors. One disadvantage to this approach is that an intermittently running, high-capacity system will be noisier than a continuously running low-capacity system.

Understanding Ventilation

If a ventilation system is considerably oversized, it will be more expensive to purchase. However, there is rarely much additional cost to modest oversizing, and it offers other advantages. For example, if you ever add on to the house, there will already be enough ventilation capacity for the new rooms. If the actual installation of the system varies from the plans (*i.e.* extra elbows or longer duct runs), the ducts may have more resistance to airflow, so the actual operating airflows may not be quite as much as the design calculations indicated. Thus, oversizing allows for a margin of safety for the designer and installer. However, a good designer should rely on careful calculations, and a good contractor should measure the capacity of the installed system to ensure it meets the design requirements, rather than relying on oversizing.

Pollutants will definitely build up when a ventilation system is shut off, so an approach with long off cycles only works if a house doesn't have any major pollution sources in it. Continuous ventilation may still be needed to flush out VOCs if they are constantly being released by building materials even when the house is empty of people. (Of course, it is always more effective to deal with non-people pollutants by using inert materials in the first place, than by diluting them with ventilation air.) Office buildings sometimes have a "Monday morning effect" if the ventilation is shut off during the weekend. Maintenance crews figure that because the building is unoccupied on Saturday and Sunday, they can forget about ventilating the place. But various pollutants are released by the building materials and furnishings over the weekend, and when the employees return to work on Monday morning, they are greeted by poor-quality indoor air. This effect can be minimized by purging the air in the building prior to the employees arriving on Monday morning.

Even if a house is built, furnished, and maintained with nonpolluting materials, an intermittently running general ventilation system having a long off cycle only works if the doors between rooms are kept open to allow the people pollutants to disperse throughout the house. If doors are kept closed between rooms, by-products of metabolism such as CO_2 and water vapor can build up in one part of the house. For example, if a general ventilation system is only operated during the day and is shut off at night, and the occupants leave their bedroom door open at night, any contaminants released into the air of the bedroom will disperse into the rest of the house, so it is unlikely that the bedroom will become stuffy. When the general ventilation system is turned on in the morning, the house will be flushed out quickly. But if the bedroom door is kept closed, the various by-products of metabolism will not be able to disperse into the rest of the house so they will build up in the bedroom during the night. A small bedroom containing two adults can easily become stuffy before morning. If a house has a forced-air furnace or air conditioner, it could be relied on to circulate and mix the air throughout the house even if doors are closed—but only in the winter and summer when heating or cooling are needed. Another way to circulate air in a house when a ventilation system is shut off is to use room-to-room fans that mount in the wall between two rooms and blow air from one room to another (Figure 12–1). Manufacturers of room-to-room fans include **Broan Mfg. Co., Inc., Grainger, Inc., Penn Ventilating Co., Inc.,** and **Reversomatic Htg. & Mfg. Ltd.** (It should be kept in mind that closed doors can have a bearing on the effectiveness of continuously-running ventilation systems, as well as intermittently-running ones, but the magnitude generally isn't as significant.)

In a study in the Pacific Northwest, it was found that over half the occupants operated their HRV ventilation system less than 30% of the time.(Lubliner) In this study, many people operated their central-exhaust ventilation systems for less than 2 hours per day. If occupants don't run their ventilation systems continuously, builders and designers need to seriously consider oversizing. Of course, it is possible that the systems in this study were already oversized and were simply too noisy. Noise is probably the chief reason why people don't operate ventilation systems.

It is also helpful to factor in how many hours a day a house is actually occupied. In a situation where both spouses work outside the home and the children are at school during the day, a house may only be occupied for 12 hours a day. Therefore, if it contains no major indoor pollution sources, there may only be a need to ventilate when the house is actually occupied. Keep in mind that even though no people are at home, there may be pets that need fresh air and that add odors to the house. One small pet certainly wouldn't need nearly as much ventilation air as an adult human being, but a house containing a dozen cats is another story.

In cold climates there is a potentially serious disadvantage to shutting a ventilation system off during unoccupied times of the day. This is because many people turn their thermostat down when they are not at home, especially in houses with expensive electric heat. After all, why heat a house if no one is home? When the temperature goes down during the day, the RH rises (see Chapter 8). Without ventilating during the unoccupied times, the moisture in the high-RH air gets absorbed slowly over the day by the structure and by furnishings. Then, when the occupants return and turn the temperature back up, they add more moisture to the air from their breath and activities, and some of the absorbed moisture is released. If the ventilation system is shut off during the day, the structure and the furnishings can absorb more moisture during the high-RH, unventilated, unoccupied, times than they release during the lower-RH, occupied, ventilated times. As a result, over a number of days, the structure and the furnishings absorb more and more moisture during the day. This is a principle cause of mold and mildew growth in electrically heated apartments in some Canadian climates.

There is no doubt that a ventilation system operating continuously will be more effective at distributing fresh air and removing stale air than a system that is operated intermittently. Because its capacity can be less, it will also be quieter, and it will have a lower installed cost. But there are times when an intermittently running ventilation system has advantages, especially with hypersensitive people.

Dilution is always a factor

With a general ventilation system, there is always some mixing of fresh incoming air with existing stale air. Therefore, the air being exhausted won't be composed entirely of stale air, and the air remaining indoors won't be composed entirely of fresh air—they are both a diluted mixture of some fresh and some stale air. To put this another way, a 150 cfm ventilation system will always introduce 150 cfm of fresh outdoor air, but because of mixing and dilution, it will always remove less than 150 cfm of stale air. It will indeed expel 150 cfm of air, but part of it is stale air and part of it is fresh air that moves through the living space without benefiting the occupants. This means that 1 ACH doesn't actually change all the air in one hour. You can maximize the amount of fresh air benefiting the occupants and minimize the amount of fresh air leaving the house by designing a good distribution pattern (see Chapter 17).

This is further complicated by how a ventilation system interacts with natural ventilation (see Chapter 7). Sometimes the mechanical-ventilation rate and the natural-ventilation rate can be added together to give you the total ventilation rate—but sometimes they can't. For example, if you operate a balanced ventilation system rated at 1 ACH, you can add the mechanical rate to the natural rate to estimate the total rate. But if you operate a ventilation system that either pressurizes or depressurizes a house, the total rate will always be less than the sum of the natural rate and the mechanical rate.

To see how big a factor dilution is, consider a house containing 8,000 cubic feet (cu. ft.) of stale air. If a ventilation system is powerful enough to introduce 8,000 cu. ft. of fresh air quickly, the incoming fresh air will mix with the stale air in the house as 8,000 cu. ft. of air is simultaneously expelled to the outdoors. It is possible to use an engineering formula to calculate that, with "perfect mixing," approximately 37% stale air will remain in the house after the 8,000 cu. ft. of air was introduced, or just under 3,000 cu.ft. Of course, if the ventilation system remains operating, the second 8,000-cu.-ft. batch of incoming air will dilute the remaining 37% even further. But the second batch of fresh air will also need to dilute any new pollutants that were generated in the meantime. The point of this is this: A general ventilation system never removes all the pollutants in a house. That is part of the reason we need considerably more air per person in a house than a scuba diver needs. A scuba diver's lungs get the benefit of 100% fresh air directly from his or her tanks.

If a little ventilation is good, is more better?

While it might seem that too much air is better than too little air, there are some definite disadvantages

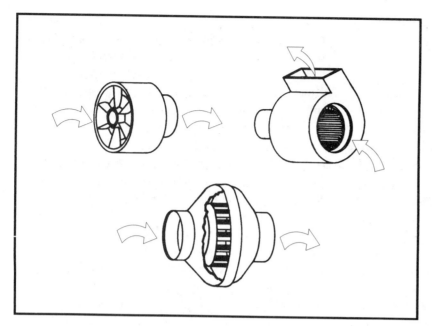

Figure 12–2. There are two types of ventilating fans—axial fans (top left) and centrifugal fans (top right). In-line fans (bottom) are a type of centrifugal fan.

to overventilating a house. If a central ventilation system has a very high capacity and it is operated for very long periods, the house can end up being too dry in the winter, too humid in the summer, or drafty, as well as wasteful of energy and costly.

The law of diminishing returns also applies. The more you increase the rate, the less benefit you receive. Three ACH will improve the indoor air quality more than 1 ACH, but the first air change is generally the most important. The laws of physics that deal with dilution theory state that if you double the ventilation rate, you will have half as much pollution. However, that is only true if emission rates are constant. With some VOCs, the outgassing rate will actually increase with the ventilation rate. With formaldehyde there will be a doubling of the outgassing rate when the ventilation rate is increased six-fold.(Godish 1993) What happens is that the formaldehyde (*e.g.* particle board) senses that the concentration in the air is lower, so it outgases faster. This is a *rate* increase, so there will still be a reduction in the formaldehyde concentration with increased ventilation—it just won't be a one-to-one relationship.

It is also important to remember that an increase in ventilation sometimes means an increase in indoor pollution. If a ventilation system creates negative pressures that cause pollutants such as radon or lawn chemicals to be sucked indoors, then the indoor concentration of those pollutants might be higher when the ventilation system is running than when it is shut off. And if the "fresh" outdoor air contains exhaust gases or biological pollutants, or if the ducts themselves are contaminated, then the air entering won't be very healthy. For example, the original outbreak of Legionnaire's Disease in Philadelphia would have been worse if the ventilation rate was increased because the bacteria were living inside the ducts.

Factoring in infiltration

Natural and accidental ventilation occur in houses with mechanical ventilation systems as well as houses that are not mechanically-ventilated. All houses, thus, are partially ventilated in a random manner. Sometimes this random infiltration and exfiltration reduces indoor pollution, sometimes not; sometimes it causes moisture problems, sometimes not; sometimes it causes backdrafting or spillage, sometimes not. Ideally, you would want a very tight house with a mechanical ventilation system exchanging all the air. But that is not very realistic because most houses just aren't built tight enough. So, what should you do?

First, you should fix any problems that exist. For example, you should weatherize for energy efficiency and comfort, then analyze and solve any pressure-related problems related to radon, moisture, or backdrafting. Once all the problems are taken care of, you will probably have tightened the house, so the amount of air that is infiltrating and exfiltrating will be less pronounced and it will not likely cause any problems. At this stage, the annual average natural ventilation rate can be estimated by performing a blower-door test. (This was covered in Chapter 11.) It is important to keep in mind that you will be working with an estimate that is an *average* rate—the actual natural ventilation rate will be higher at some times, lower at others, and sometimes it is zero. In other words, natural ventilation is like an intermittently running ventilation system—it has both on and off times. But the on/off cycle is variable, so you don't know how long the off times are. Also, remember that estimating an average annual ventilation rate is a fairly imprecise science. Still, once you know the average annual natural ventilation rate, you can start to determine how much air a general ventilation system needs to provide mechanically.

Let's say you have decided that a house needs 75 cfm of continuous ventilation, and you have estimated the average annual natural ventilation rate to be 35 cfm.

In this situation, you will need to provide 40 cfm mechanically to get the annual average rate up to 75 cfm. If you are going to use a balanced ventilator (with or without heat recovery), the rates can be added together, so you will need to select a ventilator with a 40 cfm capacity to end up with a total average rate of 75 cfm (35 + 40 = 75). But if you plan to use an unbalanced ventilator (either central-supply or central-exhaust), the natural and mechanical rates aren't additive. (You should review the three rules governing the interactions between natural ventilation and mechanical ventilators found in Chapter 6.)

When you have a low natural ventilation rate (less than half of the mechanical rate), the total rate will be the same as the rate of an unbalanced ventilator. So, in our example where the natural rate is 35 cfm and you want a total of 75 cfm, you will need to use a 75 cfm ventilator. When the natural ventilation rate is low, an unbalanced ventilator will overpower the natural airflows, sometimes totally.

When the natural ventilation rate is higher (more than half the mechanical rate), the total rate will be the natural rate plus half the rate of an unbalanced ventilator. So, if the natural rate is 45 cfm, you will need to install a 60 cfm ventilator to get a total average rate of 75 cfm (45 + ($1/2$ x 60) = 75).

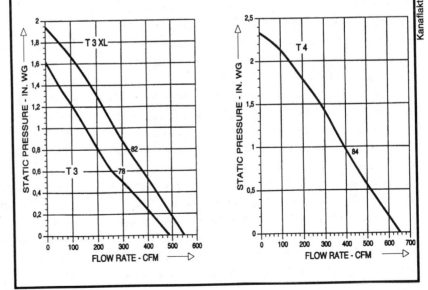

Figure 12–3. *Most fan manufacturers will provide performance charts showing the capacity of their equipment at different static pressures.*

As was discussed in Chapter 11, the average natural infiltration rate is an imprecise estimate—it could be as much as 50% too high or 50% too low. And it has on/off times that are never regular. To be conservative and to account for these variables, you could assume that the estimate is too high by 50%. For example, if the estimated annual average ventilation rate is 35 cfm, you could assume that the rate is really 17.5 cfm and use that number in your calculations. You run the risk of overventilating, but not by very much. Or, you could stick with the estimate and assume that any underventilation will be made up for by accidental ventilation (*e.g.* clothes dryer) or intermittent local ventilation (*e.g.* bath fans, range hood).

Ventilating fans

Fans provide the pressure necessary to move air through a ventilation system. There are two basic fan types used in residential ventilation equipment: axial and centrifugal (Figure 12–2). Axial fans have propeller-type blades. They work well at moving large volumes of air as long as they don't have to work against very much resistance. Window fans are axial fans. Centrifugal fans are often called blowers and they resemble the "squirrel cages" used to let pet hamsters get their exercise. Forced-air furnaces and air conditioners use centrifugal fans. In-line fans are a type of centrifugal fan to which you can attach a tube or duct to each end. (They are often called tube fans.) Many radon-removal contractors use in-line fans. In general, centrifugal fans are capable of moving more air against more air resistance than axial fans and they are often somewhat quieter than axial fans.

A variety of motors are available to power fans but permanent-split-capacitor motors and shaded-pole motors are the most common. Shaded-pole motors are fairly inexpensive to purchase but many aren't made for continuous operation, so they don't last very long. Permanent-split-capacitor motors are more expensive initially but are longer lasting, more reliable, and more economical to operate. They are typically twice as efficient, so they only use half as much electricity as shaded-pole motors. Some high-efficiency furnace/air conditioners are now being equipped with variable-speed DC (direct current) motors (see below) that operate on even less electricity.

A few pieces of ventilation equipment have their motors mounted outside the cabinet, outside any airstream. This results in the motor running hotter (the airstream helps cool it) and a hotter motor will have a shorter life span. If a motor is in one of the airstreams of an HRV, the heat released by a motor will be partially recovered by the heat-recovery process.

Many ventilation fans are meant to operate for extended periods of time, sometimes for 24 hours per day, every day, for several years. Long-running motors should be designed for *continuous use*, and they should have bearings that will last for many hours without failing. Sometimes continuous-use motors are rated by the average number of hours they will operate before failure. For example, a motor rated at 50,000 hours should last for almost 6 years (considering there are 8,760 hours in a year). Many bath and kitchen fans aren't rated for such long-term continuous operation. *Continuous duty* is not the same thing as continuous use. It is an electrical safety rating that means a motor can run for a couple of weeks without failure. Most motors on the market (elevator motors are the main exception) are continuous-duty motors, but they may or may not be rated for continuous use.

Sensitive people are sometimes bothered by odors given off by a fan motor. These odors originate from the insulation, lacquer, or lubricating oil. A ventilating fan having a motor in the fresh airstream will contaminate the fresh air with these odors. In the majority of cases, this is a minor source of pollution and it probably won't even be noticed by most people, but sensitive people must often specify that a fan motor be located outside the fresh airstream. A central-exhaust system would be a good choice for them, and some HRVs are made with the fan motor in the exhaust airstream.

Fan capacity

A fan is designed to do only so much work. It can either move a certain amount of air through a single short duct, or it can move less air through a series of very long ducts. In other words, the capacity of a ventilating fan is determined by how much air resistance it has to work against. The amount of air resistance that a fan must overcome to deliver a certain volume of air is called the static pressure. Most manufactures of good-quality ventilation equipment will provide a table or graph showing different airflow rates at different static pressures. (Figure 12–3) While you can roughly select the capacity of a ventilation system by looking at the figures in one of these performance charts, you won't be able to precisely determine the capacity of a particular ventilation system until the entire duct system has been designed. See Chapter 17 for information about sizing a duct system.

Static pressure exists because of the friction and turbulence generated as air moves through a duct. A certain amount of noise is also created. Static pressure (abbreviated s.p.) can be measured in several different units, such as Pascals, inches of water, etc. In ventilation systems it is usually measured in inches on a water gauge (" w.g.), but Pascals is sometimes used. It isn't unusual for a duct system to have enough resistance to result in 0.1" and 0.4" w.g. of static pressure. (In water pipes even higher pressures are involved, so feet on a water gauge or pounds per square inch (psi) is more appropriate when measuring water pressure.) If you attach a duct to a 100 cfm fan and measure a pressure difference between the inside of the duct and the outside of the duct of 0.2" w.g., you can say that the fan delivers 100 cfm at a static pressure of 0.2" w.g.

Many HRV manufacturers rate the flow of their products at 0.4" w.g. of external static pressure. (*External* refers to anything added to the HRV itself—ducts, grilles, dampers, etc.—that creates resistance.) In other words, these systems will deliver the rated flow when they are connected to a duct system that has enough resistance to generate 0.4" w.g. of static pressure. Many central-exhaust system manufacturers rate the flow of their products at 0.2" w.g. of external static pressure. It is usually safe to choose the capacity of a system based on these static pressures. Then, it is important to design the duct system with a resistance that will generate static pressures below these figures.

When purchasing an individual fan (*e.g.* an in-line fan), it is important to select one that will deliver the correct amount of air at the installed static pressure. Sometimes, manufacturers of low-cost fans don't list a static pressure along with a capacity. They occasionally give a capacity based on *free flow* which is the amount of air the fan will deliver at zero static pressure—with no fittings whatsoever attached. Once installed, all fans must work against some amount of static pressure, so a capacity at free flow is meaningless. In general, if a chart or graph showing cfm ratings at different static pressures isn't available, the fan is probably of low quality and should be avoided.

Motor and fan efficiency

A form of energy efficiency that promises to become important in upcoming years is motor and fan efficiency. Permanent-split-capacitor motors are commonly used in both forced-air heating/cooling systems and ventilating systems. They are typically 30–60% efficient. Shaded-pole motors, on the other hand, are usually only 10–40% efficient. Variable-speed *electrically commutated motors* (ECM) are now being used in some forced-air furnace/air-conditioners and they have efficiencies of up to 70%. Permanent-split-capacitor motors and shaded-pole motors operate on alternating current so they can be hooked directly to the electrical system of the house. ECM motors operate on direct current so they are supplied with a device that converts alternating house current to direct current. ECM motors are somewhat more expensive than permanent-split-capacitor or shaded-pole motors of a similar size. Unfortunately, ECM motors are not yet available for many residential application—but they are being used in some top-of-the-line furnaces.

One of the disadvantages of permanent-split-capacitor and shaded-pole motors is the fact that they lose about half their efficiency if they are run slower than their design speed. So, a typical shaded-pole motor, running at half-speed, may be only 20% efficient. The efficiency of an ECM motor is not dependent on its speed. If a fan is going to run for extended periods at a low speed, an ECM motor will be considerably more efficient than a permanent-split-capacitor motor. This can be an important (and money-saving) factor when a forced-air furnace/air conditioner is going to be running continuously to distribute ventilation air. Most forced-air heating/cooling fans aren't on all the time—they only run when the thermostat calls for warm or cool air. But if the heating/cooling ducts are used to distribute ventilation air, as well as heated or cooled

Understanding Ventilation

air, the heating/cooling fan needs to run whenever ventilation air is needed. In many of these installations, the fan operates continuously, but on a lower speed. (High speed really isn't needed because ventilation typically requires less air movement than heating or cooling.) When you run a permanent-split capacitor motor at low speed, it's efficiency may be only 30%, but when you run an ECM motor at a low speed its efficiency remains high—about 70%—so it will consume much less electricity. The bottom line: If a ventilation system is going to be tied to a forced-air heating/cooling system, it is often a good idea to use an ECM motor.

Actually, motor efficiency is only a part of fan efficiency. In calculating the overall efficiency of a fan, several efficiencies come into play—things such as the efficiency of the motor, the efficiency of the fan blades, the efficiency of the housing, the efficiency of the belt drive, etc.—and they must be factored together. For example, a 20% efficient motor coupled to a 20% efficient fan will have a combined total efficiency of only 4%. For more information about fan efficiency, see *The Energy Efficiency of Residential Ventilation Fans and Fan/Motor Sets*(White 1991b) and *Efficient and Effective Residential Air Handling Devices*(Allen) in Appendix C. This recent research has determined that most residential fans—not only those used in ventilation equipment, but also furnace fans, computer fans, hair dryer fans, etc.—are extremely inefficient. When the amount of energy necessary to move a given amount of air is compared to the amount of electrical energy consumed by a fan, most residential fans are less than 10% efficient. Many industrial quality fans, on the other hand, are 60–70% efficient. When a fan is hooked to a system of ducts, the efficiency of the distribution system is also a factor. In this context, efficiency has to do with the amount of energy required to move air from one place to another. In Chapter 23, we will discuss efficiency in another context—how efficiently HRVs transfer heat from one airstream to another.

For very small residential ventilating fans, or those that are only operated intermittently, there may not be a great deal of energy to save, so increasing a fan's efficiency may not result in tremendous monetary savings. A Canadian study found that an energy-efficient fan in a kitchen-range hood would have a $3 annual operating cost compared to $22 for a conventional inefficient range hood.(Sheltair) For a whole-house ventilation fan, the savings are more substantial ($11 vs. $147) because it is operated for longer periods. Even though an ECM motor can cost several hundred dollars more than a permanent-split-capacitor motor, energy savings like this can mean a quick payback. As fan efficiency data becomes more widely known, manufacturers will no doubt begin to redesign their equipment so they utilize electricity more efficiently, but today there is a great deal of room for improvement. Fan operating costs are covered in more depth in Chapter 13.

A useful way of comparing the efficiencies of two fans that have different capacities is to divide the motor's wattage by the cfm rating. This will give you the number of watts/cfm. The lower the number, the more efficient the fan/motor combination. Currently, the most efficient bath fan consumes less than $1/5$ watt/cfm. Most HRVs contain two fans and use a total of about 1 watt/cfm. If a fan uses 2 or more watts/cfm, it is relatively inefficient in moving air.

The bottom line

A modest amount of ventilation (15 cfm per person or .35 ACH) works well for most people in relatively unpolluted houses, but unusually sensitive people often require more to adequately dilute the indoor pollutants in the air. However, too much overventilation can result in poor humidity control and expensive utility bills. In any case, it should be remembered that these ventilation-rate recommendations will not guarantee you will have healthy air. The only way to do that is to build and maintain a house with inert products and to prevent pollutants from being sucked indoors.

While there are advantages to an intermittently running system (*e.g.* controlled by an automatic timer on a 15 min. on/off cycle), a ventilation system that is designed to run continuously can be of smaller capacity. Therefore, it will have a lower capital cost and will be quieter. If costs are a significant concern, operating costs (energy usage of the fan and the tempering costs) should also be considered. The cost of ventilation will be discussed in Chapter 13.

Chapter 13

The cost of ventilation

Whenever the subject of ventilation comes up, one of the first questions asked is, How much does it cost? If the cost of a ventilation system sounds like too much money, many people will decide they don't need it. After all, they have gotten along most of their lives in a home without a ventilation system, why do they need one now? At this point it becomes the responsibility of the ventilation equipment installer or supplier to educate the home buyer, the general contractor, and perhaps the bank or lending institution that a ventilation system should be a requirement, not an option, and that it is money well-spent.

Prospective home buyers rarely ask how much the electrical wiring system costs. They almost never consider what they are paying for a septic tank, or plumbing lines, or doors, or windows. When people ask how much a hot tub costs, fancier kitchen cabinets, a security system, or a built-in wet bar, they usually opt for high-quality long-lasting products or amenities that will make their lives more comfortable and easier. But a ventilation system is often not given high priority because it isn't considered necessary. This is unfortunate, but the situation will no doubt continue to a certain extent until building codes require controlled ventilation in all houses.

Our building codes will certainly require ventilation systems in houses in the future, but today many don't even mention the subject, so houses continue to be built without adequate fresh air in them—simply because a ventilation system costs money to buy and money to operate.

Granted, money is always an important factor, and there is definitely a cost associated with ventilating houses. Actually, it doesn't matter if air passes through a house in a natural, accidental, or controlled manner—all forms of ventilation cost something. But some forms of ventilation cost more than others and it would be foolish to spend more than you have to in order to get a

Understanding Ventilation

good system. So, when evaluating a ventilation system, it is important to consider all the costs, then evaluate whether one particular system makes more economic sense than another. For example, with controlled ventilation you need to purchase and install some equipment; with accidental and natural ventilation you don't. If a ventilation system has a fan, it takes electricity to run the fan and electricity costs money. In addition, when outdoor air is brought into a house by any means, it will need to be heated, cooled, humidified, or dehumidified—depending on the outdoor climate—and this adds to the house's energy bill.

It is surprising to many people that, when all the factors are considered, controlled ventilation in a tight house typically provides better temperature control, better air quality, and increased comfort, at a lower cost than uncontrolled infiltration and exfiltration in a leaky house.

Capital costs

The actual cost of an installed ventilation system can vary considerably. This is the capital cost and it includes the price of the ventilation equipment itself, incidentals, and labor. But it should be remembered that capital costs only need to be paid for once. (While no piece of mechanical equipment will last forever, high-quality ventilation equipment should last as long as a furnace.)

Capital costs are often the only costs prospective homeowners consider when they decide to reject a ventilation system from their new house. While these up-front costs are important, there are comfort benefits to having a ventilation system that you really can't put a price on. A ventilation system will also make a house more healthy, and how do you put a dollar value on your health?

By controlling excessive indoor moisture, a ventilation system will minimize mold growth and rot, so a house will be longer-lasting. Major structural repairs due to moisture-induced rot can easily cost many times as much as a ventilation system.

When you consider the health and comfort of the occupants and the health of the structure, the cost of a ventilation system is negligible. In a harsh climate, there can be monetary savings to specific ventilation strategies, something that will be covered below in the section *Operating costs*.

Equipment cost

The cost of ventilation equipment can range from less than $30 for an inexpensive bathroom exhaust fan that will probably be fairly loud and inefficient, to a little over $100 for a long-lasting, quiet, high-quality exhaust fan.

The most expensive residential balanced ventilator with heat recovery runs in the neighborhood of $1,000. And, of course, there are many options between these price extremes. For example, many HRVs can be purchased for $400–500 and central-exhaust fans are available for even less.

Many people buy an inexpensive piece of equipment, believing they are saving a great deal of money. However, there is no economy if the equipment doesn't last very long and needs to be replaced after 2–3 years.

Ventilation equipment that is expected to run for extended periods of time must contain high-quality components, and high-quality components cost more than cheap components. As with a heating/cooling system, only a quarter to a third of the installation cost of a ventilation system is for the major equipment—the bulk of the expense is for incidentals, such as the distribution system, and for labor, so you shouldn't skimp on the main piece of equipment. For example, even if you are only considering something as simple as a bathroom exhaust fan, get a unit that is well-made and quiet, rather than a bottom-of-the-line model.

Incidental costs

All ventilation systems are made up of a variety of pieces, not just one major piece of equipment. There may be a network of ducts and fittings, grilles, several through-the-wall vents or a single inlet and outlet, duct tape or mastic to seal the joints, electrical wiring, and one or more controls. There may also be a duct heater, plumbing drain, and perhaps a filter package. The incidentals involved in installing a simple exhaust fan may be minimal, but with a fully ducted heat-recovery ventilator, they can be substantial.

Chapter 13 The cost of ventilation

The cost of the incidentals can be estimated by a competent installer once the ventilation strategy has been chosen and the system has been designed.

Labor costs

Inexperienced contractors don't often have a good understanding of what a ventilation project entails. As a result, they generally underestimate what a job will cost. They then end up taking an excessive amount of time to install the system or even install it incorrectly. This can mean cost overruns and perhaps a system that doesn't do what it is supposed to. An experienced ventilation contractor, on the other hand, will often be able to perform his work more quickly and efficiently.

Heating and air-conditioning contractors are often relied upon to install ventilation systems because they have a knowledge of ducts and air movement. However, this is sometimes like asking an automobile-engine mechanic to work on an airplane engine—yes, they are both engines, but there are significant differences that only a specialist will understand. As it turns out, the knowledge necessary to ventilate a house is different from that needed to heat or cool it, and it requires more skill than simply connecting ducts together. Many heating and air-conditioning contractors are well-qualified to install ventilation systems, but some aren't. Homeowners can sometimes check with ventilation equipment manufacturers to locate installers they have personally trained. (The *Home Ventilating Institute* and some equipment manufacturers offer a great deal of information about ventilation equipment installation in training classes and manuals.) A competent installer should be able to estimate his labor costs quickly and accurately.

Adding up the installation costs

Most houses have some form of local-exhaust ventilation, such as a bathroom fan or kitchen range hood. While there can be extra cost associated with installing a quieter or more powerful fan, the additional cost usually isn't exorbitant. However, most houses don't have a central ventilation system and these systems can represent a significant investment. While a top-of-the line, fully featured, energy-conserving central ventilation system can be costly, there are lower-cost alternatives.

A late 1980s study of 152 ventilation systems in the Pacific Northwest by the Residential Construction Demonstration Project, and funded by the *Bonneville Power Administration*, calculated the following average total capital costs (equipment, incidentals, and labor):(Lubliner)

Central-exhaust heat-pump water heater	$2,857
Balanced HRV, fully ducted with a duct heater*	$1,935
Balanced HRV, fully ducted	$1,325
Balanced HRV connected to furnace return-air duct	$1,049

*duct heaters really aren't a good idea with HRVs

A smaller (24 installations) separate Residential Construction Demonstration Project survey compared the material and labor cost of HRV systems to central-exhaust systems. The non-heat-recovery systems were considerably less expensive to install—but they have a higher operating cost.

Balanced HRV	$1,015
Upgraded bath fan for central exhaust, with fresh air entering furnace return-air duct	$562
Upgraded bath fan for central exhaust, with through-the-wall vents for inlets	$393

Actually, HRV costs are typically higher than the figures given above. Though systems with simple duct layouts will be cheaper, it wouldn't be unusual for a first-class fully ducted HRV installation to run as much as $2,500–3,000, or an HRV connected to a forced-air furnace/air conditioner to be in the $1,500 range. It is important to keep in mind that costs for the same system can easily vary from one part of the country to another. However, what these figures show is that there is a central ventilation system for most budgets. The lower-cost systems may not have all the "bells and whistles," and they may have some disadvantages in a particular climate, but they can definitely provide the occupants with fresh air on a regular basis, and that is the most important function of a ventilation system.

Operating costs

The operating cost of a controlled ventilating system includes the maintenance expenses, the cost of electricity to run the ventilating fans, and the cost of

Understanding Ventilation

tempering (warming, cooling, humidifying, or dehumidifying) the incoming air. The cost of maintenance is generally minimal, but the electricity and tempering costs must be paid every month. When natural and accidental ventilation cause air to randomly infiltrate and exfiltrate through the cracks and gaps in a house, there are no installation costs, no maintenance costs, and no fan operating costs, but there is a cost associated with tempering the incoming air. However, this tempering cost is usually hidden as a part of the heating or air-conditioning bill, so we don't see it directly. How much of the heating and air-conditioning bill is attributable to tempering the infiltrating air depends on the climate and the amount of air entering the house.

It is the tempering of incoming air that can make the day-to-day operation of a ventilation system expensive, especially in very harsh climates. It makes little economic sense to purchase a low-cost fan instead of a more expensive HRV if it means high operating costs over the life of the equipment. That would be like buying a cheap automobile that only gets 8 miles per gallon. If you plan to keep such a car any length of time, the operating expenses will quickly eat up any savings in the initial purchase price. Surprisingly, the cost of tempering incoming air isn't as high as many people believe. In moderate climates it is actually quite low. In fact, in many parts of the U.S. the operating costs associated with mechanical ventilation are less than $100 per year. If the operating costs are quite low to begin with, it would be foolish to invest a great deal of money in an energy-efficient HRV because there simply wouldn't be much money to save.

Broan Mfg. Co., Inc., a producer of residential ventilation equipment, has made the following operating-cost estimates for three different pieces of their 100-cfm ventilation equipment for residences in Chicago, IL.(Wolbrink 1993b) It is assumed that the equipment is operating 60% of the time and the house is heated with an 85%-efficient gas furnace.

Central-exhaust ventilation	$73.91 per year
Balanced ventilation without heat recovery	$103.45 per year
Balanced ventilation with heat recovery	$33.61 per year

Broan's estimates are based on a 100-cfm ventilator operating 60% of the time. This may or may not be adequate—it depends on a variety of factors. But the point is this: Even $100 per year isn't very much to spend on something is so vital to our health, comfort, and well-being as fresh air.

Maintenance costs

While routine maintenance expenses must be considered, they will usually be fairly low. They include the cost of cleaning (*e.g.* filters, grilles, inlets, HRV core, ducts, etc.) and periodic lubrication. This type of work can be performed by a furnace or air-conditioner technician at a nominal cost when he services the other mechanical equipment. However, much of this type of maintenance is not difficult and can easily be done by the homeowner. Duct cleaning usually must be done by a professional, but it is generally only needed after every several years of operation.

Mechanical equipment does not last forever. Therefore, occasional fan-motor replacement should also be considered, even though it is not considered a regular expense. A high-quality fan should last for 10 years or more if it is maintained regularly. However, with a low-cost piece of ventilation equipment, the fan motors may not last nearly as long, and cheap fans tend to be noisy. In some cases, cheap fans don't even last a year. Because ventilation equipment often operates for extended periods, sometimes continuously, it is important to select high-quality long-lasting equipment, no matter what type of ventilation strategy is chosen. This will minimize both the cost of operation and of fan replacement.

Cost of electricity to run fans

The cost of the electricity to run a ventilation fan depends on three factors: the amount of time the fan runs, the wattage of the fan, and the cost of electricity. As an example, let us consider a typical HRV with fans that require 100 watts (W) of power. If the HRV is operated for 10 hours per day, it would use 365,000 watt-hours (Wh) of electrical energy per year (100 x 10 x 365). Dividing this number of watts by 1,000 gives us 365 kilowatt-hours (kWh) of energy. To determine the annual cost of running the fan, multiply the number of kWh used by your local electric rate. In the U.S., the cost per kWh ranges from a low of about 4¢ to a high of over 16¢, with an average of about 9¢ (1994 prices). If we use the average electric rate, the annual fan operating cost in our example would be $32.85 (365 x $0.09). This is only $2.74 per month (32.85 ÷ 12), or about 9¢ a day (32.85 ÷ 365)—not a great deal of money. Even if this size fan is operated continuously day and night, it would still only cost about 22¢ a day. The formula for calculating the annual cost of electricity to run a fan is:

Annual fan operating cost = (wattage of fan) x (fan run time in hours/day) x (365 days) ÷ (1,000) x (cost per kWh of electricity)

If a ventilation system is coupled with a forced-air furnace/air-conditioning system, with the larger system being used to distribute the air in the house, the additional cost of operating the furnace/air-conditioning fan should also be considered. These installations often operate for 24 hours a day. However, the furnace/air-conditioner would occasionally be running part of that time anyway for heating and cooling purposes—perhaps 33% of the time in an average year. Thus, 67% of the operation would be a ventilating cost and 33% would be a heating/air-conditioning cost. A multispeed furnace/air-conditioner fan might use 200 W on low speed. (It would typically operate at a higher speed when heating or air conditioning are called for than when only ventilation is needed, and at high speed a fan could use 500 W or more.) To calculate the annual cost of running the furnace/air-conditioning fan, first multiply the wattage (in this case 200 W) by the running time attributable to ventilation (24 hours x 365 days x 67%), then divide by 1,000 to convert to kWh, then multiply by the electric rate (we'll use 9¢ again). In this example, the additional cost to operate the furnace/air-conditioner fan to distribute ventilation air would be $105.65 per year (200 x 24 x 365 x 67% ÷ 1,000 x .09), or about 29¢ a day. A high-efficiency furnace/air conditioner with an ECM fan motor may only use less than half as much energy, so it will cost less than half as much to operate. (While it is usually a minor factor, operating a fan continuously in the winter will result in a slightly lower heating bill because the heat given off by the motor will help warm the house. On the other hand, the heat released by the motor will cause an air conditioner to work a little harder in the summer.)

Keep in mind that all installations are unique. These examples are typical, but fan wattages can vary considerably, so if operating costs are a real consideration, it is important to base your calculations on the actual wattage of the equipment you plan to use, the local electric rate, and an accurate estimate of the running time. It is also important to take into consideration a loss of air due to duct leakage.

Cost of tempering incoming air

When cold outdoor air is brought indoors in the winter, it will need to be warmed up. When hot outdoor air is brought indoors in the summer, it will need to be cooled down. If the incoming air in the summer is very humid, some of the moisture will need to be removed to prevent the house from feeling muggy or clammy. Tempering means warming, cooling, or dehumidifying the incoming air—in other words, making it comfortable. There is always a cost associated with tempering incoming air, even if the air enters by way of infiltration, but with infiltration the cost is usually hidden in the heating and air-conditioning bills. When a great deal of tempering is necessary, utility bills will be high; when only a little tempering is needed, utility bills will be somewhat lower.

Calculating the approximate average cost associated with heating or cooling the incoming air takes a little time, but it is relatively easy to do. However, determining the cost of dehumidifying incoming humid air in the summer can be somewhat more involved. To calculate the energy cost needed to heat or cool incoming air, you need to know 1) how much ventilation air is entering the house from the outdoors, 2) how much energy is needed to heat or cool that air, and 3) the cost of energy. Actually there are a number of other less significant variables involved, but it is beyond the scope of this book to explain how to perform extremely complex calculations.

What follows should yield a conservative tempering cost. If all the possible variables were factored into the equations, the actual cost would almost always be lower than the figures calculated in the next few paragraphs. In some cases, the actual cost of tempering could be as much as 30–40% lower than the estimated cost as determined below.

Understanding Ventilation

If a house has air entering through both a mechanical ventilation system and because of random leakage (accidental and natural ventilation), the total tempering cost will include the costs associated with heating/cooling both the controlled ventilation air and the uncontrolled infiltration air. The tempering cost of random infiltration can be roughly estimated by first using a blower-door to estimate the average annual infiltration rate (See Chapter 11). Calculating the cost of tempering controlled ventilation air is more accurate because the amount of air delivered by a fan is more uniform and regular. Because you can't do very much to control the amount of infiltrating air (except tighten the house), the cost of controlled ventilation is often of more interest.

When a house has a controlled ventilation system that either pressurizes or depressurizes the house, the total amount of air exchange is less than the sum of the controlled rate plus the uncontrolled rate. Only when a house has a neutral-pressure (balanced) ventilation system can you add the controlled airflow to the uncontrolled airflow to obtain the total amount of air moving into and out of a house. This is because of the way pressures and flows interact with each other (See Chapter 7 for a discussion of this). Because of these interactions, the actual tempering costs will be lower than the costs calculated below if the ventilation system either pressurizes or depressurizes the house.

Calculating the volume of air entering the house

If air enters a tight house through a controlled ventilation system, it can be relatively easy to calculate the total number of cubic feet of air entering the house during a day. You simply multiply the cfm capacity of the ventilating fan by 60 (minutes per hour) by the number of hours the fan runs.

For example, a range hood moving 200 cfm and operated 2 hours per day would move 24,000 cubic feet (c.f.) in one 24-hour period (200 x 60 x 2). A continuously operating 120-cfm central ventilation system would move 172,800 c.f. in one 24-hour period (120 x 60 x 24). The formula is:

Cubic feet of air entering a house in one day =
(cfm of fan) x 60 x
(number of hours per day the fan runs)

For outdoor air that infiltrates as a result of natural ventilation, the cfm capacity and the amount of time air is entering will both be highly variable, so an accurate calculation can be difficult. There are computer programs available that can be used to estimate aver-

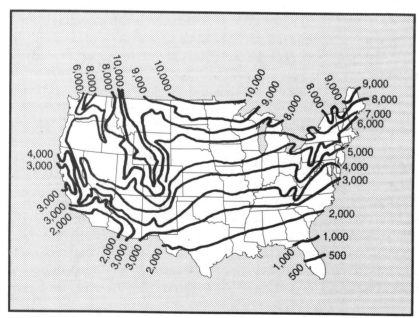

Figure 13-1. Heating degree days (DDs) in the continental United States (Base 65°F).

Chapter 13 The cost of ventilation

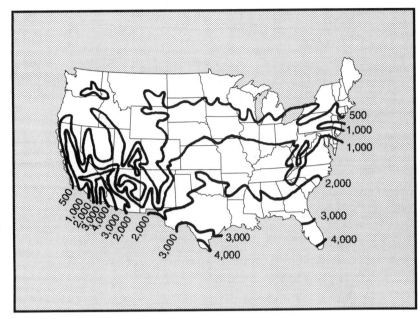

Figure 13–2. Cooling degree days (DDs) in the continental United States (Base 65°F).

age infiltration rates,(HOT2000) but blower-door testing can also help you arrive at a rough idea of the heating and cooling cost when natural ventilation is the source of fresh air.

In order to estimate the cost of natural ventilation, you can divide the CFM_{50} by 20 (or another number that the blower-door technician feels is appropriate to your locale) to get an average cfm rate, then assume that average rate would be in effect all day. In a fairly leaky house where a blower door has measured 6,000 CFM_{50}, there would be 423,000 c.f. of air entering per day under average conditions (6,000 ÷ 20 x 60 x 24). The formula is:

Cubic feet of air infiltrating into a house in one day = (CFM_{50}) ÷ 20 x 60 x (24 hours per day)

Calculating heating and cooling energy requirements

In order to calculate how much energy is needed to heat or cool the incoming air, you need to know how hot or cold the outdoor air is, and how long it is that hot or cold. This can be determined by knowing how many degree days (DDs) there are in a particular locality. A DD is a unit of measurement based on indoor/outdoor temperature differences that can be used to estimate energy consumption. Heating DDs are used to estimate heating costs and cooling DDs are used to estimate air-conditioning costs.

But what is a DD? There is one heating DD for every day in which the average outdoor temperature is a degree less than 65°F. (Engineers often use 65°F as an indoor design temperature because there are enough miscellaneous heat sources in a house (*e.g.* lighting, the stove, the refrigerator, people, etc.) to raise the indoor temperature from 65°F to 72°F.) For example, if the average outdoor temperature is 40°F over a 24-hour period, there will be 25 heating DDs (65 – 40 = 25) for that particular day. If the outdoor temperature averages 50°F the next day, then that day will have 15 heating DDs (65 – 50 = 15). By adding up all the heating DDs for an entire year, you will have a good idea how cold it is and for how long. Figure 13–1 shows a map of the U.S. with the annual number of heating DDs plotted. By locating your town on this map, you can determine the number of heating DD in your locale. For example, Indianapolis has about 5,000 DDs per year, Miami has less than 500, and there are 10,000 DDs in northern Minnesota. (There are reasons to use heating DDs based on a temperature lower than 65°F, especially in very well-insulated houses, but to be conservative, we will use data based on 65°F.)

Understanding Ventilation

In the summer, cooling DDs can be used to calculate the cost of cooling incoming warm air. There is one cooling DD for each day in which the average outdoor temperature is a degree above 65°F. Figure 13-2 shows a map of the U.S. with cooling DDs plotted. Miami has about 4,000 cooling DDs, but Indianapolis doesn't have a very long hot summer, so it only has about 1,000 cooling DDs. (Again, there are reasons to use another base temperature, but by using 65°F we are being conservative.)

Energy is measured in Btus (British thermal units), and it takes 0.018 Btus to change the temperature of one cubic foot of air by one degree Fahrenheit. (There is about one Btu of energy given off by a kitchen match.) You must add 0.018 Btu to each cubic foot of air for each degree you want to raise its temperature, or extract (with an air conditioner) 0.018 Btus from each cubic foot of air for each degree you want to lower its temperature. For example, to raise the temperature of 120 cubic feet of air from 50°F to 66°F, you must add 34.6 Btus (120 x (66 − 50) x 0.018). To lower the temperature of 160 cubic feet of air from 90°F to 72°F, you must remove 51.84 Btus (160 x (90 − 72) x 0.018).

Following are the two formulas that can be used to calculate the annual amount of energy needed to heat or cool incoming air. They are actually the same formula—the only difference is that the first uses heating DDs and the second uses cooling DDs.

Annual amount of energy (in Btus) needed to heat incoming air in the winter = (number of cubic feet of air entering the house per day) x (number of heating DDs in your locale) x (0.018)

Annual amount of energy (in Btus) needed to cool incoming air in the summer = (number of cubic feet of air entering the house per day) x (number of cooling DDs in your locale) x (0.018)

Calculating heating and cooling energy costs

Once you know how much energy is needed to heat or cool the incoming air, you can calculate how much it will cost to do so. Energy costs depend on the price of the fuel used to heat (*e.g.* oil, gas, electric) or cool (*e.g.* electric heat pump) the house. Fuels are purchased in different units (*e.g.* gallons of oil, hundreds of cubic feet (ccf) of natural gas, kilowatt-hours of electricity), so you need to know how much energy is available in each units. Following is the amount of energy that can be extracted from common heating fuels:

1 kWh of electricity = 3,413 Btus of energy
1 ccf of natural gas = 100,000 Btus of energy
1 gal. of fuel oil = 138,000 Btus of energy

ANNUAL HEATING AND COOLING COST FOR 15 CFM OF CONTINUOUS VENTILATION WITHOUT HEAT RECOVERY						
	Miami	Los Angeles	Nashville	Boston	Minneapolis	Fairbanks
80% efficient natural gas heating @ $0.60/ccf	$1	$5	$11	$17	$29	$42
100% efficient elect. resistance heating @ $0.08/kWh	$2	$17	$33	$53	$91	$131
80% efficient fuel oil heating @ $1.12/gal.	$1	$7	$14	$23	$39	$57
Heating with 10.0 SEER heat pump @ $0.08/kWh*	$1	$6	$11	$18	$25	$45
Cooling with 10.0 SEER heat pump @ $0.08/kWh	$13	$2	$5	$2	$1	$0

Figure 3–3. There is a cost associated with ventilation in all climates, but the costs are most significant in harsh locales. The costs shown in this chart are for heating and cooling 15 cfm of incoming air for an entire year. According to ASHRAE, 15 cfm is the amount of air needed by one person. The figures in this chart have been rounded to the nearest dollar.

*These figures do not take into account the fact that heat pumps use supplemental resistance heaters in very cold weather, thus the actual tempering cost of ventilation air when using a heat pump for space heating could be higher than the figures shown.

You also need to know the efficiency of your furnace, boiler, or air conditioner because you need to factor in how efficiently the fuel is consumed. This information is usually available in the owner's manual. If not, the manufacturer can supply it. Keep in mind that mechanical equipment is not always as efficient as the label states because a furnace or boiler may be out-of-tune, a heat pump or air conditioner could be low on refrigerant, filters could be clogged, etc. Duct leakage also has an effect on how well a fuel is utilized, and equipment efficiencies will vary depending on the outside temperature. Typical fuel utilization efficiencies are:

```
Electricity, resistance heating _____ 1.0
   (Always the same for resistance heating)
Electricity, heat pump or air conditioner ___ 2.0–4.9
   (This is the SEER# ÷ 3.413)
Gas-fired boiler or furnace _____ .65–.96
   (Usually expressed as a %)
Oil-fired boiler or furnace _____ .60–.95
   (Usually expressed as a %)
```

All the above information can be combined to calculate the cost of heating or cooling incoming air with the following two formulas.

Annual heating cost = (number of Btus used for heating) ÷ (number of BTUs per unit of energy) ÷ (fuel utilization efficiency) x (cost per kWh, per ccf, or per gal.)

Annual cooling cost = (number of Btus used for cooling) ÷ (number of BTUs per unit of energy) ÷ (fuel utilization efficiency) x (cost per kWh, per ccf, or per gal.)

Let's calculate the heating and cooling cost of ventilating a house with an 80-cfm fan that operates 24 hours per day in a climate having 6,000 heating DDs and 800 cooling DDs (keep in mind that some ventilation systems don't run continuously, for 24 hours a day). The number of cubic feet of air entering the house through the fan in one day = 80 x 60 x 24 = 115,200 c.f. The amount of energy needed to heat this air in the winter would be 115,000 c.f. x 6500 DD x 0.018 = 13,455,000 Btus. The amount of energy needed to cool this air in the summer would be 115,000 c.f. x 880 DD x 0.018 = 1,656,000 Btus.

If we heat in the winter with a high-efficiency (95%) gas furnace, and gas prices are $0.60/ccf, the cost to heat the ventilation air will be 13,455,000 Btus ÷ 100,000 ÷ .95 x $0.60 = $84.98. If we cool in the summer with a heat pump having an efficiency of 3.0, and the electricity rate is $0.09/kWh, the cost to cool the ventilation air will be 1,656,000 Btus ÷ 3,413 ÷ 3.0 x $0.09 = $14.56. Therefore, in this example, the total annual heating and cooling cost for ventilation will be $99.54 ($84.98 + $14.56). This is the cost of heating and cooling the ventilating air *without* using an HRV. If we ventilate at the same rate using a 70% efficient HRV, we will conserve 70% of the energy; therefore, we will save 70% of the cost. So, with a 70%-efficient HRV, the heating and cooling cost will be only $29.86 ($99.54 x 30%). To get a total operating cost, we would need to add to these figures the cost of the electricity needed to run the fan itself. With a 50-watt fan @ $.09/kWh, this would be an additional $39.42 (50 x 24 x 365 ÷ 1,000 x $0.09) per year. A 100 watt fan would cost twice as much, or $78.84 (100 x 24 x 365 ÷ 1,000 x $0.09) to operate for a year.

While there is always a cost associated with ventilation, it is often not exorbitant. The table in Figure 13–3 lists the cost of providing continuous ventilation (24 hours per day), without heat recovery, for one person (15 cfm), in several locales. Note how the cost varies significantly with different heating fuel types. The costs will change further if fuel costs are higher or lower, or if furnaces or air conditioners with other efficiencies are used. These figures are only for the cost of heating or cooling the incoming air. They do not include the electricity cost of running any fans.

Calculating the cost of dehumidifying incoming air

Unfortunately, it is somewhat more difficult to accurately calculate the cost of dehumidifying ventilation air than it is to determine the cost of heating or cooling that air. Without getting into complex calculations, you can get a rough idea of the cost of dehumidifying incoming air by multiplying the cooling cost by 30%. As you can see from the estimates in Figure 13–3, Miami has a cooling cost for ventilation of approximately $13 per person, so it would have a dehumidification cost of about $4 per person. The formula is:

Annual dehumidification cost of ventilation air = (0.30) x (cost of cooling ventilation air)

Understanding Ventilation

Some HRVs are capable of transferring moisture from one airstream to the other. In an air-conditioning climate, this can save energy. With these particular HRVs, the humid incoming air passes through the core and transfers a percentage of heat *and* moisture into the outgoing airstream. Therefore, less moisture is brought indoors. This means both the cooling and dehumidification costs will be less. In heating climates, these HRVs can prevent a house from becoming excessively dry, so they can help conserve on humidification costs as well as heating costs.

Is an HRV cost-effective?

The cost of electricity to actually power a balanced ventilating system with heat-recovery compared to a central-exhaust system will depend on the size of the electric fan motors. But usually a balanced system will consume more electricity than a central-exhaust system. This is because the balanced system generally has two motors compared to one motor in a central-exhaust system. If the balanced system has heat recovery, the motors are larger because they must overcome extra resistance to airflow due to the heat-recovery core. The cost of tempering the incoming air will also vary. This is because with any non-heat-recovery ventilation system, the air entering the house will be at whatever temperature it is outdoors and the air leaving the house will be at room temperature. In climates where the outdoor temperature doesn't vary from 70°F very much, the temperature of the incoming ventilation air won't have much of an effect on heating or air-conditioning bills simply because those bills aren't very high to begin with. The three factors that have the greatest impact on tempering costs are severity of the climate, cost of fuel, and capacity/running time of the ventilator.

In very cold climates, the cost of tempering incoming air can be significant. In the northern parts of the U.S. and in Canada, heating costs in the winter can be especially high, and ventilation can add significantly to the heating bills. So, it is in these climates where heat-recovery ventilation makes a great deal of economic sense.

HRVs can also conserve energy in air-conditioning climates, especially if they are capable of transferring moisture between airstreams. How much energy they conserve depends both on how hot the climate is, and on how humid it is. This is because it takes energy for an air conditioner to remove humidity from the air as well as cool the air. In moderate climates, an HRV might not offer sufficient energy savings in either the winter or the summer to justify the higher equipment and labor costs when compared to a simpler system without heat recovery.

The price of fuel is a significant factor in tempering costs. Because electricity is generally more expensive per Btu than natural gas, an HRV in an electrically heated house may be more cost-effective than an HRV in a gas-heated house.

If a high rate of air exchange is needed, heat recovery will be more cost-effective than when low flows are sufficient. For example, ventilating a swimming pool or spa enclosure can require extra capacity to remove evaporated moisture, making heat recovery an economically viable option compared to a non-heat-recovery ventilator.

A ventilator that runs continuously will cost more to operate than a similar-sized ventilator that runs only a few hours per day. This is why local-exhaust fans (*e.g.* range hoods and bath fans), which don't operate for very long periods of time, rarely have a tempering cost high enough to justify heat-recovery. For example, the annual tempering cost of running a 50-cfm bath fan two hours a day in most parts of the U.S. is less than $15.

For an HRV to be cost-effective, it should conserve enough energy to pay for itself within a few years. In order to determine if an HRV is cost-effective when compared to a non-heat-recovery ventilator in a particular locale, you need to calculate the cost of tempering the incoming air with and without heat recovery and compare the two figures. For example, let's say you have calculated the cost of tempering ventilation air to be $170 per year without heat recovery, and you have calculated the cost of tempering with an HRV to be $50 per year. That means your annual tempering cost will be $120 less (170 – 50) with the HRV. To determine if this is cost-effective, you also need to compare cost of running the fans and the capital costs. Suppose the non-heat-recovery option will cost $570 and the HRV will run $480 more, or $1,050. If you determine a simple payback by dividing the extra capital cost by the an-

Chapter 13 The cost of ventilation

ANNUAL COST OF 80 CFM OF CONTINUOUS VENTILATION WITH AND WITHOUT HEAT RECOVERY*						
	Miami	Los Angeles	Nashville	Boston	Minneapolis	Fairbanks
Total ventilation operating and tempering cost without heat recovery	$133	$82	$135	$150	$188	$269
Total ventilation operating and tempering cost with heat recovery	$69	$54	$70	$75	$86	$110
Annual savings	$64	$28	$65	$75	$102	$159

Figure 13–4. This chart shows the total yearly operating costs (fan operating cost plus tempering cost) for a typical 80 cfm balanced ventilation system in a house that is heated and cooled with an average-efficiency natural gas furnace and central air conditioner at typical fuel prices.

*rounded to the nearest dollar.

nual savings (480 ÷ 120), you can see that the HRV will pay for itself in 4 years. After the first 4 years, you will save $120 a year.

An HRV isn't going to be cost-effective in all climates. If we use the same equipment in a mild climate we might calculate that the cost of tempering without heat recovery is $45, and the cost of tempering with an HRV is only $14. There will still be an annual savings, but it will only be $31 (45 – 14). The up-front costs are the same ($570 vs. $1,050), so a simple payback will take much longer—about 15.5 years (480 ÷ 31). When you factor in the expense of running the fans (an HRV usually has higher-wattage fans than other strategies), a non-HRV option will often be more cost-effective in a mild climate.

Whether an HRV is cost-effective for a particular situation depends on several factors, as well as how long you wish to wait to recoup your investment. Keep in mind the fact that most mechanical ventilation equipment has a useful life of 10–20 years. While ducts will typically last longer than that, motors, controls, etc. will need to be replaced periodically. To be cost-effective, an HRV should pay for itself before it wears out. Because climates vary considerably throughout the U.S., because fuel costs also vary significantly, and because ventilation capacities and run times depend on a particular design, it is always important to perform a few quick calculations before determining if an HRV is cost-effective or not. To calculate the number of years it will take to pay for an HRV, use the following formula:

Number of years required to pay for an HRV = [(Cost of HRV) – (Cost of a non-heat-recovery option)] ÷ [(non-HRV tempering cost) – (HRV tempering cost)]

The table in Figure 13–4 shows the estimated cost for 80 cfm of continuous balanced ventilation in several U.S. cities with and without heat recovery. These figures include all tempering costs (heating, cooling, and dehumidifying) and the cost of running the fan. They are based on the calculation methods given in this chapter (so the tempering costs could be as much as 30–40% too high) and several assumptions: heating with natural gas @ 80% efficiency, cooling with a 10.0 SEER air conditioner, operating a 60 W fan continuously (an HRV fan might have a higher wattage but a central-exhaust fan would typically have a lower wattage), natural gas costing $0.60/ccf, electricity costing $0.08/kWh, dehumidification cost being 30% of cooling cost, and an HRV having 70% efficiency. While all the figures will change with different energy costs, fan run times, efficiencies, etc., they can give you a rough idea of how much money heat recovery can save. Based on these figures, you can see that there is a savings in all climates, but it is most significant in cold locales. The greatest savings will be in the hottest and coldest climates—Arctic and equatorial regions.

The cost estimates derived from the above calculations are valid if you compare a balanced ventilator without heat recovery to a balanced ventilator with heat recovery. It is important to note that they don't take infiltration into consideration. As a result, they are a little misleading if you try to compare a balanced system to a system that pressurizes or depressurizes a house. This is because a ventilation system that pressurizes or depressurizes a house will overpower some (or all) the natural infiltration and exfiltration. On the other hand, the flow of a balanced ventilator (with or without heat recovery) can be added to the flow attributable to natural ventilation to arrive at the total amount of air exchange in a house. As a result of the way mechanical ventilators interact with natural ventilation, the total (natural plus mechanical) ventilation rate in a house with a balanced ventilator will be higher that the total (natural plus mechanical) ventilation rate in a house where a ventilator either pressurizes or depressurizes a house (assuming the ventilators have the same capacity). This is covered in more depth in Chapter 7 but the following example should be helpful: There are two houses of equal tightness (let's say the average annual natural ventilation rate is 40 cfm in each house), one house has a 100-cfm HRV, and the other has a 100-cfm central-exhaust ventilator. The total average ventilation rate in the HRV-ventilated house will be 140 cfm, but the total average ventilation rate in the central-exhaust-ventilated house will be 100 cfm. So, if you compare tempering costs based only on the controlled-ventilation capacity (100 cfm in both cases) you won't be getting a true picture because one house has a higher overall ventilation rate than the other. To obtain the most accurate cost comparison between different ventilation strategies, you should calculate both the natural and mechanical rates and combine them appropriately. Actually, the best estimates use daily averages—but that is a job for a computer program.

In harsh climates, the cost-effectiveness of an HRV may hinge partially on the cost of the heating or air-conditioning equipment. For example, adding a ventilation system in a cooling climate can mean increasing the load of the air conditioner when compared to an unventilated house. If the use of an HRV capable of removing moisture from the incoming air will mean you can use a smaller air conditioner, the air-conditioner equipment savings will help to offset the cost of the HRV. In a heating climate, an HRV will place less of a tempering load on the furnace than a non-HRV option. So, you might be able to get by with a smaller (slightly less expensive) furnace if you use an HRV.

Altech Energy has estimated that in a southern U.S. climate, an HRV capable of transferring both moisture and heat between the airstreams can save about $1/3$ ton of air-conditioning capacity per 100 cfm when compared to a ventilation system without heat recovery.(Lossnay)

Is ventilation worth the cost?

Suppose we ignore the fact that HRVs have a lower tempering cost than other types of ventilators. Is ventilation itself worth the cost? The only way to answer this question is to put a value on the health of the building and the health and comfort of the occupants, then compare that to the capital and operating costs of the ventilation system. This is very difficult to do. After all, who can say how much health and comfort are worth. Is your health and comfort worth $20 a year? $100? $500? $1,000? $100,000? What if you have a house with no ventilation system, and it develops a hidden moisture problem that causes several studs and some sheathing to rot. How much damage would you be willing to accept? $500? $1,000? $5,000?

Let's consider what a ventilation system actually costs. A top-of-the-line, expensive, fully ducted HRV installation might have a capital cost of $3,000. While it is possible to install a very simple central-exhaust ventilation system for considerably less, or a do-it-yourselfer could avoid labor costs, we will consider a worst-case scenario. Three thousand dollars would add $24.14 to your monthly mortgage payment (based on a 30-year loan at 9% interest). In most climates in the United States, the annual operating cost will be less than $100 for such a system. So, the total yearly ventilation cost would be $389 (($24.12 x 12) + $100), or just over a dollar a day. A simple ventilation system, consisting of a small, quiet, continuously running, central-exhaust fan might have a total cost (capital and operating) of less than one third that much.

Yes, a dollar a day (or 33¢ a day) represents one more expense that many of us can ill afford. But we are already probably spending considerably more than that on health insurance and homeowners insurance. Insurance only pays for itself after we get sick or our house has been damaged. Most of us would agree that we would be better off not getting sick in the first place and not having to repair a rotted house.

Ventilation is like preventive medicine—it provides the benefits of comfort, a healthy body, and a healthy structure. The bottom line: Most of us would rather not spend the money, but we would all agree that ventilation is worth at least as much as insurance. And it generally costs somewhat less.

Chapter 14

Controlling a ventilation system

There are many different approaches to controlling a ventilation system, ranging from the simple to the complex. Basically, controls fall into two broad categories: automatic and manual. Automatic controls require little or no human input, while manual controls are operated by the occupants.

Whatever type of control is used, it must be user friendly—that is, it must be easy to understand, because even automatic controls require occasional attention. All controls should be clearly labeled, and the builder should make sure the occupant has a user's manual explaining how the control operates.

A control, manual or automatic, can do one of three things: 1) It can turn a ventilating fan on or off; this is the most common use of a control. 2) A control can vary the speed of the fan, thus changing its cfm capacity. This increases or reduces the ventilation capacity of the entire ventilation system. 3) A control can activate an adjustable damper or grille to vary the volume of air flowing through one part of the system. Adjusting a grille serving a single room may change the overall capacity of the system very little—the air just tends to be distributed differently.

Several types of controls are available. Some are suited to controlling an intermittently operated local-exhaust fan, others are more applicable to a general ventilation system. If a house has more than one type of ventilation system (most houses can benefit from both central ventilation and one or more local exhaust fans), it generally needs more than one type of control. For example, an automatic carbon-dioxide detector makes more sense for a general ventilation system than it does for a local ventilation fan, but a manual timer would work well for a local-exhaust fan that is only used occasionally. On the other hand, it is possible to control a local exhaust fan in a bathroom with an automatic motion sensor, while the house's general ventilation system operates on an automatic timer.

Understanding Ventilation

Before deciding on a particular ventilation control (or several controls to be connected in combination), you should analyze just what you are trying to accomplish, then select a control strategy that the occupants can easily understand, utilize, and afford. Some controls are fairly expensive—up to several hundred dollars. Some of these can cost more than the ventilation equipment itself, but low-cost controls often work just as well.

There are some definite advantages to eliminating complex and expensive controls and running a general ventilation system continuously, 24 hours a day. (It is a good idea to install a simple on/off switch on such a system so it can be shut off when no one is home, or when the outdoor air is especially polluted, such as when the neighbors decide to spray pesticides on their fruit trees or lawn.)

When considering a complex control system, keep in mind that the average human nose is generally far more sensitive to air pollutants than any control. Therefore, allowing the occupants to use their sense of smell to determine when to turn their ventilation system on and off has a great deal of merit.

Some of the controls on the market are physically larger than a conventional on/off switch and they require a slightly deeper electrical box inside the wall. So, even if a simple control is being installed, it is a good idea to use a deeper electrical box to make it easier for a future occupant to upgrade to a more complex control. Low voltage (24 volt) controls do not require an electrical box, but it is easier to attach a cover plate if an electrical box is installed anyway.

Manual controls

Manual controls have a certain appeal because they allow the occupants to operate the ventilation sys-

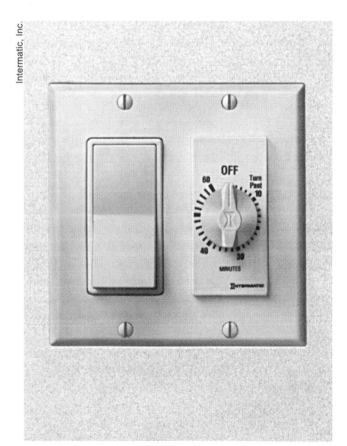

Figure 14–1. A manual timer allows the occupants to turn a ventilation system on at any time. The timer will then shut the system off automatically.

Chapter 14 Controlling a ventilation system

Figure 14–2. A delayed-off switch is a manual timer with an internal adjustment that determines how long the ventilation fan remains on.

tem. This is often the obvious choice—place people in charge of their own fresh air. After all, who else is better equipped to judge when extra air is needed. In fact, some people are more sensitive to indoor pollution than instruments and gauges. For these extremely sensitive people, a manual control can be very desirable because they often want to turn the ventilation system on and off more frequently, depending on a wide variety of environmental and weather conditions.

Unfortunately, some people are not very good at sensing poor air quality, so we might not realize when it is time to turn a ventilation system on and off. And, of course, even if a control system is very simple and easy-to-understand, if it is manually operated, there is nothing that can be done to force people to use it. Some people simply don't want to be bothered with another responsibility. For these individuals, automatic controls are the answer.

Following are the types of manual controls that are currently available. Some are incorporated by manufacturers as a part of their ventilation package, but most can also be purchased separately.

On/off switch

An on/off switch is the simplest manual control available. It is easy to understand and operate—we all know how to turn a switch on and off. This simple manual control is often used with bath fans, range hoods, and other local-exhaust fans. An on/off switch can also be used on a central ventilation system—the occupants simply turn the system on whenever they are home, then turn it off when they leave. This can be made easy to remember by locating the switch near the entry door. If the capacity of the ventilation system is oversized, the occupants can turn it off for a portion of the time they are home, depending on what the outdoor conditions are like, or if they need extra fresh air to overcome a temporary polluting indoor activity, they can let the system run continuously.

A variation of this control method would be to use multiple switches located in various rooms. For example, a pair of three-way switches could be used to control a central ventilation system from two different entry doors. If it is desirable to have switches in several locations for easy access, low-voltage switches with a series of relays can be useful. When switches are used in multiple locations, there should be an indicator light next to the switch so you will know when the ventilation system is running.

Sometimes a simple on/off switch is used to open or close a motorized damper or grille of a general ventilation system to redirect the flow of air. When this is done, the fan continues to run but the distribution pattern is changed.

Manual timers

Manual timers are activated by the occupants, then they turn themselves off automatically. Sometimes they are called interval timers or countdown timers. Because they turn themselves off, they are sometimes considered semiautomatic controls, but we will refer to them as manual controls because they require regular human input. Spring-wound manual timers (sometimes called crank timers) are readily available with ranges of 0–5, 0–15, 0–30, or 0–60 minutes or 0–1, 0–6, 0–12, or 0–24 hours (Figure 14–1).

Manual timers are useful controls in a variety of situations, for both local ventilation and for general ventilation. For example, a 0–24 hour timer can be coupled to a central ventilation system. The occupants can twist the knob on the timer to the number of hours they plan to be in the house that day and then they don't have to worry about turning the system off, the timer will do it for them. Some timers have a useful "hold" feature. When set on hold, the timer function is bypassed and the timer remains in the on position until it is manually turned off.

A 0–60 minute timer can be used to shut off a local ventilation fan, such as a bathroom-exhaust fan, after the room has been vacated. A manual timer in a hobby room could be used to open an adjustable ventilation grille. This would provide extra ventilation to that particular room (and less to the rest of the house) while activated. Manual timers are available from many different ventilation equipment distributors as well as **Grainger, Inc.**, **Grasslin Controls Corp.**, **Intermatic, Inc.**, and **Tork**.

Broan Mfg. Co. Inc. has a manual timer called a #64 Delayed-Off Switch that is often sold for use in bathrooms to control a local-exhaust fan (Figure 14–2).

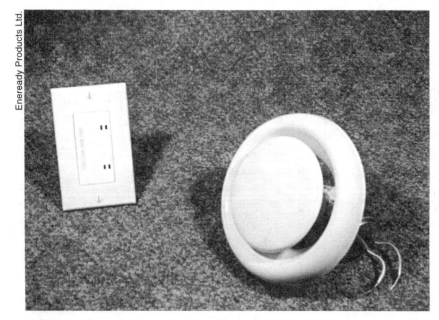

Figure 14–3. The PoshTimer and POWERGrill are designed to be used together to temporarily boost the flow of air through one zone of a central ventilation system. In this way, one system can be used for both central ventilation and local ventilation.

Chapter 14 Controlling a ventilation system

Figure 14–4. This special cover plate can transform an existing wall switch into a delayed-off timer.

These devices have three positions: on, off, and delayed-off. They can be used to turn a fan on and off like a simple on/off switch, but when switched to the delayed-off position, the internal timer is activated and they will shut off after a certain amount of time has passed. The time delay can be preset by the occupant (up to 60 minutes) by removing the cover plate. These switches can also be wired to control both a light and a fan. When installed in this way, turning them on activates the light and the fan, turning them off shuts off both the light and the fan, and turning them to the delayed-off position turns off the light immediately and the fan after the preset delay.

Penn Ventilator Co., Inc. also has a similar switch with a time delay called the Airminder ARM12. **Intermatic, Inc.** has three Interval In-Wall Timers with either a 15-minute, 1-hour, or a 4-hour delay, but they cannot be adjusted internally like the Broan or Penn timers. **Grasslin Controls Corp.** has a KLT 715 electronic timer that can be programmed to automatically shut off anywhere from one minute to 24 hours. It is designed to replace a standard wall switch.

Eneready Products Ltd. manufacturers a PoshTimer, a unique control that is both a speed control and a timer with several settings. This control is designed to be used in conjunction with Eneready's POWERGrill—a motorized Whisper exhaust grille that opens and closes when activated by the PoshTimer (Figure 14–3). The PoshTimer doesn't look like a conventional switch. The face plate has 4 LED lights and two heat sensors. To activate it, you simply wave your hand across the two sensors. Wave up for on, wave down for off. The default setting for the timer is 28 minutes but you can easily set it to remain on for 1, 4, or 8 hours.

Home Equipment Mfg. Co. has a special cover plate that transforms an ordinary wall switch into a manual timer that can be preset to shut off after 15 minutes, or 1, 2, 4, 8 or 12 hours. It is primarily marketed for turning lights on and off, but can also be used to control a ventilation system (Figure 14–4).

Speed controls

Range hoods (local ventilation) and HRVs (general ventilation) often have built-in variable-speed controls to change the motor's RPM and, as a result, the airflow capacity of the fan. These solid-state controls can be useful if ventilation requirements aren't constant. A fan running at low speed may be sufficient most of the time, but when additional air is needed for a certain period, the high speed can be temporarily activated. Simple residential speed controls are readily available. They are often sold for paddle-type ceiling fans.

Variable-speed controls and two- or three-speed controls are generally sold through local building-supply outlets, but most ventilation equipment suppliers also offer them. Variable-speed controls allow the occupants to fine-tune the capacity of a ventilation system to suit their own particular needs. Ventilation equipment manufacturers sometimes use built-in speed controls called booster switches because they boost the fan speed. Speed controls are also available from **Grainger, Inc.** (several styles) and **Penn Ventilator Co., Inc.** (Lek-Trol).

A drawback to solid-state variable-speed controls is the fact that they often cause a fan motor to have an annoying hum at low speeds. The hum is the result of a vibration within the components of the motor itself. This cannot be eliminated, but it can be minimized if the control has an adjustable low-speed stop point. With this feature, a minimum speed can be set so humming is less noticeable yet the fan will provide enough air for normal activities. Still, occupants often complain of humming when speed controllers are used.

Sometimes a fan is activated by both a solid-state speed control and a second on/off switch in such a way that the second switch can be used to override the speed control and boost the fan to high speed. This may sound like a good idea, but it often causes damage to the speed control if the on/off switch is turned on when the fan is running at low speed. The sudden change in voltage to the fan causes a surge in the variable-speed control that results in damage.

Shaded-pole motors and permanent-split-capacitor motors are usually much less energy efficient when run at slower speeds. If these motors are run at too slowly, they may just sit there consuming electricity without moving very much air. Most of these motors won't operate properly below 50% of their design speed.

Some motors, such as permanent-split-capacitor direct-drive furnace/air-conditioner motors, are specifically manufactured to operate at two (or more) speeds. These motors use electricity more efficiently, and they are much less likely to hum than if a speed control is used to control a single speed motor. ECM motors are specially designed for variable speed, energy-efficient operation, but they tend to be costly. Multispeed motors or ECM motors are primarily used with forced-air furnace/air-conditioner fans but, because of the extra expense, rarely with ventilating fans.

The purpose of a speed control is to vary the airflow, something that can be done by other means. For example, the volume of air flowing through a ducted system can be changed periodically by using an adjustable damper. When the damper is closed, or partially-closed, the resistance to airflow increases and the volume decreases. Such a damper may be activated by any type of control, such as a timer.

Automatic controls

A fully-automatic ventilation system is appealing inasmuch as it gives the occupants one less thing to worry about. The ventilation fan simply runs whenever needed without a thought from anyone. Uniform temperatures are maintained indoors automatically with thermostats, so why not maintain indoor air quality automatically? While some occupants may prefer manual controls, others may opt for an automatic control system, and some may want an automatic control with a manual option.

If properly designed and installed, any control system can function quite well. However, an automatic system is often easier said than done. This is because there are a number of variables that determine when fresh air is needed and how much is needed. For example, extra fresh air is required when the number of people indoors increases, and the occupancy rate may vary by time of day or it may be highly irregular. An automatic control system might also need to account for temporary increases in the indoor humidity, or occasional indoor pollution problems such as burnt toast or spilled disinfectant. For a completely automatic control system to account for all of the possible contingencies, it will be fairly complicated—and expensive.

If the outdoor air quality is inconsistent—sometimes acceptable, sometimes of poor quality—an automatic control must be able to measure the pollutants in the outdoor air and shut the general ventilation system off when the outdoor air quality declines. Another way of dealing with poor quality outdoor air would be to install a suitable filter to clean the incoming air. In either case, it is often desirable to have a simple manual override switch that the occupants can use to shut off an automatic control if the outdoor air is especially poor, or if they expect to be gone (perhaps, on a vacation) for an extended period of time.

Because automatic controls are designed to have little human input, occupants may have no idea how the ventilation system actually operates. Therefore, if part of the system malfunctions, they may not have enough understanding to override the automatic controls without telephoning a technician or serviceman. Controls should always be clearly labeled and the occupants should be provided with an instruction sheet or operating manual.

Demand-controlled ventilation (DCV) is another way of referring to an automatically controlled ventilation system—one that is designed to run only when the occupants need fresh air. A DCV system operates whenever there is a *demand* for fresh air by people. A DCV system turns on when people are indoors and shuts off when they leave. DCV is only a viable option when pollutants released from building materials don't build up when the system is shut off. Therefore, pollutant source control is very important with a DCV system; construction materials, furnishings, and maintenance products must be as low-polluting as possible.

A DCV system can sense occupancy in a number of ways. One study found that even though people

Figure 14–5. Automatic timers can be preset to turn a ventilation fan on and off several times a day.

Understanding Ventilation

exhale moisture in their breath, relative humidity is a poor indicator of occupancy because there are moisture sources in a house other than people. Dehumidistats do not always work very well as DCV controls for a general-ventilation system, but they can be viable for a local exhaust fan, say in a bathroom. The Aerco general ventilation system distributed by **Airex** utilizes moisture sensors and has been used successfully in France for many years. This system is designed to operate continuously, 24 hours per day, but at different rates. When the relative humidity is low, the general-ventilation rate throughout the house is low, but when the relative humidity rises, the ventilation rate increases, primarily in the rooms with high humidity. So, it also functions as a local ventilation system.

Carbon dioxide (also from exhaled breath) is a good indicator of how many people there are indoors, so carbon dioxide sensors can be viable DCV controls. They are generally used to turn a general ventilation system on and off, rather than vary the ventilation rate. A mixed-gas sensor could also be used, but they generally aren't as accurate as carbon-dioxide sensors. A motion sensor might also be a way to control a DCV system, but it would need to account for the fact that people are sometimes motionless for extended periods—while sleeping, for example.

While the concept of a DCV system is intriguing, economic analysis has shown that it is possible to achieve only a small savings in operating costs at current energy prices.(Sheltair and SAR) Because the more complicated and exotic controls are relatively expensive (several hundred dollars for a CO_2 sensor), it is doubtful if they will see widespread use in residences in the foreseeable future.

No control

Having no control means hot-wiring the ventilation system into a house's electrical system so that it runs all the time. The ventilation fan would stop running only in the event of a power failure or if the circuit breaker in the power panel is shut off. This lack of control is really the simplest form of automatic control because it requires no occupant input. Thus, it can work well for a general ventilation system if the outdoor air quality is usually good, and in most cases, the outdoor air quality is far better than the indoor air quality. Having no control is a useful option when the occupants can't be relied on to operate a manual control.

A major advantage to a ventilation system that runs 100% of the time is the fact that it can be sized quite small to exactly meet the daily needs of the occu-

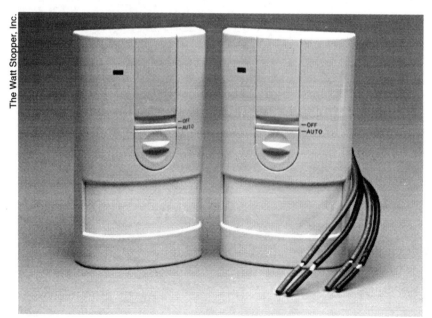

Figure 14–6. Motion sensors can be used to activate a ventilation system when there is movement in a room.

Chapter 14 Controlling a ventilation system

Figure 14–7. A dehumidistat can be adjusted to turn a ventilation system on automatically when the relative humidity is too high indoors.

pants. A small-capacity ventilation system will be low in cost. As an example, if it is known that there will only be two people living in a house for 24 hours a day, and there are no major pollution sources in the house (except for the people), then a small 30-cfm fan running continuously may be all that is necessary to maintain good indoor air quality—if the 30 cfm is distributed evenly throughout the living space. Small-capacity, continuously running ventilation systems can work very well, but they are often undervalued by people who are in a position to make a profit by selling a more-complicated and expensive ventilation system.

The big disadvantage to a continuously running, low-capacity ventilation system is that it allows for no extra fresh air when conditions call for it, nor is it designed to be shut off easily. However, the extra capacity can easily be provided by a separate local-exhaust fan. To shut such a system off easily, you can install a simple on/off switch that is labeled "This switch controls the house's ventilation system. This system is designed to run 24 hours a day."

Automatic timers

An automatic timer both turns on and off a ventilation system (or opens and closes an adjustable damper or grille) at specific times of day. These are more complicated than the manual timers described above. Sometimes they are called time clocks or time-of-day timers. In a typical application, if no one is home between 7 AM and 6 PM, the timer can be set to turn a general ventilation system on at 5 PM (to allow an hour to flush out any pollutants that have built up indoors during the day) and off at 8 AM (to allow an hour to flush out the last of the pollutants generated during breakfast).

Traditionally, automatic timers have been mechanical devices with a single on and off setting, but many of the newer electronic models can be programmed to turn fans on and off several times a day. For example, you could turn a general-ventilation fan on and off every 15 minutes. Time clocks are available in a variety of styles through many ventilation-equipment suppliers and most electrical-supply houses. They can also be obtained from **Grainger, Inc., Grasslin Controls, Inc., Intermatic Inc.,** and **Tork**. Some styles are somewhat commercial-looking, but residential models are readily available (Figure 14–5).

Motion sensors

Motion sensors (also called occupancy sensors) can be used to control a whole-house ventilation system, but they are most suitable for use in a single room, perhaps for a local ventilation fan. They are often used to turn lights on and off, but they can also be used to control a vntilating fan. Some are manually turned on, then they turn off automatically. Others turn on and off whenever there is motion in the room. Many motion sensors can be adjusted so that they remain on for a certain time period after activity has ceased. This is important because people are often motionless for several minutes at a time.

Though their purpose is to detect the movement of people, motion sensors that turn on automatically can also be triggered by pets or a rustling curtain. Besides simply turning a fan on or off, motion sensors can also be used to activate the high speed of a fan or to open (or close) a motorized grille or damper.

Airex distributes an Opto X infrared motion sensor that fits in the end of an exhaust grille. When motion is sensed in the room, the Opto X opens a damper to allow extra air to flow through the grille. The size of the opening can be preset to correspond with different occupancy levels. These are most often used in commercial applications, such as offices, but they can also be adapted to residences.

Broan Mfg. Co., Inc. has a SensAire combination light and fan for use in bathrooms that incorporates a motion sensor. When it detects movement in the bathroom, the fan is activated and it remains running for a certain period after the motion stops. The amount of time the fan remains on can be set internally for up to 60 minutes after the motion ceases.

Lightolier offers several types of Insight motion-detecting devices in different wattages. Ceiling-mounted or wall-mounted sensors are available as well as on/off wall switches that resemble conventional designer switches. The switches are turned on manually, then after motion in the room stops (there is an adjustable delay), they turn themselves off. With the Insight sensors, you

Figure 14–8. A carbon dioxide controller can activate a ventilation system when carbon dioxide from the occupant's exhaled breath builds up in the indoor air.

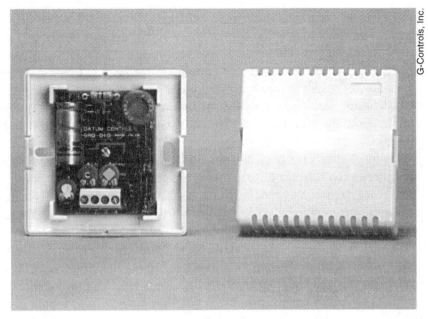

Figure 14–9. A mixed-gas sensor will detect a variety of different gases in the air and turn a ventilation system on when concentrations increase.

have a choice of manual-on/automatic-off or automatic-on/automatic-off.

The Watt Stopper, Inc. has a wide variety of infrared and ultrasonic motion sensors. Designed primarily for controlling lighting, these devices can also be used for controlling ventilating systems. Some units mount on the ceiling, some can replace a wall switch, and some have adjustable time-delays and coverage patterns so they can be adapted to a variety of different situations (Figure 14–6).

Tork also has a series of infrared motion sensors that can be used to automatically turn ventilation equipment on and off, and **Honeywell, Inc.** has an infrared motion sensor that can be set to turn a fan (or light) off 15 sec. to 15 min. after motion stops. Honeywell's sensor is designed to replace a conventional wall switch.

Humidity sensors

Humidity sensors are popular ventilation controls. A dehumidistat (sometimes incorrectly called a humidistat) can be set to activate a ventilation fan (or motorized grille or damper) when the relative humidity rises to a certain level (Figure 14–7). They are available from most ventilation equipment suppliers. Bathroom local-exhaust fans are often controlled by dehumidistats because these are rooms where excess moisture is likely to be generated. Dehumidistats are sometimes used to control a central ventilation system by sensing the overall humidity in a house. When this approach is used, it is still important to remove moisture quickly from high-humidity rooms such as the kitchen and bathroom, using local exhaust fans.

A big problem with using a dehumidistat as a control is the fact that many occupants find them difficult to understand. For example, to have increased ventilation, they often turn a dehumidistat up instead of down. Turning it up means it will not come on until the humidity is even higher; thus, you have less ventilation. It is also not unusual for a homeowner to set a dehumidistat at 70%, thinking they are setting a thermostat at 70°. And it is not unusual for an electrician to connect the wires to a dehumidistat backwards.

Another disadvantage lies in the fact that both outdoor and indoor relative-humidity levels vary with the seasons. For example, it wouldn't be unusual for a house to have an indoor RH of 20% in the winter and 60% in the summer. The outdoor RH can also vary by 30 percentage points or more from one season to another. Thus, what is considered a high RH in one season may not be considered high in another. This means that a dehumidistat will need to be readjusted season-

Understanding Ventilation

ally—sometimes they need to be readjusted every month—in order to function optimally.

Dehumidistats are sometimes located inside ducts containing exhaust air. When the humidity in the duct goes down—a sign that the air in the house is drier—the dehumidistat deactivates the ventilation system. Some HRVs are equipped with dehumidistats already installed in their cabinets that increase the fan speed when they sense an increase in humidity. This strategy may need careful adjusting to account for variations in seasonal or day-to-day humidity and still provide adequate running time.

Dehumidistats are widely available from many different manufacturers. They can be obtained from most ventilation-equipment suppliers, as well as through local electrical- and building-supply outlets.

The Aerco central-exhaust ventilation system distributed by **Airex** uses an automatic humidity-sensing exhaust grille to vary the amount of air leaving the house through the grille. These devices are generally located in moisture-prone rooms and open up as the indoor humidity rises to allow more air to pass outdoors from rooms with high humidity. The grilles also contain a unique manual override to temporarily boost the airflow. The override consists of a wall-mounted pneumatic squeeze bulb. You poke the flexible bulb with your finger and it opens the grille for a preset time interval of 15–90 minutes, depending on the model.

Broan Mfg. Co., Inc. has found a way around seasonal humidity variations by using a dehumidistat in its SensAire bath fans that don't sense humidity *per se*, but rather a change in humidity. A SensAir will activate the bath fan when the humidity rises quickly (for example, when you take a shower), yet will not be affected by a gradual seasonal rise in humidity. This is a good control for a local exhaust fan, but not for a general ventilation system.

Tamarack Technologies, Inc. offers a series of Humitrak solid-state dehumidistats that come with manual override switches that can be used to keep a fan running for a preset time before returning the system to automatic operation. They also have a built-in indicator light so you know when the fan is operating.

Carbon-dioxide sensors

A high level of carbon dioxide (CO_2) is not necessarily toxic itself, but it is often used as in indicator that the overall indoor air quality is poor. We all exhale CO_2 with every breath, but if the air in a house is being diluted regularly with fresh outdoor air, the CO_2 won't have a chance to build up. On the other hand, poorly

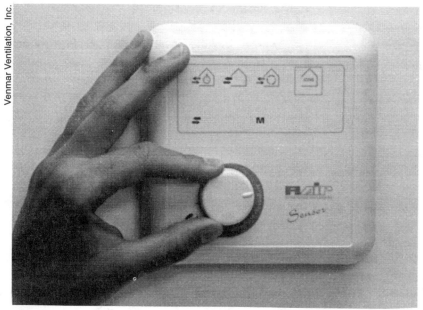

Figure 14–10. Many ventilation-equipment suppliers are offering multipurpose controls, such as this one that senses relative humidity and allows you to choose different modes of circulation.

ventilated houses often have high CO_2 levels. When indoor CO_2 levels are below 400 parts per million (ppm), the indoor air quality is generally considered to be quite good, 600 ppm is good, 800 ppm is adequate, and 1,000 ppm is often considered minimally acceptable. (About 80% of the population are satisfied with the level of body odors in a house having 1,000 ppm CO_2, however 1,000 ppm CO_2 is certainly not toxic.) While these numbers can be used as a rule of thumb to estimate indoor air quality, they are by no means an absolute guarantee of unpolluted air because there may easily be other pollutants in the air that can negatively affect health. This can be especially true if an occupant is very sensitive. See *Understanding ASHRAE's recommendations* in Chapter 12 for a discussion of the limitations of using CO_2 as an indicator of air quality.

Though they aren't commonly used in residences, CO_2 monitors are readily available that can be used to activate a general ventilation system based on CO_2 levels in the house (Figure 14–8). They can be set to turn a fan on (or open a motorized grille or damper) when CO_2 levels reach 400, 600, 800, or 1,000 ppm. CO_2 sensors are manufactured by several companies such as **Telaire Systems, Inc.**, and are also distributed by **Johnson Controls, Inc.**, and **Honeywell, Inc.** under their own brand name. Since CO_2 sensors are relatively expensive when compared to other control strategies, it is doubtful if they will become popular. After all, at over $500, they cost more than some entire ventilation systems. They also consume a certain amount of power at all times and, in order to function optimally, they must be recalibrated on a regular basis by a trained technician.

Mixed-gas sensors

A mixed-gas sensor is an electronic device that can detect a variety of different gases and vapors that are found in the indoor air.[Meier][Gas] When stale air passes through one of these sensors, a tiny heat source oxidizes the gaseous air contaminants onto a special material. The oxidized gases change the electrical conductivity of the material, which then generates an electrical signal that can be used to control a ventilating fan (or a motorized grille or damper).

Mixed-gas sensors are designed to detect a wide variety of gases that are oxidizable, so they will detect everything from tobacco smoke and cooking odors, to VOCs and bad breath. Therefore, a mixed gas sensor can be used to activate a general ventilation system whenever detectable gases are present—even if no people are in the house. They could even be used to detect pollutants in the outdoor air and turn a ventilation system off when the outdoor air quality declined.

Mixed-gas sensors are useful controls in industrial applications, where concentrations of pollutants are often high and vary widely. The big drawback to using one in a residence is the fact that they are not sensitive enough to detect the small, subtle variations in air quality that can occur in a house. For example, the difference between acceptable air and polluted air in a house may be too small for a mixed-gas sensor to detect. Therefore, they are more suitable in applications where dramatic short-term changes in pollution occur.[Schell]

Starting at about $200, mixed-gas sensors are considerably less expensive than CO_2 sensors. **G-Controls, Inc.** (Figure 14–9) and **Staefa Control System, Inc.** offer duct-mounted mixed-gas sensors that are used primarily in commercial installations, but can be adapted to residences. **Vent-Axia, Inc.** has a residential wall-mounted mixed-gas sensor that can be used with fans or dampers that draw less that 1 amp of current, and **Nutech Energy Systems, Inc.** and **Venmar Ventilation, Inc.** offer a mixed-gas sensor to control their HRVs. Though mixed-gas sensors can be used to control residential general ventilation systems, it is doubtful if they will prove sensitive enough to be viable, except in very specialized applications.

Air-pressure controllers

It is also possible to automatically control a ventilation system by sensing the air pressure indoors in relation to the outdoor air pressure. For example, an air-pressure sensor could open a motorized damper connected to an inlet whenever the house experiences a negative pressure indoors due to a powerful local-exhaust fan. This can be useful to prevent negative-pressure-related problems such as pollutants being sucked indoors or backdrafting.

Devices called *differential-pressure transducers* sense air-pressure differences and convert that information into a low-voltage electrical signal. The electrical signal can then be used to control either a fan's speed or a motorized damper in order to maintain a desired pressure (or a lack of pressure) in a house. Manufacturers of differential pressure transducers include **AirPro, Inc.**, **Dwyer Instruments, Inc.**, **Econsys, Inc.**, **Modus Instruments, Inc.**, and **Setra Systems, Inc.**

It is possible to use an air pressure controller to maintain a slight positive pressure in a building to keep out radon or other soil gases. It is also possible to control pressures in one part of a building, relative to another part of the same building. For example, a low-pressure area could be maintained in the kitchen to keep cooking odors from migrating into the rest of the house. It is also possible to use a differential-pressure transducer to sense negative pressures in a house and open a make-up-air damper to prevent a backdrafting problem. While all of this is possible, because of the technical know-how required to correctly design an air-pressure-control system, differential-pressure transducers are rarely used in residences. Most of the time, the air pressures in houses are fairly small, so an air-pressure controller would need to be very sensitive to detect them. But the more sensitive the controller, the more easily it will be affected by the day-to-day or season-to-season fluctuations in pressure resulting from wind or stack effect. For example, the wind could easily cause the pressures in different parts of a house to fluctuate by several Pascals.

AirPro, Inc. has a system specifically designed to sense the pressure difference between the soil and the living space of a house, and is controlled by a fan that slightly pressurizes the basement. This prevents radon and other soil gases from getting into the house.

Combined control strategies

Any of the above manual or automatic controls can be combined in a given house to control either local or general ventilation. The way in which they are wired and adjusted will determine which controls have priority over others. As an example, a general ventilation system may be set up to operate at low speed (say 60 cfm) when activated by a programmable timer. A CO_2 sensor in the master bedroom could be adjusted to boost the fan to high speed (say 120 cfm) when CO_2 levels reach 1,000 ppm—a level that could be reached in a few hours when the bedroom door is closed. A dehumidistat could open a motorized grille in the bathroom when the relative humidity in the bathroom reaches 60%, thereby diverting more ventilation capacity to that room. And finally, a motion sensor could activate the high speed of the ventilator whenever the pet cat uses its litter box. A much less complicated general ventilation system might operate at a low speed when activated by a 24-hour manual crank timer, or at high speed when activated by a simple switch.

Some manufacturers of ventilation equipment are producing some fairly sophisticated and easy-to-use control systems. These are either incorporated into the cabinet of the ventilator or wall-mounted. For example, **Venmar Ventilation, Inc.** offers a wall-mounted control that has a built-in dehumidistat and three different circulation modes (Figure 14–10). Built-in speed controls and dehumidistats are quite common in HRVs.

Controlling a ventilation fan and a heating/cooling fan simultaneously

When a general ventilation system is connected to the ducts of a central forced-air furnace/air-conditioner, it is often necessary for the furnace/air-conditioner fan to operate whenever the ventilating fan is running. This is because heating/air-conditioning ducts are larger than ducts that only carry ventilation air, and heating/air-conditioning fans move considerably more air than ventilation fans. To ensure proper distribution of the lower volume of ventilation air through the larger system of ducts, the higher-capacity fan is needed. During operation, the lower-volume fresh ventilation air is mixed with the higher volume of recirculated air from the living space and the more-powerful fan distributes the mixture throughout the house. In most cases the more-powerful fan can be operated at a lower-than-normal speed when only ventilation, and not heating or cooling, is called for. However, some fan motors aren't very energy-efficient when operated at low speed.

If a low volume of fresh air is deposited into a forced-air heating/cooling duct system, and the furnace/air-conditioner fan is *not* operating, the fresh air will

Chapter 14 Controlling a ventilation system

take the path of least resistance through the duct system. Generally, it will all pass through the nearest grille and into the living space. This means that most (or all) of the fresh air enters only one room, resulting in little or no fresh air to other rooms. So, with the furnace/air-conditioner fan off, there is much less effective distribution. When you have less effective distribution, you need much more ventilation capacity to dilute the air in all of the rooms. Running both fans simultaneously provides effective distribution of fresh air throughout the house, so you need less ventilation capacity.

To control both a furnace/air-conditioner fan and a ventilation fan simultaneously, three control strategies are possible.

First, you can let the heating/air-conditioning thermostat turn on the ventilation fan whenever it activates the furnace/air-conditioner fan. This is the easiest to hook up but it means that you will only be ventilating whenever heating or cooling are called for—at times when you have the most natural ventilation due to stack-effect pressures. In mild weather there may be no need for either heating or cooling, so the thermostat won't turn anything on. This control strategy results in no mechanical ventilation in mild weather—just when you need it most because there is very little natural ventilation due to stack-effect pressures.

Second, you activate the furnace/air-conditioner fan with either the thermostat (when heat or cooling are called for) or with a separate ventilating system control. With this setup, the furnace/air conditioner functions normally by turning on and off according to the temperature settings on the thermostat. In addition, a separate control for ventilation air is wired so that it turns on both the ventilation fan and the furnace/air-conditioner fan. This control doesn't activate the heating or cooling cycle—it only turns on the fan—so the house doesn't get heated or cooled every time ventilation is called for. If the thermostat has already turned on the furnace/air-conditioner fan, then the ventilation control will only activate the ventilation fan—because you can't turn the furnace/air-conditioner fan on twice.

Third, you can run the furnace/air-conditioner fan continuously and turn the ventilation fan on and off as required. You might even run both continuously. Some furnaces and air conditioners, and some thermostats, have a switch that will allow the fan to run continuously, even when heating or cooling aren't called for. However, not all furnace or air-conditioner fans are quiet enough, or energy-efficient enough, for the occupants to put up with continuous operation. Many new high-efficiency furnaces and air conditioners use variable ECM speed motors that are specially designed for energy-efficient, continuous operation at a variety of speeds, making them more economical to operate than other motors at quieter, slower speeds. Actually, if you plan to operate a furnace/air-conditioner fan continuously for any reason, an energy-efficient ECM motor should be considered mandatory.

Chapter 15

Air-filtration equipment

Air filters cannot remove humidity, they cannot bring fresh air indoors, and they cannot generate oxygen, so they are not a substitute for a ventilation system. As a rule, the most effective way to deal with air pollutants are source control, separation, and ventilation. In evaluating the benefits of air filters for people with allergic respiratory diseases, the greatest benefit is generally derived from source control.(Nelson, Harold) Still, filters can help improve air quality, so incorporating one into a house can have certain benefits. Unfortunately, many people have the impression that an air filter will solve all their indoor air-quality problems. They definitely aren't a cure-all, but they can help, especially when used in combination with other pollution-reduction strategies.

All air filters will remove something from the air. But not all filters are created equal—some are better than others in removing pollutants. And different types of filters remove different types of contaminants. To understand what a filter can and cannot do, it is helpful to place all the various air pollutants into two broad categories: particles (more correctly called particulates) and gases. Particulates are very small solid (or liquid) substances that are light enough to float in the air. Typical particulates include mold spores, pollen, house dust, dead insect parts, dust mite feces, animal dander, fibers from clothing and furnishings, fragments of larger materials, and dirt. In urban areas, minute bits of tire rubber and asbestos fibers from brake linings are common. Gases include combustion gases, by-products of human and animal metabolism, formaldehyde, and the hundreds of other VOCs floating in the air. Tobacco and wood smoke are composed of both gases and particulates.

Filters that remove gases will do almost nothing to remove particulates, and vice versa. In order to remove both categories of pollutants, you need two types of filters: a particulate filter for particulates and an adsorption filter for gases. Note that adsorption is spelled with a "d" because it is derived from the word *adhere*.

Understanding Ventilation

There are literally hundreds of filter manufacturers, distributors, and suppliers in the U.S. A few of the larger manufacturers are mentioned in this chapter whose products are widely available, but there are similar products made by other companies that are of equal quality. A wider selection is listed in *The Energy Source Directory*[(Iris)].

It should be emphasized that all filters have a limited life and must be maintained. Some filters must be replaced periodically. If not, they can cause excessive resistance to airflow and an increase in electricity consumption. (The use of any filter generally results in some energy usage—either to operate, as with an electrostatic precipitator, or in fan energy to overcome the resistance to airflow.) Other filters (*e.g.* electrostatic precipitators) must be cleaned occasionally. (Maintenance is something that many occupants ignore, so it is a good idea for the filter installer to place brief filter-maintenance instructions in a conspicuous place where they will be noticed regularly.)

Efficiency and resistance

Many residential ventilating devices, as well as forced-air heating/cooling systems, already contain filters. Thin aluminum-mesh filters and thin throwaway fiber filters are common. Usually, the primary purpose of these inexpensive filters is to protect the fan motor from relatively large particulates, or in the case of a kitchen-range hood, to capture particles of grease. Occasionally, thin ($1/4$") activated carbon adsorption filters are used in ventilation systems, but they usually don't contain enough carbon to be very effective. This is a problem with many filters: They really don't remove very many pollutants from the air. So, why not just replace an inexpensive filter with a better one? That is often easier said than done. To understand why, let's look closer at how filter efficiency and resistance to airflow are related.

Efficiency

Air pollutants come in different sizes, ranging from those fairly large that can be seen with the naked eye, to tiny, invisible, microscopic particles. As a rule, large pollutants are easier to remove than small ones. Typical allergens such as mold spores and pollen are much larger than the microscopic particulate components of smoke, so they are fairly easy to filter out of the air. In fact, many medium-efficiency filters do a very good job with the larger particulates. Because smoke

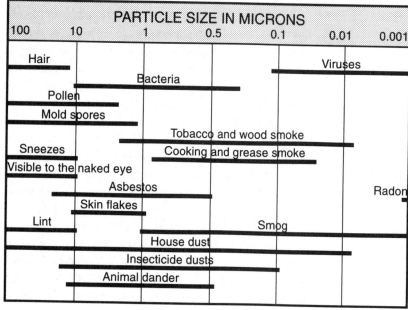

Figure 15–1. Air pollutants come in a variety of sizes. Larger pollutants are easier to remove from the air than smaller pollutants.

Chapter 15 Air-filtration equipment

ATMOSPHERIC SPOT DUST EFFICIENCIES	
10%	Good for capturing lint. Somewhat helpful for ragweed pollen. Not very good for smoke and staining particles.
20%	Fairly good at capturing ragweed pollen. Not very good for smoke and staining particles.
40%	Good at capturing pollen and airborne dust, some smudging and staining particles. Not very good for tobacco smoke particles.
60%	Very good for all pollens and most particles that cause staining and smudging. Partially helpful for tobacco smoke particles.
80%	Very good at removing smudging and staining particles, coal dust, oil smoke particles, and tobacco smoke particles.
90%	Excellent protection for all particles.

Figure 15–2. Filters vary in their efficiency. Those with higher efficiency generally are more expensive and often have more resistance to airflow.

particles are so small, you need a more efficient filter to remove them, but if there is no smoke in a house, a high-efficiency filter may not be necessary. A small particulate often attaches itself to a larger particulate, allowing a less efficient filter to remove both at once.

The size of a pollutant is measured in microns—a very small unit of measurement. The period at the end of this sentence measures about 400 microns in width. A very fine human hair might be 40 microns in diameter. Most people usually can't see anything smaller than about 10 microns. Your nose, sinuses, and windpipe are designed to filter out particulates down to 3–5 microns in size. Particulates smaller than this can find their way deep into your lungs. Figure 15–1 shows a variety of pollutants by size.

A filter's efficiency is a measure of how well it removes pollutants from the air. But how efficient is good enough? What does it mean when a filter advertisement says "removes 88% of the pollutants from the air?" Is an 88% efficient filter a good one? The answer depends on how the filter was tested—something that leads to a great deal of confusion. There are three different ways of measuring filter efficiency that are in common use: the weight-arrestance test, the atmospheric-spot-dust test, and the DOP-smoke-penetration test. All three tests measure a filter's efficiency at capturing particulates. Unfortunately, there is no standard test to determine the efficiency of a filter to adsorb gases.

Weight-arrestance test

An efficiency based on arrestance will tell you how well a filter performs at removing very large and heavy particles, something most filters do reasonably well. This test is described in *ASHRAE*'s Standard 52–76,(American Society 1976) and it is only good for measuring low-efficiency filters. For example, a standard 1" thick furnace filter doesn't do a very good job of capturing most common pollutants, but when tested for weight arrestance, it will be about 80% efficient. Because this test only measures large and heavy particulates (it was originally developed to measure efficiency at capturing "Arizona road dust" over 8 microns in size), it offers no indication of how a filter will do at capturing the smaller particulates typically found in houses and easily inhaled into the lungs. So, a high percentage on a weight-arrestance test doesn't mean very much.

Atmospheric-spot-dust test

The atmospheric-spot-dust test (also described in *ASHRAE*'s Standard 52–76) is a much more useful way of measuring filter efficiency than the weight-arrestance test because it measures particles between 0.3 and 6

Understanding Ventilation

microns in size. A standard 1" thick furnace filter will only be 3–5% efficient when measured by an atmospheric-spot-dust test. A filter rated at 80% on an atmospheric spot dust test is a very efficient filter. So, if a filter advertisement says "88% efficient by *ASHRAE* Standard 52–76," keep in mind that both of these tests are contained in *ASHRAE* Standard 52–76, so you need to look for the words "atmospheric-spot-dust" or "arrestance" to know if it is a very good filter or not.

While the atmospheric-spot-dust test is a more accurate way to evaluate a filter than the weight-arrestance test, it does have some drawbacks. For example, it is not very good at measuring how good a filter is at capturing very small particulates (less than 0.3 microns)—a size that many small respirable particles (*e.g.* cigarette or wood smoke) fall in. A filter can be 95% efficient on the atmospheric-spot-dust test, yet only capture 50–60% of these tiny particles. Figure 15-2 shows what spot-dust-test percentages mean in terms of what they are capable of removing from the air.

DOP-smoke-penetration test

This test is only used for very-high-efficiency air filters—those that are rated above 98% efficiency on the atmospheric-spot-dust test. The most-efficient filters available are called HEPA filters and are generally over 99% efficient when measured using the DOP-smoke-penetration test. HEPA is an acronym that stands for *high efficiency particulate air*. While this is a very accurate way of measuring the efficiency of an air filter for all size ranges of particles, it is only used for filters that are inherently good at capturing most small respirable-sized particles, so only extremely efficient filters are rated with this test. The test gets its name from an inert smoke-like gas called di-octyl phthalate (DOP) that is used in the test procedure. This test is often called a Military Standard 282 test because it is described in that U.S. military document.(U.S. Dept. of Defense)

Resistance to airflow

At first, it may seem logical to select a filter with the highest efficiency possible—as long as it isn't too expensive. After all, more efficiency usually costs more money. Ideally, superefficiency may be a good idea, but from a practical standpoint, there is a major drawback to many of the high-efficiency filters. They can have a great deal of resistance to airflow, so it is difficult for a fan to force air through them. This means the fan works very hard, but doesn't move much air. The precise amount of reduction in airflow depends on the particular characteristics of the filter, the fan, and the duct sys-

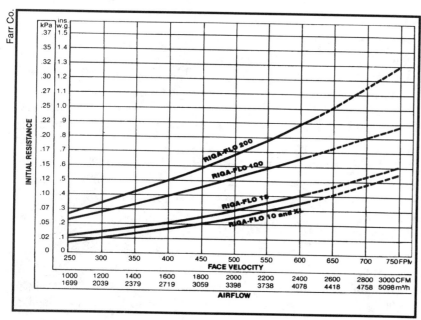

Figure 15–3. Manufacturers generally supply a chart or graph showing the amount of resistance (static pressure drop) their filters have at different airflows.

tem. Very few residential furnace/air-conditioning fans or ventilating fans are powerful enough to be used with a filter having a great deal of resistance. The only type of filter that has very little inherent resistance to airflow is an electrostatic precipitator (see below).

To determine a filter's resistance to airflow, a pressure reading is made both before and after the filter while there is a certain amount of air flowing through the filter. This is usually done at several flow rates. The difference in the two pressures is called the *static pressure drop* and it is usually measured in inches on a water gauge (" w.g.). There will be a greater static pressure drop (more resistance) at a high flow rate, and a lower static pressure drop (less resistance) at a low flow rate. Filter manufacturers often supply a chart showing the relationship between resistance (static pressure drop) and airflow (Figure 15-3).

A standard 1" thick furnace filter and most other low-efficiency air filters will have almost no resistance to airflow. In contrast, medium-efficiency filters often have 0.2–0.3" w.g. static pressure drop at their rated airflow and HEPA filters may be rated at 1.0" w.g. or more of resistance. These typical static pressure drops will increase as the filter becomes dirty because a clogged filter has more resistance to airflow.

In order to couple an air filter to a forced-air system (either a ventilation system or a heating/cooling system), the static pressure drop across the filter must be compatible with the fan. Filters are available in a wide variety of sizes and it is usually possible to match a filter to a given fan, but it is not always practical. This is because high-efficiency filters can have a relatively large physical size, depending on their flow and static pressure specifications. This can make it difficult to fit them into an existing cramped mechanical room. In new construction, it is important to make sure everything will fit into the allotted space.

Most residential ventilating systems have relatively small capacity fans (compared to forced-air heating/cooling fans), and there are a limited number of filters specifically made for use with these fans. However, it is possible to combine a small-capacity ventilating fan with a physically larger heating/air-conditioning filter. For example, if you want to add a medium efficiency filter (40% on the atmospheric-spot-dust test) to a 100 cfm ventilation system, you might consider a filter that is designed for a 1000 cfm furnace/air-conditioning system. Because resistance decreases with less airflow, it will probably have a fairly low static pressure drop at 100 cfm and have minimal effect on the airflow of the ventilating fan. Of course, whenever one manufacturer's filter is used in conjunction with another manufacturer's fan, it is important to work closely with both suppliers to ensure that the filter's resistance to airflow won't be excessive and the combined system will function correctly.

Adsorption filters can also have a great deal of resistance to airflow. To minimize the static pressure drop, these filters often end up being too physically large to be suitable for residential applications. However, some manufacturers are now introducing relatively thin adsorption filters with less resistance. Unfortunately, they don't contain very much adsorption material so they have less ability to capture gases, and they don't last very long.

Where to install a filter

A filter can be used to clean the outdoor air before a ventilator brings the air into the house. This strategy will ensure that only good-quality air is brought indoors. Airports regularly do this to keep their terminals free from airplane fuel and exhaust odors. This approach only works with a ventilation system that has a single inlet port, because with only one inlet, all the air entering the house can be handled with a single filter. Central-supply systems, balanced systems, and heat-recovery balanced systems typically have a single inlet. Some HRV manufacturers offer medium-efficiency filters as options for their units.

It is not possible to use a filter with a central-exhaust system having several through-the-wall inlets. This is because through-the-wall vents are designed to operate passively. A filter must have a fan actively pushing or pulling air through it—air will not move through a filter passively. If you place filters on through-the-wall vents, the resistance of the filters will cause air to come in elsewhere because there is almost always a path of lesser resistance somewhere else.

Another approach is to use a better-than-average air filter in the forced-air furnace/air conditioner to clean the air that is already in the house. If the ventilation system and the forced-air furnace/air-conditioning system are not interconnected, outdoor air will come indoors through the ventilator, mix with house air, then be filtered whenever it passes through the furnace/air-conditioning filter.

By coupling a ventilator with the forced-air heating/cooling system, it can be possible to mix the fresh air with the house air inside the duct system, then filter the mixture before it enters the living space. Because not many medium-efficiency filters or adsorption filters are specifically designed for use with residential ventilation systems, coupling a filter with a forced-air furnace or air conditioner is a popular option. One of the advantages of this approach is the fact that all the air in the house will eventually pass through the furnace/air conditioner filter more than once; in fact, the air will be filtered as often as the system recirculates the air. When air is filtered multiple times, medium-efficiency filters work very well at removing many common pollutants.

A final filtration strategy would be to use room-sized portable filters to clean the indoor air. These units can sometimes filter the air in a house several times an hour. Inasmuch as their fans are considerably less powerful than a forced-air furnace/air-conditioning fan, they are generally less effective at cleaning the air than the first two approaches. However, portable filters are popular because of their lower cost. Because they cannot be tied to a ventilation system, portable filters will not be covered in this book. The various filtration strategies are shown in Figure 15–4.

It is important to remember that filtering outdoor air before it enters the living space will do nothing to remove the pollutants that originate indoors (except through dilution). For example, you must use either a portable filter or a filter attached to a forced-air heating/cooling system to deal with formaldehyde that is released from wall paneling or kitchen cabinets. (Removing strong formaldehyde sources from a house—or separating them from the occupied space—is generally more effective than trying to reduce formaldehyde levels with a filter.) In addition, many particulate sources indoors can only be dealt with by using portable filters or forced-air heating/cooling system filters. For example, the natural fibers (*e.g.* cotton and wool) that are popular today release tiny bits of fiber into the indoor air. Dust mites and mold spores also often originate indoors.

To use a forced-air furnace/air-conditioner filter to optimum advantage, the air in the house must be recirculated through the filter at least 2–4 times per hour, and the fan must run continuously. This amount of air movement will ensure that heavier respirable particulates remain airborne rather than settle onto the floor. If the particulates aren't airborne, they can't be removed from the air. When particulates settle onto the floor, they simple get stirred up by foot traffic (exposing the occupants), then settle again.

Portable filters also need to recirculate the air at least 2–4 times per hour and run continuously in order to be effective. They can work well in a single closed room, but they are much less effective as whole-house filters. Central ventilation systems rarely move enough air to be very effective at removing particulate pollutants that are released indoors.

The location of the fan with respect to the filter can have an effect on how well the entire surface of the filter is utilized. As a rule, air pulled through a filter (fan downwind) travels more slowly and uniformly through all parts of the filter than if air is blown through the same filter (fan upwind). When air is pushed (rather than pulled) through a filter, most of the air passes through the center of the filter quickly and very little of it passes through the perimeter. When air is pulled through a filter, it does a better job and lasts longer, and because the fan motor doesn't have to work as hard, less electricity is consumed.

Particulate filters

The standard 1"-thick furnace/air-conditioner filter is a particulate filter yet it is only designed to remove large, heavy pollutants from the air, and because it is less than 5% efficient (on the atmospheric-spot-dust test), it doesn't do a very good job. Actually, its primary purpose is not to clean the air for occupants, but rather to protect the fan motor in the furnace or air conditioner. It is a *furnace* filter, not a *people* filter, and a motor doesn't need a very efficient filter. Besides protecting the motor, it also prevents the cooling coil in an

Chapter 15 Air-filtration equipment

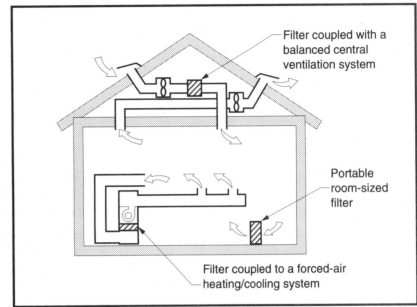

Figure 15–4. Filters are typically used in three different locations: as part of a forced-air heating/cooling system, as part of a central ventilation system, or as a stand-alone portable filter.

air conditioner or heat pump from getting clogged. People require more filtration than mechanical equipment, and many different filters on the market do a much better job: flat filters, pleated filters, bag filters, cube filters, etc. (Actually, we often protect our mechanical equipment better than ourselves. For example, the filter that cleans the air going into an automobile engine's carburetor is of higher quality than the filter that cleans the air for occupants riding in the seating compartment.)

Four kinds of particulate filters are being promoted in the residential market that are better than a standard 1" thick furnace/air-conditioner filter: medium-efficiency filters, electrostatic precipitators, electrostatic air filters, and to a lesser extent, HEPA filters. While virtually any type of industrial- or commercial-grade filter, of any efficiency, can be adapted to a residential ventilation system, we will focus our discussion primarily on these four.

Medium-efficiency and HEPA filters are often called *media* filters because they are made of materials (media) that work like a coffee strainer, capturing particulates too large to pass through the strainer. They are also sometimes called mechanical filters because they *mechanically* capture particulates. An electrostatic precipitator, on the other hand, relies on electric charges to capture particles—it *electrically* captures particles. Electrostatic air filters are a hybrid approach; they rely partially on media and partially on a static charge.

It is possible to improve the efficiency of a media filter by coating it with a sticky oil called a tackifier, but if this material has any odor at all, the odor can contaminate the airstream.

Some people have expressed concern that media filters, which are usually made of polyester or glass fibers, will deteriorate and shed some fibers into the airstream. No doubt, this occurs, but it is not believed to be a significant issue. One study (produced by a fiberglass manufacturer) found that the number of fibers shed was considerably less than the number typically found in outdoor air.[Shumate] Thus, even though a filter may shed a few fibers, it will generally tend to remove far more material from the air than it releases. The net effect is considerably cleaner indoor air than if no filter were used.

An important point to keep in mind when selecting any filter is that efficiency changes with time. With some filters, efficiency improves over time; with others it declines. As a media filter becomes partially clogged, it actually becomes more efficient because the captured debris partially plugs the strainer, resulting in smaller openings. It wouldn't be unusual for a media filter's efficiency to double when partially loaded with

Understanding Ventilation

dust. However, as the efficiency increases, so does resistance to airflow. The result is better filtration, but less air. Electrostatic precipitators, on the other hand, are most efficient when they are first installed; then the efficiency declines as they get dirty, so they become less efficient with time.

Medium-efficiency filters

For improved efficiency in a media filter, the air must pass through smaller openings. Yet, this makes it harder for air to get through the filter. Thus, media filters with higher efficiency also have more resistance to

Figure 15–5. Pleated, medium-efficiency filters are often used in forced-air heating/cooling systems and can be adapted to central ventilation systems.

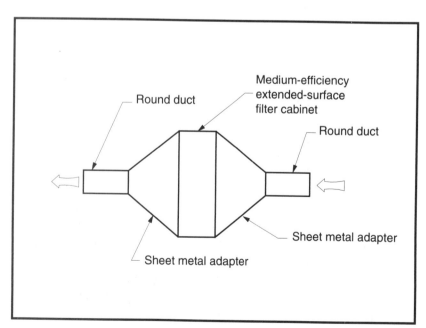

Figure 15–6. Sheet-metal adapters can easily be constructed to use a relatively large filter (designed for a forced-air heating/cooling system) with smaller ventilation ducts.

airflow than standard furnace filters. In order to compensate, higher-efficiency filters are often manufactured in a pleated accordion shape. This configuration results in much more surface area and less overall air resistance. Because they have more surface area, they are often called *extended-surface* filters. The service life of an extended-surface filter is often in the range of a year or more.

Extended-surface filters are available in a wide range of sizes and efficiencies. Some, such as the 20-20 filters by **Farr Co.**, are 1" thick and yield up to 20% efficiency on the atmospheric-spot-dust test. They are designed to replace a standard low-efficiency furnace/air-conditioner filter.

Several extended-surface filters are being offered to the residential market in the 25-45% range on the atmospheric-spot-dust test, for use with forced-air heating/cooling systems. Most heating/air-conditioning contractors are familiar with these medium-efficiency filters. Manufacturers include **Carrier Corp.**, **General Filters Inc.**, **Honeywell, Inc.**, and **Research Products Corp.** (Space-Gard brand) (Figure 15-5). These filters are several inches thick, so they cannot be used to simply replace a standard 1" furnace filter. Instead, they are supplied with a special housing that must be incorporated into the duct system. These particular medium-efficiency filters are designed for airflow ranges of 600-2,000 cfm, but they can easily be used with smaller airflows. In fact, when used with most ventilation systems, which typically move less than 200 cfm, they have only minimal resistance to airflow. They are more efficient at lower airflows. Because they are approximately two feet square, some duct adapters will need to be fabricated to use them with the smaller (6-7") round ducts typical of most ventilation systems (Figure 15-6).

Pleated filters and bag filters are available with atmospheric-spot-dust-test efficiencies up to 95%, but they are not commonly installed in residences.

HEPA filters

HEPA filters are actually a special type of extended-surface filter (Figure 15-7). They are sometimes called *absolute filters* because they are typically 97.99% efficient with either the atmospheric-spot-dust test or the DOP-smoke-penetration test. These filters were originally developed during World War II by the Atomic Energy Commission for use in nuclear laboratories to filter out deadly plutonium particles. They are often used to maintain clean rooms in the electronics industry and in laboratories. HEPA filters are now available in some portable and central residential equipment.

Figure 15-7. HEPA filters are extremely efficient at removing particulate pollutants, but they are more commonly used in laboratories than in residences because of their high resistance to airflow.

HEPA filters are more costly than other particulate filters, but they represent state-of-the-art efficiency in achieving clean air. Because their pores are so small, they have considerably more resistance to airflow than other extended-surface filters. As a result, they often require a fan motor capable of operating at high static-pressure drops to force air through them.

In most cases, even though medium-efficiency filters aren't perfect, most residences do not require the superefficiency of HEPA filters. In other words, HEPAs are overkill in most situations. (Actually, hospital operating rooms often use 95% filters through which air is circulated at a rate of 10–20 air changes per hour, rather than HEPAs.) Medium-efficiency filters are generally sufficient to remove the most bothersome particulates (*e.g.* mold and pollen) found in houses. As a compromise, an 80% filter (atmospheric-spot-dust test) will do a very good job of cleaning the air in a house, and at a lower cost than a HEPA. Still, very sensitive people may opt for a HEPA. If smoking is allowed indoors, a HEPA filter can be a good idea because smoke particulates can be quite tiny. However, it is always far more effective to ban smoking indoors than to try to reduce its effect with a filter.

There are a number of HEPA filter manufacturers, including **Airguard Industries, Inc.**, **Air Filtration Products**, and **Farr Co.** However, these companies usually only supply filters—not fans—so care must be taken to ensure that a particular ventilation system or heating/air-conditioning system is compatible with the static-pressure drop of the filter. In most cases a HEPA installation will need to be custom-designed. A few companies offer HEPA filters as a part of a central filtration system in which all the design work is already taken care of (See *Packaged combination filters* below). Because the static-pressure drops are higher with HEPA filters, extra care should be taken to ensure that the ducts are well-sealed, else there will be excessive leakage.

Figure 15–8. Electrostatic precipitators (electronic air cleaners) are quite efficient at removing particulate pollutants, but they produce small amounts of ozone (itself a pollutant). Unless they receive regular maintenance their efficiency drops substantially.

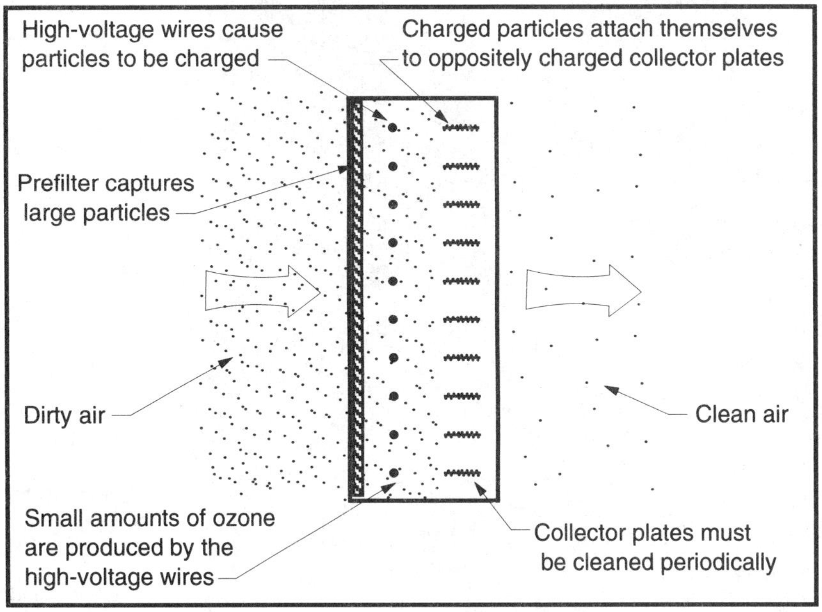

Figure 15–9. *Electrostatic precipitators place a charge on particulate pollutants which then cling to oppositely charged collector plates. If the collector plates aren't cleaned occasionally, the pollutants pass right through the filter.*

A HEPA filter should last at least a couple of years in a typical house. For optimum life, it is a good idea to use a coarser prefilter, such as a **Farr Co.** 20–20 extended-surface filter, upwind of the HEPA to capture the larger pollutants (some people recommend a 40% efficient prefilter). If the prefilter is changed regularly, the HEPA won't be wasted on the easy-to-capture large particles and, thus, will last longer.

Electrostatic precipitators

Many homeowners have heard of an electronic air cleaner. Its proper name is an *electrostatic precipitator*. These devices are generally designed to be permanently mounted into the ducts of a forced-air furnace/air conditioner (Figure 15–8). They operate by placing an electrical charge on particulates that pass through the filter. To accomplish this, they generate a high voltage (approximately 20,000 volts) that gives the dust a static-electric charge. Because opposite charges are drawn toward each other, the charged particulates are attracted to and cling to special, oppositely charged metal plates that are built into the filter (Figure 15–9). Once the plates are full of particulates and can hold no more, they must be removed and cleaned. This can usually be done at home in a dishwasher or tub of water.

Depending on how dirty the air is, cleaning may be necessary as often as several times a month. Because of their design, electrostatic precipitators have virtually no resistance to airflow.

Electrostatic precipitators are often rated at 90% efficiency (atmospheric-spot-dust test), but their efficiency declines quickly as the plates fill with dust. The dirtier the air, the sooner the plates fill, and the quicker efficiency drops off. It wouldn't be unusual for one of these filters to be only 20% efficient after only a few days of use. While they are considerably better at handling particulates than a standard furnace filter, they have one primary disadvantage: The high voltage generates a certain amount of ozone, and ozone is a well-known air pollutant. While the amount of ozone produced isn't considered significant by many people, it can be a concern to sensitive people.

In general, ozone is less of a concern with an intermittently running system because it tends to dissipate rather quickly. However, filters are more effective when the air is recirculated through them continuously, and a continuously running system can result in higher ozone levels indoors. Also, because electrostatic precipitators operate at high voltages, electrical sparks occasionally form inside them. Sparks can disturb some of the dust on the collector plates, causing a certain

Understanding Ventilation

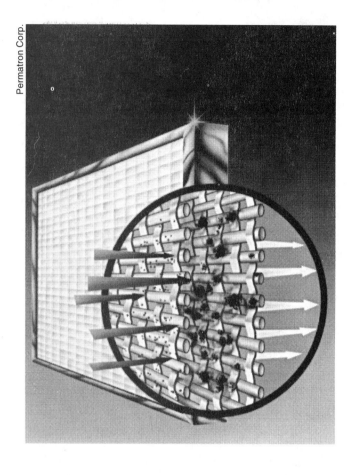

Figure 15–10. Electrostatic air filters have a charge of static electricity that captures oppositely charged particulate pollutants. Though not nearly as efficient as electrostatic precipitators, they are somewhat less expensive.

amount of it to be released from the filter. Thus, they can actually release some of the dust they have collected.

Among other things, the efficiency of an electrostatic precipitator is dependent on the air speed and on turbulence. Ideally, air should flow through them smoothly. This means that elbows or changes in direction in the duct system should be avoided near the filter. It is possible to install turning vanes inside the elbows so that exiting air has a more uniform flow pattern. However, this is seldom done in residential installations because of the increased cost.

Currently, no electrostatic precipitators are specifically designed for residential ventilation systems. However, furnace-sized models can easily be adapted. Like the medium-efficiency filters discussed above, most electrostatic precipitators are designed for 600–2,000 cfm flow rates, but when used at lower flows are even more efficient. Manufacturers of electrostatic precipitators for forced-air heating/air-conditioning systems include **Carrier Corp.**, **Emerson Electric Co.** (Electro-Air brand), **Honeywell, Inc.**, and **Trion, Inc.** To alert occupants when it's time to clean their electrostatic precipitator, **Honeywell, Inc.** offers an electronic sensor with an indicator light that comes on when the filter is dirty. They also market an extended-surface filter that can be later upgraded to an electrostatic precipitator.

Electrostatic air filters

Static electricity sometimes results when you shuffle your shoes across synthetic carpeting. Electrostatic air filters work much the same way, but instead of your shoes causing the friction, it is produced from air moving through special fibrous plastic media (*e.g.* vinyl, polyester, or polystyrene). These filters operate similarly to electrostatic precipitators, but are not hooked to the house's electrical system. The static electricity causes the filter material itself to become charged, then oppositely charged particulates in the airstream cling to the filter's fibers (some particulates are uncharged)

(Figure 15–10). Some electrostatic air filters don't even need moving air to generate a static charge; they are pre-charged with *electrets*. An electret is a plastic material that carries a permanent static charge.

While electrostatic air filters aren't very effective at capturing very small particles, they do a reasonable job with mold spores and pollen. Overall they are about 10–15% efficient on an atmospheric-spot-dust test, so they are somewhat more efficient than a standard furnace filter, but not as good as a medium-efficiency filter. They are widely available from a number of manufacturers, including **3-M Do-It-Yourself Division** (Filtrete brand), **Air Kontrol**, **Dust Free, Inc.**, **Newtron Products Co.**, and **Permatron Corp.**

Most electrostatic air filters are 1" thick and their resistance to airflow is low, so they can easily be substituted for a conventional furnace filter. Therefore, they are an inexpensive way to get a modest upgrade in filter efficiency, but overall, medium-efficiency filters and electrostatic precipitators are somewhat more efficient.

Gaseous filtration (adsorption)

Particulate filters either work like a coffee strainer—trapping the pollutants that are larger than their pore size—or capture pollutants by utilizing an electric charge. Gaseous filtration requires a totally different approach: *adsorption*. Adsorption is a process by which gas molecules adhere to a solid surface. Although adsorption is a physical/chemical process, a crude analogy would be to compare an adsorption filter to flypaper. Once an adsorption filter has captured all the gases it can hold, it must be replaced.

Like many other filters, adsorption filters can have a considerable amount of resistance to airflow. For very polluted air, a great deal of adsorption material is necessary, therefore a great deal of surface area is required to reduce the resistance to airflow for the fan to function properly. In commercial applications, trays filled with the adsorption material are arranged in an accordion shape to provide plenty of surface area (Figure 15–11). Manufacturers include **Barneby & Sutcliffe**, **Farr Co.** and **RSE, Inc.** As a rule, these kinds of adsorption filters are expensive because of their size and special fan requirements. To reduce costs in residential applications, filters containing less adsorption material and, thus, less air resistance, have become popular for use in houses. With less adsorption material, they are not as efficient, but they are much lower in cost, so they are within the budgets of many homeowners. The big drawback is the possibility that they

Figure 15–11. In commercial applications, adsorption materials, such as activated carbon, are placed in trays arranged in an accordion-like shape to minimize the resistance to airflow. These filters generally have too much resistance for use in most residential applications.

Figure 15–12. *A partial-bypass filter allows some air to pass through without coming in contact with any adsorption material. Though this reduces the efficiency, it also minimizes the resistance to airflow. These filters are available in a 1" thickness to fit in forced-air furnaces and air conditioners. (The photograph is shown about ³⁄₄ size.)*

have so little adsorption material in them, they may not be worth the effort, especially with very dirty air.

One approach to less air resistance is a *partial-bypass* adsorption filter. With these filters, some of the air passing through the filter comes in contact with the adsorption material while some doesn't. Partial-bypass filters look like a honeycomb with several granules of adsorption material in each ¹⁄₈" cell (Figure 15–12). This gives them less resistance to airflow than if the cells were packed full. If a partial-bypass filter is used in a forced-air furnace/air conditioner, the recirculating air in the house will continuously be returning to the filter, so that after several passes, most of the air should come in contact with some of the adsorption material.

A partial-bypass filter will be less effective when used on a fresh-air inlet, because it will only filter part of the air entering the house.

In a house with very contaminated air, a small to moderate-sized adsorption filter may not be big enough to provide any noticeable improvement in air quality. In a relatively unpolluted house, such a filter can be used to polish the air. Even though there are no guidelines or standards that can be used to test and compare adsorption filters, substantial adsorption material is needed if you want to adsorb a great deal of gaseous pollution. As a rule-of-thumb, about one pound of adsorption material will capture as much as ¹⁄₂ pound of gas (less for some gases). That sounds like a lot because, after all, gases don't weigh very much. But the air in a 1,500-sq.-ft. house weighs about 1,000 pounds and if filtered at a rate of .35 ACH, you would be asking the filter to process about 100,000 pounds of air per day. If the gaseous pollutants are at a concentration of 1 ppm, a filter would need to adsorb about 1.5 oz. of gas every day. A filter containing less than a pound of adsorption material certainly wouldn't last very long.

Besides containing more adsorption material, thickness greatly enhances an adsorption filter's effectiveness—gases are in contact with the adsorption material for a longer period of time in a thick filter. The best adsorption filters are several inches thick, but they have far too much resistance to airflow for most residential applications.

While adsorption filters are designed to *capture* gases, they can also *release* some of those gases back into the air. This is a particular problem with adsorption filters. For example, activated carbon, a popular adsorption material, will readily remove a variety of gases from polluted air. But the air rarely has the same degree of pollution at all times. A filter will adsorb gases during a period when the air is highly polluted and then, if the air happens to be less polluted for a while, those gases can be released into the clean air. In a study by the National Institute of Standards, researchers varied the concentration of toluene in the air flowing through an activated-carbon adsorption filter. When the toluene concentration was high, the filter adsorbed most of it, but when the concentration fell, the filter slowly released the gas it had captured. Over the 45-hour experiment, the total amount of toluene captured by the filter was about equal to the amount released.(Mahajan) So an adsorption filter actually provides you with a more uniform concentration of pollutants by reducing the peak concentrations and then releasing pollutants when concentrations are low.

A variety of materials can be used to adsorb gases. In some applications (primarily industrial) adsorbents such as zeolites, soda lime, and silica get are used. Many of these materials are specially processed to only adsorb specific pollutants. More commonly available for residential applications, activated carbon and activated alumina work well at adsorbing a wide variety of different gases. Activated carbon and activated alumina can only hold a certain amount of gas, so when they become saturated, they must be replaced, else they will start releasing pollutants. To determine if the materials are no longer capable of adsorbing anything, try to sniff an odorous substance through them.

Activated carbon

Activated carbon (often called activated charcoal) is the most common adsorption material on the market (Figure 15–13). It can be made from coal, coconut husks, wood, or a several other materials. When viewed under a microscope, activated carbon is extremely porous. This means it has a great deal of surface area for its size, offering plenty of places to which gases can adhere. Activated carbon does an excellent job of removing most gaseous pollutants from the air. Filter manufacturers often list the wide variety of gases it will adsorb. The chart in Figure 15–14 is typical, but it can be somewhat misleading. It lists many specific chemicals found in indoor air, but it also contains items such as *household smells*, *lingering odors*, *perfumes*, and *stuffiness*—all of which are impossible to categorize accurately from a chemical standpoint. Because such odors could easily vary in composition from house to house, no chemist could begin to define them, yet the chart claims that activated carbon will adsorb them quite well. Mold and pollen are also listed, yet mold spores and pollen are particulates and adsorption filters are designed for gases. While activated carbon can capture *odors* that are released from mold and pollen, it should not be considered effective at capturing the particulates themselves.

Figure 15–13. Activated carbon is probably the most common adsorption material on the market.

Understanding Ventilation

ACTIVATED CARBON PERFORMANCE CHART

Substance	Rating	Substance	Rating	Substance	Rating	Substance	Rating
Acetaldehyde	2	Dead animals	4	Indole	4	Pentanone	4
Acetic acid	4	Decane	4	Industrial wastes	3	Pentylene	3
Acetic anydride	4	Decaying substances	4	Iodine	4	Pentyne	3
Acetone	3	Deodorants	4	Iodoform	4	Perchloroethylene	4
Acetylene	1	Detergents	4	Irritants	4	Perfumes, cosmetics	4
Acrolein	3	Dibromethane	4	Isophorone	4	Perspirations	4
Acrylic acid	4	Dichlorobenzene	4	Isoprene	3	Persistent odors	4
Acrylonitrile	4	Dichlorodifluoromethane	4	Isopropyl acetate	4	Pet odors	4
Adhesives	4	Dichloroethane	4	Isopropyl alcohol	4	Phenol	4
Air-Wick	4	Dichloroethylene	4	Isopropyl ether	4	Phoagene	3
Alcoholic beverages	4	Dichloroethyl ether	4	Kerosene	4	Pitch	4
Amines	2	Dichloromonofluormethane	3	Kitchen odors	4	Plastics	4
Ammonia	2	Dichloronitroethane	4	Lactic acid	4	Pollen	3
Amyl acetate	4	Dichloropropane	4	Lingering odors	4	Popcorn & candy	4
Amyl alcohol	4	Dichlorotetrafluoroethane	4	Liquid fuels	4	Poultry odors	4
Amyl ether	4	Diesel fumes	4	Liquor odors	4	Propane	2
Animal odors	3	Diethylamine	3	Lubricating oils & grease	4	Propionaldehyde	3
Anesthetics	3	Diethyl ketone	4	Lysol	4	Propionic acid	4
Aniline	4	Dimethylaniline	4	Masking agents	4	Propyl acetate	4
Antiseptics	4	Dimethylsulfate	4	Medicinal odors	4	Propyl alcohol	4
Asphalt fumes	4	Dioxane	4	Melons	4	Propyl chloride	4
Automobile exhaust	3	Dipropyl ketone	4	Menthol	4	Propyl ether	4
Bathroom smells	4	Disinfectants	4	Mercaptans	4	Propyl mercaptan	4
Benzene	4	Embalming odors	4	Mesityl oxide	4	Propylene	2
Bleaching solutions	3	Ethane	1	Methane	1	Propyne	2
Body odors	4	Ether	3	Methyl acetate	3	Putrefying substances	3
Borane	3	Ethyl acetate	4	Methyl acrylate	4	Putrescine	4
Bromine	4	Ethyl acrylic	4	Methyl alcohol	3	Pyridine	4
Burned flesh	4	Ethyl alcohol	4	Methyl bromide	3	Radiation products	2
Burned food	4	Ethyl amine	3	Methyl butyl ketone	4	Rancid oils	4
Burning fat	4	Ethyl benzene	4	Methyl cellosolve	4	Resins	4
Butadiene	3	Ethyl bromide	4	Methyl cellosolve acetate	4	Reodorants	4
Butane	2	Ethyl chloride	3	Methyl chloride	3	Ripening fruits	4
Butanone	4	Ethyl ether	3	Methyl chloroform	4	Rubber	4
Butyl acetate	4	Ethyl formate	4	Methyl ether	3	Sauerkraut	4
Butyl alcohol	4	Ethyl mercaptan	3	Methyl ethyl ketone	4	Sewer odors	4
Butyl cellosolve	4	Ethyl silicate	4	Methyl formate	3	Skatole	4
Butyl chloride	4	Ethylene	1	Methyl isobutyl ketone	4	Slaughtering odors	3
Butyl ether	4	Ethylene chlorhydrin	4	Methyl mercaptan	4	Smog	4
Butylene	2	Ethylene dichloride	4	Methylcyclohexane	4	Soaps	4
Butyne	2	Ethylend oxide	3	Methylcyclohexanol	4	Smoke	4
Butyraldehyde	3	Essential oils	4	Methylcyclohexanone	4	Solvents	3
Butyric acid	4	Eucalyptole	4	Methylene chloride	4	Sour milk	4
Camphor	4	Exhaust fumes	3	Mildew	3	Spilled beverages	4
Cancer odor	4	Female odors	4	Mixed odors	4	Spoiled foodstuffs	4
Caprylic acid	4	Fertilizer	4	Mold	3	Stale odors	4
Carbolic acid	4	Film processing odors	3	Monochlorobenzene	4	Stoddard solvent	4
Carbon disulfide	4	Fish odors	4	Monofluorotrichlormethane	4	Stuffiness	4
Carbon dioxide	1	Floral scents	4	Moth balls	4	Styrene monomer	4
Carbon monoxide	1	Fluorotrichloromethane	3	Naptha (coal tar)	4	Sulfur dioxide	2
Carbon tetrachloride	4	Food aromas	4	Naptha (petroleum)	4	Sulfur trioxide	3
Cellosolve	4	Formaldehyde	2	Napthalene	4	Sulfuric acid	4
Cellosolve acetate	4	Formic acid	3	Nicotine	4	Tar	4
Charred materials	4	Fuel gases	2	Nitric acid	3	Tarnishing gases	3
Cheese	4	Fumes	3	Nitro benzenes	4	Tetrachloroethane	4
Chlorine	3	Gangrene	4	Nitroethane	4	Tetrachloroethylene	4
Chlorobenzene	4	Garlic	4	Nitrogen dioxide	2	Theatrical makeup odors	4
Chlorobutadiene	4	Gasoline	4	Nitroglycerine	4	Tobacco smoke odors	4
Chloroform	4	Heptane	4	Nitromethane	4	Toilet odors	4
Chloronitropropane	4	Heptylene	4	Nitropropane	4	Toluene	4
Chloropicrin	4	Hexane	3	Nonane	4	Toluidine	4
Cigarette smoke odor	4	Hexylene	3	Octalene	4	Trichloroethylene	4
Citrus and other fruits	4	Hexyne	3	Octane	4	Trichloroethane	4
Cleaning compounds	4	Hospital odors	4	Odorants	4	Turpentine	4
Combustion odors	3	Household smells	4	Onions	4	Urea	4
Cooking odors	4	Hydrogen	1	Organic chemicals	4	Uric acid	4
Corrosive gases	3	Hydrogen bromide	2	Ozone	4	Valeric acid	4
Creosote	4	Hydrogen chloride	2	Packing house odors	4	Valericaldehyde	4
Cresol	4	Hydrogen cyanide	2	Paint & redecorating odors	4	Varnish fumes	4
Crotonaldehyde	4	Hydrogen fluoride	2	Palmitic acid	4	Vinegar	4
Cyclohexane	4	Hydrogen iodide	3	Paper deteriorations	4	Vinyl chloride	3
Cyclohexanol	4	Hydrogen selenide	2	Paradichlorobenzene	4	Waste products	3
Cyclohexanone	4	Hydrogen sulfide	3	Paste & glue	4	Wood alcohol	3
Chlohexene	4	Incense	4	Pentane	3	Xylene	4

4. High adsorption capacity. One pound of activated carbon takes up about 20% to 50% of its own weight of these substances.
3. Satisfactory capacity. One pound of activated carbon takes up about 10% to 15% of its own weight of these substances.
2. Low capacity. One pound of carbon takes up less than 10% of its own weight of these substances.
1. Activated carbon cannot be used to remove these substances under ordinary circumstances.

Figure 15–14. A typical promotional chart used by activated carbon suppliers.

It is also important to remember that while activated carbon will indeed adsorb many different gases, it is not very effective at capturing gases with low molecular weights. Formaldehyde, a very common indoor pollutant, happens to be one of the gases that activated carbon doesn't adsorb very well. However, specially treated carbons are readily available that can remove formaldehyde from the air. Industrial processes use activated carbon that has been specially treated to remove other gases such as ethylene, hydrogen sulfide, and mercury compounds. According to Figure 15–14, activated carbon isn't effective for other common indoor contaminates such as carbon monoxide, carbon dioxide, ethane, and methane.

Activated carbon can be successfully used to combat ozone, but it doesn't do so by adsorption. Instead, the carbon catalyzes the ozone (which is composed of 3 oxygen atoms) into benign oxygen gas (which is composed of a pair of oxygen atoms). Many companies that manufacture electrostatic precipitators offer a thin carbon filter that can be used to handle the ozone those devices produce. Some other gases are also catalyzed by activated carbon.

Several manufacturers produce thin (1") filters containing activated carbon that can be substituted for a standard furnace filter. They include **Aero Hygenics, Inc.**, **Barneby & Sutcliffe**, **Columbus Industries** (Polysorb brand), **Dust Free, Inc.**, and **Permatron Corp.** The **E.L. Foust Co.** is a retail, mail-order source for 1" filters containing activated carbon. None of these filters contains much carbon, so they probably aren't suitable for very contaminated situations. However, they may work well in a relatively unpolluted house.

ACS Filter Division offers 1" activated-carbon filters in standard furnace/air-conditioner sizes as well as in sizes to fit Electro-Air, Emerson, and Honeywell electrostatic precipitators for ozone control.

Activated alumina

Carusorb and Purafil are two popular adsorption materials made of activated alumina impregnated with potassium permanganate (Figure 15–15). These products are manufactured by **Carus Chemical Co.** and **Purafil, Inc.**, respectively.

Activated alumina is generally more costly than activated carbon. Although it doesn't have as much adsorptive power as most activated carbons, it does have the ability to catalyze some low-molecular-weight gases, such as formaldehyde and ethylene, into relatively-harmless water vapor and carbon dioxide. (The gases are adsorbed onto the surface of the activated alumina

Figure 15–14. Activated alumina is pink when fresh and turns dark brown as it adsorbs gases. By crushing a pellet to see the color of its center, you can determine if it is still effective. A pink center means it is still capable of adsorbing gases.

and then the potassium permanganate oxidizes them into water vapor and carbon dioxide.) If chlorinated hydrocarbons are present in the air moving through the filter (and often they are, being released from some chlorinated cleaning products), they can be converted to hydrochloric acid when they are oxidized. This new pollutant can then be released into the air.

Over time, activated alumina becomes depleted to the point that it can no longer react with gases. When fresh, it is a bright pink color, but as it ages it gradually changes to dark brown, signifying that it should be replaced. Sometimes activated alumina is combined with activated carbon in the same filter to enhance the gas-removing ability.

Miscellaneous air-cleaning strategies

A number of miscellaneous approaches are available to clean the air. Because they tend to be cheaper, they have their applications— but the above-mentioned approaches are generally more effective.

Occasionally, air fresheners are sold as air cleaners. In reality, they do absolutely nothing to clean the air. Instead, they add fragrance to cover up the odor of pollutants. The pollutants are still there; they just aren't as immediately noticeable. In reality, the fragrance actually adds to the pollutant load.

Negative-ion generators

Negative-ion generators can be used to clean indoor air. Some models simply spew out negatively charged ions (electrons) into the air. These charged ions attach themselves to dust particles, causing the dust to be negatively charged. Then, the charged dust is attracted to and clings to positively charged surfaces such as walls and ceilings. Rather than have dust float through the air, the walls and ceilings act like a giant dust collector. The result can be dirty walls and ceilings. After a while, some of the particles lose their charge and reenter the air. Some negative-ion generators contain a built-in collection filter to capture the charged particles. They work much like the electrostatic precipitators described above, and they too can produce a certain amount of ozone.

Researchers have determined that when there are negatively charged ions in the air, some people may have a sense of well-being.[Yates] Conversely, people can feel bad when the air contains positively charged ions.[Giannini] While this effect has nothing to do with cleaning the air, many people have installed negative ion generators for this very reason.

Figure 15–16. In the 1980s, the National Aeronautics and Space Administration (NASA) found that common house plants were capable of reducing indoor pollution. Later studies have determined that certain bacteria living in the soil around specific plants can consume some gaseous pollutants, such as formaldehyde, but the pollutant reduction in a typical house isn't very significant.

Chapter 15 Air-filtration equipment

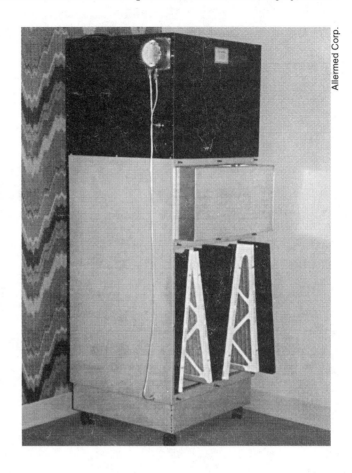

Figure 15–17. This whole-house filter contains both a HEPA filter and a bank of adsorption filters. Marketed primarily to highly sensitive individuals, it can be used as a stand-alone system or in conjunction with a forced-air heating/cooling system.

Ozone generators

Ozone is a highly reactive gas, composed of three oxygen atoms, that can react with soft tissues in the body and cause breathing difficulties. Yet, ozone generators are being sold to *improve* indoor air quality. Even though ozone is a well-known pollutant, it has the unique ability to react with VOCs in the air and convert them into harmless water vapor, carbon dioxide, and oxygen. The use of ozone can also cause an *increase* in the level of VOCs in a house. This occurs when ozone reacts with nonvolatile or semivolatile VOCs and converts them into VOCs. Ozone will also kill mold. But even though ozone can kill mold, allergic individuals react to dead mold spores as well as live ones. In addition, too much ozone in the air can easily react with a person's eyes, skin, and soft respiratory membranes, and as a result, damage them.

Portable ozone generators are currently being marketed to purify the air in homes. Many experts point out that while these devices sometimes reduce pollution levels, they are potentially a very dangerous way to clean the air because the ozone itself can impact the occupants in a negative way. Even though some of the portable ozone generators are only designed to emit a minimal amount of ozone, it is possible that in a small room they can produce unhealthy ozone levels. The U.S. Food and Drug Administration allows ozone generators to produce up to 0.05 ppm of ozone, but sensitive people often react to levels as low as 0.001 ppm. Individuals with respiratory problems such as asthma are also more vulnerable.

Officials in at least two states, Minnesota and North Carolina, have expressed serious reservations about the use of ozone generators in residences. In October 1991, the Minnesota attorney general went to court and won a case against an ozone-generator manufacturer under the state's consumer-fraud statutes.(Summary) In 1992 the North Carolina Department of Environment, Health, and Natural Resources issued a bulletin saying

Understanding Ventilation

Figure 15–18. The Scrubber filter package is designed to work in conjunction with an HRV. Different options are available for the individual inserts.

it does "not support exposing people to a toxic air pollutant to reduce concentrations of other air pollutants."(Etheredge)

Ozone generators are available from **Kleen-Air Company, Inc.** that are designed to be permanently mounted in furnace/air-conditioning ducts or in ventilation ducts. With these devices, ozone is generated only when air is flowing through the duct and is diluted quickly. This can be a good way to distribute small amounts of ozone throughout a house. While improper installation can certainly result in excessive indoor levels of ozone, it is possible to carefully size these units so only a minimal amount of ozone is produced. However, the actual ozone concentration in a room will depend on several factors: the volume of the room, the rate ozone is generated, the dilution rate (due to ventilation), and the rate at which ozone reacts with pollutants. With so many variables it can be difficult to control the ozone concentration at all times. Thus, a small bedroom might have a higher concentration than a large living room, and both rooms might experience even higher concentrations when the ventilation exchange rate is low.

One of the best uses of ozone in residences is by contractors who specialize in fire and flood restoration work. In this application, high concentrations of ozone are used to neutralize the odors associated with smoke and water damage. This can only be done when no one is in the house. All in all, as a method of controlling indoor air pollution on a day-to-day basis, ozone is probably not a good idea.

House plants

In the mid 1980s the National Aeronautics and Space Administration (NASA) sponsored some pioneering research that evaluated the use of house plants as air-cleaning devices.(Wolverton) In their early work, spider plants in particular were shown to remove formaldehyde from the air (Figure 15–16).

While NASA's original studies were done in a controlled laboratory setting, more recent research has attempted to determine if the same holds true in residential applications. Unfortunately, it has been found that, while many different indoor plants can indeed reduce pollutant levels in a house, the amount of reduction isn't nearly as significant as NASA's original research implied.(Godish 1986) In fact, it is believed that the reduction is due, not to the plants metabolizing the pollutants, but rather to soil bacteria consuming them.(Godish 1988) Sometimes activated carbon is added to the soil of house plants to store excess air pollutants until the microbes can digest them.

Having several plants in a house can raise the indoor relative humidity, resulting in increased populations of biological pollutants, such as mold or dust mites, that thrive at higher RH levels. For people with sensitivities to these microorganisms who wish to have house plants, cacti or succulents need less water, so they generally don't contribute to a higher RH. While they don't have the potential to reduce indoor pollution, they do add greenery to a house.

Pollution-reduction strategies such as ozone or house plants offer simple solutions to a relatively complex problem—and simple solutions have great appeal to many people. However, complex problems are rarely solved by simple solutions.

Packaged combination filters

A few companies produce residential filtration equipment that combines a high-efficiency filter with an adsorption filter. These units are sold as packaged systems pre-engineered to work in residential applications. Some are designed specifically for use with ventilation systems. Those that are primarily designed for use with central furnace/air-conditioning systems can

Figure 15–19. This HEPA/activated-carbon filter can be installed adjacent to a forced-air furnace/air conditioner in a partial-bypass configuration, or as a stand-alone central filtration system.

Figure 15–20. The manufacturer of this compact HEPA/activated carbon filter uses more powerful fans in its HRV to overcome the resistance of the filter.

be useful when a ventilation system is coupled with the heating/cooling system. Since many of these systems utilize filters that have a considerable amount of resistance to airflow, they generally require more fan energy than simpler, less-efficient filters. Because the operating cost can be significant, it should be estimated accurately during the planning phase.

Allermed Corp. specializes in portable and whole-house air filters for hypersensitive individuals. The company makes a CS-2000 filtration unit that can be used in conjunction with a forced-air furnace/air conditioner or with its own system of ducts as a stand-alone central filtration system. This unit consists of an insulated cabinet, a built-in high-efficiency fan motor, a prefilter, 40 pounds of adsorption material (several different products are available), and either a 90–95% (atmospheric-spot-dust test) or a 99% HEPA (DOP-smoke-penetration test) final filter (Figure 15–17).

Nutech Energy Systems, Inc. offers a Scrubber filter system designed for use in conjunction with an HRV central ventilation system. The standard model consists of a metal cabinet with eight slots containing a 1" prefilter, six 1" thick polyester media filters impregnated with activated carbon (Polysorb), and a 2" pleated filter. As an option, either an electrostatic precipitator or an electrostatic air filter can be substituted for one of the carbon filters, or other more efficient filters can be used. The amount of air resistance depends on the specific filters selected. Because this unit does not contain its own internal fan, the airflow is dependent on the particular ventilation system with which it is used. When coupled with a typical HRV capable of delivering 170 cfm, a Scrubber containing eight filters could have enough resistance to reduce the airflow by about a third (Figure 15–18).

Pure Air Systems, Inc. produces a 600 HEPA Shield that contains a prefilter, a 2" activated carbon filter, and a 99+% HEPA (DOP-smoke-penetration test) filter, all housed in a metal cabinet. Instead of the HEPA, the company's 600 OdorAdsorber contains a 16"-thick activated-carbon (or activated-alumina) filter for optimum gas removal. Either unit can be fitted with a single-speed or an optional two-speed $1/3$-hp. fan capable of delivering 600 cfm. For higher-capacity situations, a 2,000-cfm model is available. Designed primarily for use in conjunction with a forced-air furnace/air conditioner, these filters can also be used as stand-alone whole-house recirculating units, or adapted for use with a fresh-air ventilation system (Figure 15–19).

Raydot, Inc. has a Purepak filter module that can be added to its HRVs to clean the incoming air before it reaches the living space. A Purepak consists of a 14" x

Chapter 15 Air-filtration equipment

14" x 16" insulated metal cabinet with 10" inlet and outlet collars. Inside the cabinet is a Pre-sorber pleated activated-carbon prefilter and a 95% efficient HEPA filter (DOP-smoke-penetration test). The HEPA should last 3–5 years. Raydot's HRV fan is powerful enough to pull air through this filter, but other HRV fans are not, so this filter can't always be adapted to other systems (Figure 15–20).

Therma-Stor Products produces a Filter-Vent that consists of a metal cabinet, a built-in 110-watt fan, a seven-day timer with two-hour intervals, a high-efficiency particulate filter, and a thin adsorption filter. There are two options for the high-efficiency filter: either a 90–95% efficient bag filter or a 98% efficient pleated filter (both are atmospheric-spot-dust ratings). The adsorption filter contains a $1/2$"-thick polyester media impregnated with activated carbon (Polysorb). The Filter Vent is rated at 300 cfm. During operation, a certain amount of the airflow is pulled in from the outdoors (77 cfm of incoming air is standard, but optional amounts of 100 cfm or 124 cfm can be specified), and the remainder of the 300 cfm is recirculated through the house. This unit can be used as either a stand-alone central-supply ventilation system (resulting in slight house pressurization) or in conjunction with another ventilation strategy (Figure 15–21).

The cost of filtration

Filter costs vary widely. To install an electrostatic precipitator, a medium-efficiency filter, or a HEPA filter often runs several hundred dollars. Complicated installations might cost $2,000 or more. The actual installed cost will depend on whether it is a new or existing installation, engineering design requirements, system capacity, labor rates, precise filtering requirements, etc. Relying on readily available medium-efficiency filters, electrostatic precipitators, electrostatic air filters, or the packaged filters mentioned above is generally more cost-effective than creating a complete new filter system from scratch.

The annual media replacement costs also vary widely—from a couple of dollars for a standard furnace filter, $30–40 for a medium-efficiency filter, $40–50 for an electrostatic air filter, up to $100–200 for a HEPA replacement. Raw adsorption material sells in bulk for $1.50–3.00 per pound, and 1" adsorption filters are often in the $50 range.

Electrostatic precipitators have no annual media replacement cost, but because they are electric, have an operating cost of $30–50 per year. If any filter installation requires a more powerful fan (which may or may

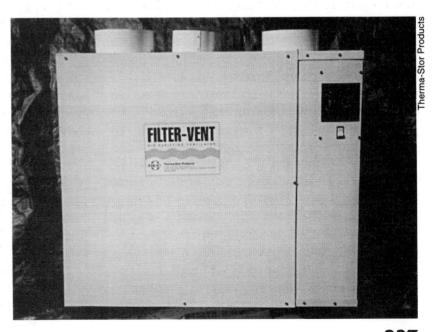

Figure 15–21. The Filter-vent can be used as a stand-alone central-supply filtration system or with another ventilation strategy.

Understanding Ventilation

not be an energy-efficient fan), or if it requires a forced-air furnace/air-conditioning fan to operate continuously, any extra electricity needed to run the fan should be factored in as a part of the operating cost.

Filtration for sensitive occupants

Many sensitive people (see Chapter 4 for a discussion of MCS) are bothered by air filters. Sometimes this is because of captured contaminants that are later released by the filter. But more often it is because, despite the fact that filters are designed to remove pollutants from the air, they can also *generate* tiny amounts of air pollution.

Most particulate filters (*e.g.* extended-surface filters and HEPA filters) are made of fiberglass or polyester fibers held together by some type of synthetic resin. This resin can sometimes outgas enough to bother hypersensitive individuals. The amount of outgassing is extremely small, and isn't considered a problem for most people, but there are individuals who can be bothered by it. In fact, there are numerous cases of well-meaning contractors installing high-efficiency filters for sensitive people, only to have the filter itself become a problem for this very reason.

It is sometimes possible to reduce the outgassing from a media filter by "baking" it on an oven at 200°F for a couple hours. If this is done, be sure the sensitive person is out of the house, the range hood is used, and the kitchen and oven are thoroughly aired out after the baking has been completed. It is a good idea to check with the filter manufacturer before baking a filter because the high temperature can sometimes cause the resin to deteriorate and the filter to fall apart.

Electrostatic precipitators contain no resin, so there is no outgassing, but during normal operation they do generate ozone. In most cases, not enough ozone is produced to worry about, but some sensitive people are bothered by very low levels—levels that are considered safe by many experts. Electrostatic air filters don't generate ozone, but because they are made of plastic, may bother someone sensitive to synthetic materials.

If an adsorption filter is placed downwind of a particulate filter, it can either adsorb the outgassing of the resin from a media filter or catalyze the ozone from an electrostatic precipitator. This is a good idea, but it has its drawbacks. Adsorption materials, such as activated carbon, are generally very dusty and can release this fine dust into the air. They are often placed upwind of a particulate filter so the particulate filter can capture any dust released. If you want to place an adsorption filter downwind of a particulate filter, the dust can be minimized by thoroughly vacuuming the adsorption filter prior to installation.

A variety of adsorption materials are available on the market, and sensitive people sometimes react to one but not another. For example, activated carbon made from coconut husks may be tolerable but that made from coal may not be. Occasionally, an extremely sensitive person will be bothered by any kind of filter.

For most people, filters can improve the quality of the air. However, with sensitive people it is important to determine if they will be affected by the filter itself before anything is installed. The best solution for many sensitive people is to relocate to an area where the outdoor air is of high quality, then rely on a ventilation system without filters to bring clean air indoors. Of course, this will do little to reduce the concentration of airborne pollutants that are released from materials indoors such as fibers from cotton or wool furnishings, and clothing. The best way to effectively deal with these particulates is to use a portable air filter, an air filter coupled with a forced-air heating/cooling system, provide regular house cleaning and vacuuming, or dilute the pollutant concentration with a ventilation system.

Chapter 16

Fresh-air inlets and stale-air outlets

Houses with ventilation systems have *deliberate* openings, called *inlets* and *outlets*, through which air passes between the indoors and the outdoors. Air enters a house through an inlet. Air leaves the house through an outlet. As has been discussed, air also can enter or leave a house through *random* holes in the structure.

An inlet or outlet can directly connect the indoors to the outdoors (*e.g.* through-the-wall vents) or they can be connected to ducts to move air to or from specific locations (see Chapter 17 for a discussion of distribution). Balanced ventilation systems (with or without heat recovery) generally have a single inlet and a single outlet. Central-supply ventilation systems can have one of each, or one inlet and several small outlets in the form of through-the-wall vents. Central-supply ventilation systems can also have one inlet for air to enter a house and, instead of outlets, air could leave through the random holes in the structure. Many central-exhaust ventilation systems have one outlet and several through-the-wall-vents for inlets, but those connected to a forced-air heating/cooling system only have a single inlet. It is also possible for a central-exhaust ventilation system to utilize one deliberate outlet and the random holes in a house instead of deliberate inlets.

Inlets and outlets can be either *active* or *passive*. If a fan pushes or pulls air directly through an inlet or an outlet (*e.g.* the fan and the inlet or outlet are connected together with a duct), the inlet or outlet is active. If air moves through an inlet or outlet because of pressure changes in a house (by any type of pressure: natural, accidental, or controlled), the inlet or outlet is passive. As an example, a window fan actively blows air out of a house through a window (the window is an outlet, a deliberate opening) and at the same time an equal volume of air enters the house passively either through random holes or other deliberate openings. If a house has an active inlet or outlet, and the fan is off, air can move through the opening passively as a result of

Understanding Ventilation

some other pressure. For example, if a clothes dryer is blowing air outdoors, make-up air could enter an inlet or an outlet passively when the ventilation fan that is connected to the inlet or outlet is turned off.

No matter how many inlets a system has, it is important that they be located where the outdoor air is clean so only unpolluted air is brought indoors. It is also important to expel stale air to the outdoors where it won't be blown in someone's face or be pulled back indoors through a nearby inlet.

When outdoor air passes through an inlet, it will either be cooler or warmer than the indoor air—depending on the season—so it will often need to be tempered. Tempering means changing the temperature (or humidity) of the incoming air to more closely match the indoor conditions. If fresh air reaches the occupants before it is tempered, it can feel drafty or uncomfortable.

Deliberate openings vs. random holes

Air can enter and leave a house either by way of unplanned random holes in the structure, or through deliberate openings (inlets or outlets). In most houses, air moves through a combination of random and deliberate openings. As was discussed earlier in Chapters 8–10, air movement through random openings can cause various moisture- or pollution-related problems.

If a house has an exhaust ventilation system, make-up air will enter through any opening it can find when the house becomes depressurized. In a very tightly built house, there is no choice: You must provide deliberate openings, such as through-the-wall vents, so the make-up air can enter. If the house is loosely built and has through-the-wall vents, some make-up air will enter through the inlets and some will enter through the random holes in the structure. It can be difficult to predict exactly how much air will enter the inlets and how much will infiltrate randomly, but it is often proportional. If a third of the openings in the house are through-the-wall vents and two-thirds are random holes, then approximately a third of the air will enter through the inlets and two-thirds will enter randomly.[Heller] Consider a moderately-tight house that has a total of 50 sq. in. of random holes and also has 10 through-the-wall vents, each of which has a 5 sq. in. opening. If this house is uniformly pressurized, approximately half of the air will exfiltrate through the random holes, and half will escape through the deliberate openings. Actually, the precise proportion depends on several factors.

The combined area of all the random holes in a house is often expressed as an Effective Leakage Area or ELA. (In Canada it is called Equivalent Leakage Area. See also Chapter 10.) Predicting precisely how much air will pass through deliberate openings compared to that passing through random holes can get complicated. When you analyze the mathematics of airflow at different pressures, it can be shown that through-the-wall vents often perform differently than the random holes as estimated by an ELA. For example, when you estimate an ELA, you must base the estimate on a house sensing a certain amount of pressure. If the actual pressure on the house is *less* than the pressure used to estimate the ELA, proportionally more air will flow through the vents than through the random holes. If the pressure on the house is *more* than the pressure used to estimate the ELA, proportionally less air will flow through the vents than through the random holes. In other words, if the area of the through-the-wall vents is equal to the ELA, more air will flow through the through-the-wall vents (in fact, up to twice as much) than through the random holes—at the low pressures common in houses.

In spite of the difficulty in predicting where air enters, one rule is certain: If you want to minimize the amount of air infiltrating through random holes in order to ensure that most of it enters through the deliberate inlets, then you must tighten up the house. In other words, you need to seal up the random holes.

The same reasoning holds true for central-supply ventilation systems having through-the-wall vents as outlets. If the house is tightly constructed, most of the air will escape through the vents when the house is pressurized. But if the house is loosely constructed, air will leave through both the deliberate openings (the through-the-wall vents) and the random holes. To maximize the amount of air leaving through the deliberate openings, you must tighten the house.

Because most of the air will pass through random holes, through-the-wall vents are not very effective in anything but a tightly-constructed house.

Consider a house of average tightness with a central-exhaust ventilation system and through-the-wall vents for inlets. If only 20% of the make-up air enters through the inlets and 80% enters through the random holes, you aren't getting much benefit from the inlets. However, small rooms, such as bedrooms, may not have as many random holes as large rooms. Therefore, a small bedroom will often have less make-up air entering than a large living room, even though it needs just as much fresh air when occupied. So, bedrooms sometimes benefit from through-the-wall vents, even in loose houses.

Large houses of average tightness often have so many small random holes in them that if you install through-the-wall vents, a great deal of air will still tend to flow through the random holes. Smaller houses—say those less than 1500 sq. ft.—are inherently tighter because they have fewer square feet of exterior walls, floors, and ceilings. Apartments are often tightly constructed for privacy and sound control, so they often benefit from the use of through-the-wall vents. Everything else being equal, air movement will be more likely to occur through deliberate inlets and outlets in a small house than a large one simply because it is tighter. Of course, the more you tighten up any size house—even a large one—the more effective you will be at forcing air to enter and leave where you want it to.

Inlets and outlets are simply holes in a house

It should be remembered that inlets and outlets are simply holes in a house, and air can move through them in either direction, depending on the indoor/outdoor air-pressure differences. You may plan for an opening to be an outlet, but it may actually become an inlet under some conditions. The direction air moves through an inlet or outlet is affected by natural, accidental, and controlled pressures. If, for whatever reason, the indoor air pressure is greater than the outdoor air pressure in the vicinity of an inlet or outlet, air will tend to exit through that opening. If the indoor air pressure is less than the outdoor air pressure in the vicinity of an inlet or outlet, for whatever reason, air will tend to enter through that opening.

What happens if the ventilation system is shut off?

Air can move through an inlet or outlet passively even when a central ventilation system is turned off—if there are any other pressures acting on the house.

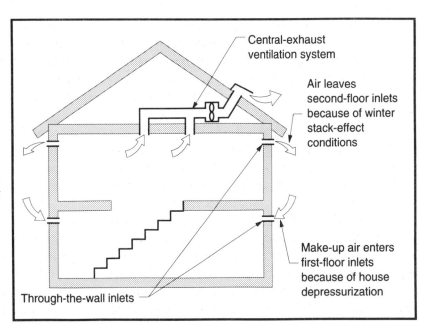

Figure 16–1. Just because a through-the-wall vent is called an inlet doesn't mean it can't function as an outlet under some circumstances. House pressures can easily result in air moving in different directions at different times of the day or year. The situation in this illustration can be minimized if the depressurization resulting from the central-exhaust system is greater than the strongest pressures expected due to stack effect.

Understanding Ventilation

Figure 16–2. Spring-loaded backdraft dampers can be used to prevent air from moving backwards in a duct, but the system's fan must be powerful enough to overcome the strength of the spring.

Because of the many natural, accidental, and controlled pressures a house can experience, this is always a possibility. For example, during the winter, air can leave the house through an inlet located high on a sidewall because of stack effect. Or the through-the-wall vents on the lower level of a two-story house could serve as inlets, while the through-the-wall vents on the upper level act as outlets. An outlet on the windward side of a house can become an inlet whenever the wind is blowing. A clothes dryer can sometimes cause enough depressurization to cause make-up air to enter passively through both inlets and outlets no matter where they are located.

What if the ventilation system is running?

When a ventilation system is running, air will most likely enter through inlets and exit through outlets—but not always. Even though air flowing the wrong way through inlets and outlets is more likely when the ventilation system is off, it can also occur when a ventilation system is in use.

For example, consider a two story house in the winter with a central-exhaust ventilation system and through-the-wall inlets on each level. Even when the exhaust fan is operating, it is possible for air to only enter through the first-floor inlets. This is because the outward pressure due to stack effect on the second floor works against the inward pressure generated by the exhaust fan. The net result can be a certain amount of air leaving through the second-floor inlets (Figure 16–1). Interactions such as these can occur whenever a natural or accidental pressure competes with the pressure generated by a ventilation fan. Such interactions may not cause any particular pressure-related problems, but they can result in a different airflow pattern than you planned.

Central-exhaust and central-supply ventilation systems are often designed to generate more depressurization than typically occurs because of natural pressures. For example, in cold North American climates, a central-exhaust system that generates a 10-Pa. depressurization will be stronger than any normal pressures generated by wind and stack effect.

Minimizing unwanted air movement

We already know unwanted air movement through random holes can be minimized by building a tight house. But what can you do to guarantee that air only enters (and doesn't leave) through inlets, and only exits through outlets? One solution is to equip all the deliberate openings with motorized dampers (either low-voltage or 110-volt) that only open when the ventilation fan is activated. Motorized dampers will prevent air from moving in either direction when they are closed. As an example, you could connect a motorized inlet damper and a local exhaust fan to the same control switch, so that when the control activates the fan, it also opens the inlet damper. If there is more than one exhaust fan in a house (*e.g.* a central-exhaust fan and a local-exhaust fan), they may each need an inlet with a motorized damper, or you could use a single damper and wire it to open when either exhaust fan operates. It is even possible to activate a motorized damper with a differential-pressure transducer so it will open automatically whenever a fan causes a significant pressure change in the house. (Actually, the equipment isn't yet readily available to do this.) These options are only feasible for controlling air movement through a single inlet (or outlet). Because each opening would require a costly motorized damper, they are not suitable for a house having multiple through-the-wall vents. (See also Chapter 14, *Controlling a ventilation system.*)

Backdraft dampers can also help, because they operate like a one-way valve and only let air move in one direction. Many ventilation equipment manufacturers, such as **Artis Metals Co.**, **American Aldes Corp.**, and **Jenn-Aire Co.**, offer inlets and outlets with built-in backdraft dampers. A backdraft damper doesn't prevent all unwanted air movement (as a motorized damper does); it only prevents the air from moving in the wrong direction. In other words air could still enter an inlet passively when something other than the ventilation system causes a negative pressure in the house.

Some backdraft dampers require very little air pressure to open them. Spring-loaded butterfly dampers that mount inside ducts are popular (Figure 16–2), but low-powered fans sometimes don't generate enough pressure to open them.

Dampers must be located where they can be cleaned occasionally because dust and lint can build up and prevent them from operating smoothly. Though it may seem obvious, it is important to make sure backdraft dampers are installed correctly; it is not unusual for them to be mounted backwards. An incorrectly installed backdraft damper may have no air flowing through it when the fan is operating. Dampers are also subject to freezing shut in the winter, especially if humid air is passing through them.

Another way to minimize unwanted air movement through inlets and outlets is to choose a ventilation strategy that has the fewest number of deliberate openings between the indoors and the outdoors. For example, a balanced system only requires two openings in a house (one inlet and one outlet), but a central-exhaust system could have a single outlet and several through-the-wall vents as inlets. The fewer the number of openings, the fewer chances there are of air entering (or leaving) through the wrong opening when the ventilation fan is off.

Actually, air entering an outlet (or leaving through an inlet) isn't necessarily bad. It is only undesirable if it causes the house to become more polluted. For example, if fresh air enters an outlet, it often only means more fresh air for the occupants. However, if air enters an outlet that is located near an outdoor pollution source, such as an idling automobile, it could bring some of the exhaust gases indoors. Similarly, if air enters an outlet that is located near an indoor pollution source such as a gas kitchen range, it could spread that pollution throughout the house. Both of these scenarios can be avoided simply by using backdraft dampers. (Of course, any pollution released by an indoor source should always be exhausted outdoors, or else it will disperse into the rest of the living space even if the inlets are equipped with dampers.)

Inlet and outlet locations

Inlets and outlets are usually visible on the outside of a house. To look less unsightly, they are often located in inconspicuous places. For example, many commercial buildings have fresh-air inlets on the back of the building, often near the loading dock. This can result in delivery-truck exhaust being pulled indoors and distributed throughout the building. It is important that the outside of a house looks uncluttered, but it is more important that inlets and outlets be placed where they won't compromise indoor air quality.

Understanding Ventilation

There are a number of ways to minimize the visual impact of inlets and outlets besides locating them at the rear of a house. For example, some through-the-wall vents are made to mount inconspicuously in window sashes. Inlets and outlets can also be painted to match the color of the siding, or they can be located in soffits or on roofs.

Fresh-air inlet locations

Wherever air enters a house, it will bring with it outdoor pollutants that are in the vicinity of the entry point. While it is not possible to control the locations of random holes in a loosely built house, deliberate inlet openings can be placed where the outdoor air is the

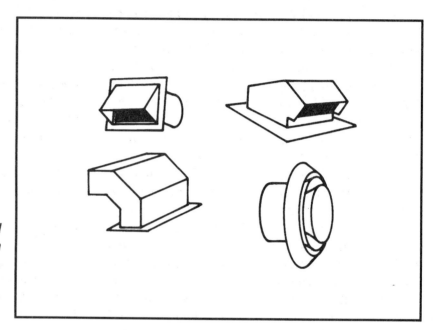

Figure 16–3. A variety of inlets and outlets are available for use on roofs, soffits, and walls.

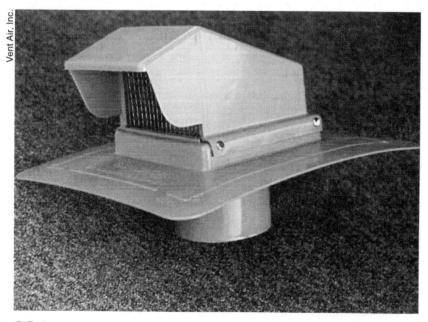

Figure 16–4. This roof outlet has a built-in collar on the bottom to which a duct can easily be attached.

cleanest. For example, a fresh air inlet should not be near a driveway where the exhaust from an automobile might be brought indoors, near a patio where a barbecue is located, or near a plumbing vent.

Fresh-air inlets should not be located near outlets from clothes dryers, central vacuums, bath fans, HRVs, or range hoods. Otherwise polluted air leaving the house will simply be drawn back indoors. Locations near chimneys, garbage cans, animal cages, etc. also should be avoided. Inlets near the ground can be blocked by snow or leaves and can pick up mold spores or lawn-mower exhaust more easily than inlets high on a wall. Inlets on a roof can also be blocked by snow unless they are raised up in the air, or they can allow asphalt odors (from hot fiberglass shingles in the summer) to enter the house. Because wind flowing around the corner of a house causes a certain amount of turbulence that can affect the amount of air entering an inlet, inlets should be located at least three feet from house corners. (Active inlets are less affected by wind turbulence than passive inlets.)

When planning the location of a fresh-air inlet, look around the outside of a house carefully to find a spot where the air is usually the cleanest. If the prevailing wind direction can be determined, inlets should be upwind of outlets to minimize drawing pollutants back indoors. While reintroduction can be minimized, it can't be prevented 100% of the time because non-prevailing winds tat can occasionally cause air from an outlet to move toward an inlet.(Lepage) Ten feet of space between an inlet and an outlet is probably a minimum separation distance, but the best defense against reintroduction is to locate inlets as far as possible from outlets.

It is helpful to locate inlets where they can be easily reached for occasional cleaning. If an inlet becomes partially blocked by leaves, lint, debris, etc., the operation of a ventilation system can be compromised, sometimes severely. It is often a good idea to label an inlet with a statement such as "Fresh-Air Inlet, Keep Clean" as a reminder to future occupants.

Stale-air outlet locations

Stale polluted air leaves a house through outlets, so they should not be placed near a door, porch, or patio where people are likely to be standing or sitting. Because outlets expel both pollutants and moisture from the living space, care must be taken to locate them so the moisture doesn't cause any problems. For example, locating an outlet near a skylight or window should be avoided, because the moist air could fog them at certain times of the year. In a cold climate, an outlet terminating through a roof should be installed with an ice membrane under the shingles to prevent the outgoing moisture from freezing and backing up under the shingles. Also, don't locate an outlet above an electric meter where moisture could create an electrical hazard.

Sometimes local-exhaust fans are installed so that they simply blow stale air into an attic with the assumption that the air will eventually make it outdoors by passing through roof vents. This is a mistake. An exhaust duct can be passed through an attic and connected to a roof-mounted outlet, but you should never blow exhaust air directly into an unheated attic, crawl space, or basement (especially in a cold climate) because the moisture in the air can cause serious condensation problems.

Other considerations

Inlets and outlets are often covered with screen, wire mesh, or louvers to prevent birds or rodents from entering the house. Sometimes it is tempting to select a fine screen to keep out insects. However, fine screen is easily clogged with lint, cottonwood fiber, leaves, etc. Screens with openings less than $1/8"$ are occasionally accidentally painted over—in effect sealing them shut. Screen or mesh with $1/4"$ holes is usually a good compromise. Keep in mind that any restrictions can partially block the airflow to a certain extent. To minimize this, select inlets and outlets based on their *net free area* (nfa). This refers to the area that allows unobstructed flow. Most good-quality inlets and outlets have such a rating. However, this is generally only important when designing a ventilation system from components of different brands because manufacturers who supply inlets and outlets with their ventilation equipment generally take such restrictions into account.

Aluminum or plastic inlets and outlets will be less likely to rust than steel ones. Typical products on the market include wall caps, roof caps, soffit caps (Figure 16–3). These can often be purchased through local building-supply stores, but they are also available from

Understanding Ventilation

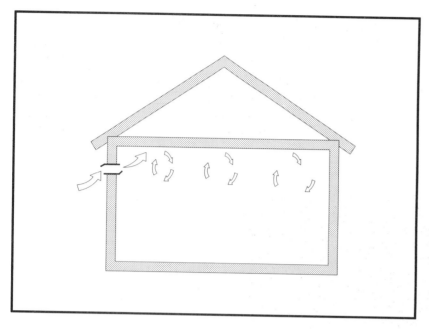

Figure 16–5. When outdoor air enters a well-designed through-the-wall vent, it will mix with room air above the headspace before entering the occupied zone.

most ventilation-equipment suppliers. In addition, ventilation-equipment suppliers often have specialty products such as round grilles and through-the-wall vents. The Roof-Cap Vent by **Vent Air, Inc.** works very well as a roof-mounted outlet. It is available in 4–6" sizes and has a round collar on the bottom to which a duct can easily be attached (Figure 16–4).

Tempering the incoming air

In the winter, incoming fresh air can be quite cold. In the summer it can feel hot and humid. Either situation can be a source of discomfort, but it is in cold climates where complaints are most likely to occur. As was stated earlier, tempering involves changing the temperature (or humidity) of the incoming air to more closely match indoor conditions. In mild climates where outdoor temperatures aren't severe, tempering is often not very important.

If a ventilation system is carefully designed and laid out, occupants rarely complain of uncomfortable drafts. The key can be as simple as mixing the incoming air quickly with the more temperate house air. For example, with a central-exhaust ventilation system, if outdoor air enters by way of through-the-wall vents, the incoming air will usually mix quickly with room air above the heads of occupants. With an HRV, occupants rarely complain of drafts because the incoming air is preheated by the exhaust air inside the core of the HRV.

Tempering the incoming air is also important to prevent damage to a furnace. If too much cold outdoor air is brought directly into a forced-air furnace through a fresh air duct, it can damage inside the furnace. For example, the sudden shock of cold air on a hot surface can cause a heat exchanger to crack. Or if the humidity and temperature conditions are right, moisture can condense out of the air and lead to corrosion. Some reports recommend that air reaching a furnace's heat exchanger be at least 15°C (55°F) to minimize damage.(Geddes, Union, et al.) For damage to occur, a great deal of very cold air generally must come in contact with the furnace's heat exchanger. Fortunately, in many climates where outdoor temperatures aren't severely cold, the incoming air mixes thoroughly with the warmer air inside the ducts before entering the furnace itself, thus avoiding a sudden shock of cold air to the furnace. Also, damage is unlikely if only a modest amount of cold air is brought in.

Over the years, a number of methods have been employed to temper incoming air. Some work very well, but others have potentially serious health-related draw-

Chapter 16 Fresh-air inlets and stale-air outlets

backs. The two most popular tempering methods are through-the-wall vents and HRVs.

Through-the-wall vents

Using through-the-wall vents with a central-exhaust ventilator is a good way for outdoor air to passively enter directly into the living space. When properly designed and located, they cause the outdoor air to mix quickly with the indoor air before it reaches the occupied zone (Figure 16–5). This prevents the occupants from sensing sudden temperature or humidity changes. Still, through-the-wall vents work best in milder climates. As outdoor temperatures get below

Figure 16–6. An electric duct heater can be used to warm incoming air in the winter.

Figure 16–7. To prevent cold incoming outdoor air from damaging the furnace in the winter, the cold air can be mixed with a small amount of warm air from the supply side of the furnace to warm it up. However, this approach requires careful measuring of airflows and pressures to work well.

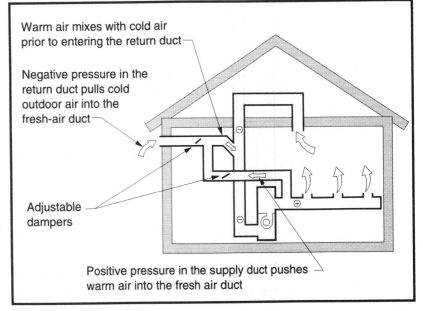

237

freezing, the occupants will be more likely to notice a temperature change when near such vents. In extremely cold climates, it is sometimes desirable to employ a method of preheating the cold air before it passes through a vent and enters the occupied space. In Sweden, where outdoor temperatures can be quite cold in the winter, through-the-wall vents are sometimes located low on a wall behind radiators so that the incoming air will be warmed by the heaters before it reaches the occupied space.

Through-the-wall vents manufactured for use with residential ventilation systems are relatively small. They are not designed to allow a great deal of air to enter. This is so that, in the winter when the outdoor air is cold, the occupants do not feel a chilly draft when they are near an inlet. Most people would complain if 100 cfm at 30°F entered their living room through a single inlet, but 20 cfm entering through each of five different inlets in separate rooms is rarely noticed, especially if the inlets are located high on the walls. The same total amount of air is entering the house, but it is distributed into several rooms so there are no cold drafts.

Through-the-wall vents are most effective in small, tight houses. They are especially suitable in apartments that have been tightly constructed for sound control and privacy. Loose houses simply have too many random holes for air to pass through. If a house is very loose and also has through-the-wall vents, most of the air will pass through random holes. (See *Specific openings vs. random holes* above.) In a tight house with through-the-wall vents, the air will enter through the vents because there are no other pathways.

Because they are passive openings, through-the-wall vents work best when a ventilation system is designed to run continuously. Otherwise, when the ventilation system is off, air will move through them anyway—perhaps when you don't want it to. Some through-the-wall vents can be closed manually, but this

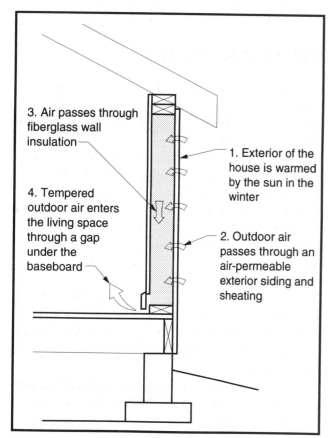

Figure 16–8. *A south-facing permeable wall can warm incoming air in the winter, but there is concern about the negative health effects of breathing bits of fiberglass insulation. A permeable wall is impossible to clean.*

Chapter 16 Fresh-air inlets and stale-air outlets

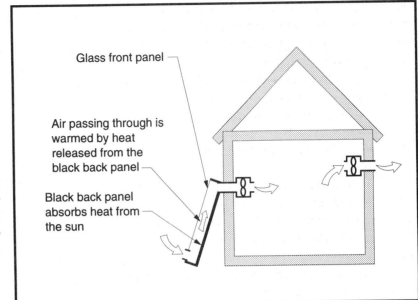

Figure 16-9. A solar panel can be used to preheat cold incoming air in the winter. In the summer, it can be covered up or temporarily disconnected.

can become tedious. A continuously running system ensures a uniform rate of air movement at all times.

Sources of through-the-wall vents include **Airex, American Aldes Corp., Broan Mfg. Co., Inc., Therma-Stor Products, Thinking Vents, Inc., Titon, Inc.,** and **Vent-Axia, Inc.** See Chapter 19 for descriptions of these products.

Tempering with an HRV

An HRV can be one of the best ways to temper incoming air, especially in harsh climates, because a good percentage of the heat that would normally be exhausted in the winter can be captured within a heat-recovery core, then the core can warm the incoming cold fresh air. In hot, humid climates the incoming air can be both cooled and dehumidified with some models. HRVs work very well at tempering the incoming fresh air for occupant comfort, energy conservation, and protection of the furnace from damage.

Because most HRVs are balanced ventilation systems, they offer the advantage of having two fans: a fresh-air fan and a stale-air fan. This makes it possible to distribute the air easily to various rooms.

While an HRV can temper air in either the summer or in the winter, it is usually more cost-effective in the winter in cold climates (see Chapter 13). The major disadvantage of an HRV is its high capital cost compared to some of the other tempering methods.

Electric duct heaters

If outdoor air enters through a single fresh-air duct, an electric-resistance heater can be placed inside the duct to warm the incoming air (Figure 16-6). These consist of an electric heating coil (2-3 kilowatts is typical) permanently mounted inside the duct. The heater is often controlled by the house's main thermostat so the duct heater comes on when the inside of the house is cold. A better control method would be to place a thermostat inside the duct where it can sense the outdoor temperature. In this way the thermostat can be adjusted to turn the heater on only when the outdoor temperature is low.

To prevent a fire hazard, it is important to install an electric duct heater so it only comes on when a fan is blowing air through the duct. In other words, electric duct heaters should only be used with active inlets, not passive inlets.

Electric duct heaters typically raise the temperature of the incoming air 35-50°F whenever they are on, but they can be difficult to control accurately; they of-

Understanding Ventilation

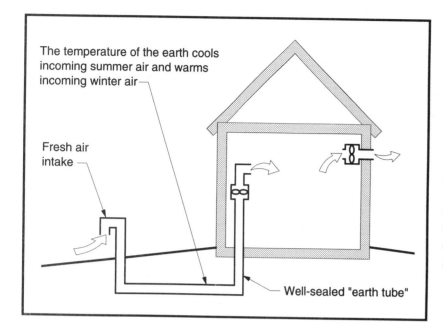

Figure 16–10. An earth tube can warm incoming air in the winter and cool incoming air in the summer—but it can also become a breeding ground for microorganisms.

ten either overheat or underheat the incoming air, depending on the outdoor temperature. Because they are simply electric heaters, they can be expensive to operate, especially if they are operating for long periods of time. HRVs are often somewhat more cost-effective in the long run.

Duct heaters are available from many different manufacturers of ventilation equipment, including **Broan Mfg. Co., Inc., Conservation Energy Systems, Inc., Nutech Energy Systems, Inc.,** and **Reversomatic Htg. & Mfg. Ltd.**

Hydronic duct heaters

Sometimes, a water coil or radiator that contains recirculating hot water is placed inside a duct to warm the incoming air. A thermostat is can be used as a control to activate a pump which will circulate water between a hot water storage tank or central boiler and the radiator. This makes it a more complex and costly installation when compared to an electric duct heater. However, if the fuel used to heat the water is inexpensive (*e.g.* natural gas), this may be cheaper to operate than an electric heater.

Keep in mind the fact that the connecting pipes could be prone to freezing if they are run through unheated crawl spaces or attics. In addition, if the system that heats the water is ever shut off (or is temporarily broken), cold incoming winter ventilation air could freeze the pipes. A safeguard against freezing would be to use an antifreeze-filled loop of piping between the hot water storage tank and the ventilation system, or use electric heat tape on the pipes that will turn on under freezing conditions.

Mixing with warm air

If you mix incoming cold air with warm air quickly enough, it will be tempered before it reaches the occupants. This is basically what through-the-wall vents do. But mixing can also be done with a fresh air duct connected to a forced-air furnace. **Air-King Ltd.** has a Fresh-air 970 thermostatically controlled damper that does just that. It allows a small amount of heated air from the supply side of the furnace to mix with the cold incoming fresh air before it enters the return side of the furnace (Figure 16–7). This can be useful when cold fresh air is brought directly from the outdoors into the furnace system. The Air-King system only tempers the air when the furnace is putting out heat, so you don't get any outside air unless the furnace is running. When no heat is needed (*e.g.* when the fan is running, but the thermostat isn't calling for heat), the damper remains closed. **Skuttle Mfg. Co.** has a similar product called a Make-Up Air Control.

Environment Air Ltd., **Nutech Energy, Inc.**, and **Venmar Ventilation, Inc.** use this mixing approach to tempering with their non-heat-recovery ventilators. They all mix a certain amount of house air with the incoming air to temper it before it is distributed into the living space. See Chapters 19 and 22.

Permeable walls

Permeable walls have been used experimentally in a few demonstration houses in cold climates.(Canata 1992) This approach utilizes a central-exhaust (depressurization) ventilation system and allows make-up air to pass through the structure of the insulated walls. Air is pulled through a permeable sheathing, through fiberglass insulation (which acts as a filter to remove pollen and mold spores), then the make-up air enters the living space through a narrow gap under the baseboard. As the air passes through the wall, it is slowly warmed by the heat escaping through the wall to the outdoors (Figure 16–8). Solar radiation falling on a south-facing wall can also significantly raise the temperature of the incoming air.

A permeable wall will do the best job of warming incoming winter air when it faces south. It is less practical for a north-facing wall. Because outdoor air only enters the rooms that are adjacent to the permeable wall, there will likely be some rooms that don't receive enough incoming air (unless there is an air circulating system in the house). Therefore, a permeable wall is usually not very effective in distributing air throughout a house.

This approach is sometimes called a *dynamic wall* and has some drawbacks as far as indoor air quality is concerned. Fiberglass insulation can contaminate the incoming air with either tiny fibers of glass or formaldehyde from the resin binding the fibers together. Proponents say that the air moves much too slowly for fibers to be pulled loose and enter the house, and the resin isn't a major outgassing source. But fiberglass insulation installers are urged to wear respiratory protection when working with this product, and it makes little sense to deliberately expose occupants to it on a regular basis. With the insulation acting like a giant filter, it will eventually become contaminated with outdoor pollutants which could eventually be released into the house. (Of course, this is exactly what happens when air infiltrates into a house through random openings.) A permeable wall is virtually impossible to clean.

Solar preheaters

A solar preheater consists of a solar panel through which incoming air passes before entering the house. A simple unit is shown in Figure 16–9. Air moves between the black panel and the glass cover, then enters the house. The black panel is warmed by the sun, so it tends to warm any air flowing over it—as long as the sun is out. A solar preheater will not warm incoming air at night. The panel must be constructed of inert materials so it doesn't contaminate the incoming air (*e.g.* black paint might outgas when warmed by the sun). The panel must also be constructed so it can be opened periodically for cleaning.

Conserval Systems, Inc. offers a Solarwall system that is designed to mount on the south side of a house where it will be warmed by the sun. Incoming ventilation air is warmed as it passes through the Solarwall prior to entering the living space. This is an all-metal system that uses no glass or plastic.

Earth tubes

Bonneville Power Administration used an *earth tube* in some demonstration houses in 1992 to temper the fresh air entering a house(Earthtubes) (Figure 16–10). The idea was to use the temperature of the earth itself to raise the temperature of the incoming air in the winter. During most of the year, the ground temperature several feet below the surface is about 50°F. By pulling cold outdoor air through a buried tube in the winter, the air will be warmed up before it enters the house. The system apparently does warm the incoming air. In some of Bonneville's installations, when the outdoor air temperature was 39°F, the air entering the house was 48°F. This approach can also be used to cool incoming air in the summer.

There are a number of disadvantages to earth tubes. Because a more powerful fan is needed to pull

Understanding Ventilation

the air through a long tube than to simply pull the air directly from the outdoors, electricity costs can be slightly higher. In the summer the inside of the tube will be cool, so hot humid air entering the house can condense inside it. Because of condensation, there is the possibility for microbial growth inside the tube. In fact, earth tubes have the potential to become seriously contaminated with microorganisms. Occasionally, they become inhabited by small rodents and snakes. If the joints between sections of the tube aren't well-sealed, radon or other soil gases could be pulled indoors.

When an earth tube gets contaminated, it must be thoroughly cleaned—which sometimes can be difficult—or sealed and abandoned.

Chapter 17

Distribution: moving air around a house

Distribution involves moving air around in a house: moving air from one room to another. There are two different types of air-movement systems that typically do this: forced-air heating/cooling systems and ventilation systems—and both have different functions. Forced-air heating/cooling systems are designed to maintain a uniform temperature throughout the living space. Ventilation systems are designed to provide clean, fresh air to occupants. Because of their different functions, these two systems should be designed with different criteria in mind. However, the differing functions can sometimes be combined into a single system. This chapter deals primarily with ventilation distribution systems, but it will also refer to forced-air heating/cooling distribution systems.

To distribute ventilation air to the best advantage means: 1) removing stale air as close to a pollution source as possible so it doesn't contaminate the rest of the house, and 2) supplying fresh air to those parts of the house where the occupants are most likely to be located. When this is done well, the distribution system is said to be *effective*. An effective system needs less capacity than an ineffective system, so capital and operating costs will be less.

To distribute air in a house (either ventilation air or heating/cooling air), a difference in air pressure is necessary to push the individual air molecules along. When air is distributed through ducts, a fan actively creates the air-pressure difference that causes the air to move through the ducts. After leaving a duct, air usually passes through a *supply grille* as it enters a room. (Air leaves a room through an *exhaust grille*.) Grilles are often decorative, but their primary function is to direct the flow of air.

Air can also be distributed passively from one room to another without ducts by changing the air pressure between rooms. For example, the central-exhaust ventilation system shown in Figure 17–1 will create a

Understanding Ventilation

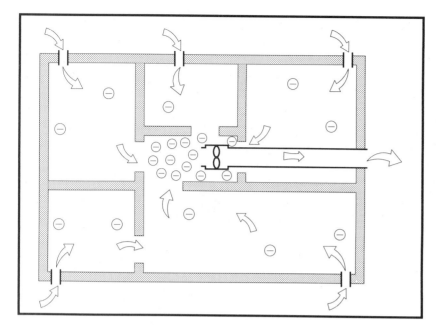

Figure 17-1. A centrally located exhaust fan will depressurize a tight house so make-up air will enter through-the-wall vents.

low-pressure area in the central hallway. Make-up air will enter the house passively by means of through-the-wall vents (or random holes) because the whole house is under a negative pressure. It will mix with stale room air, then the mixture of fresh/stale air will passively move into the low-pressure area in the hallway where it will be actively blown outdoors. Most ventilation systems use a combination of active and passive air movement to distribute the air in a house.

In many cases, ventilation air is not supplied to *and* exhausted from each room. If it is supplied to a room, it is usually allowed to find its way to an exhaust grille in another room by passing through rooms, halls, under doors, etc.

It was pointed out in Chapter 6 that closed doors can cause rooms to become pressurized or depressurized. This is more pronounced with forced-air heating/cooling systems than with ventilation systems, because ventilation systems generally move much less air. (Most residential ventilation systems are typically rated at less than 200 cfm, but heating/cooling systems usually move 900–2,000 cfm.) Therefore, undercutting doors by $1/2$" to 1" is usually enough to relieve the imbalanced pressures created by a ventilation system when doors are closed. If a high-volume ventilation system is installed, problems related to either pressurization or depressurization can be avoided by using one of the methods described in Chapter 6.

Ventilation effectiveness

Ventilation effectiveness has to do with how well a system removes stale air from where pollutants are produced and how well it introduces fresh air where people need it. When a supply grille is located next to a person's head, a ventilation system will obviously be more effective at providing that person with fresh air than if the grille is on the other side of the room. When an exhaust grille is located very close to a toilet, the ventilation system will be effective at removing odors. The more effective a ventilation system is, the less capacity it needs; therefore, it will be less costly to install and operate.

Category A rooms and category B rooms

To determine which parts of a house require exhaust grilles and which require supply grilles, the rooms are sometimes grouped into two categories. Category

A rooms are considered general-occupancy rooms. They include bedrooms, living rooms, family rooms, dens, etc. These are rooms that are most likely to have people in them. Category B rooms are those that have regular moisture or pollution sources—bathrooms, laundry rooms, kitchens, home workshops, home offices, etc. Category B rooms are often called service rooms.

It is generally more effective to supply air to a Category A room and exhaust air from a Category B room. In fact, many Category B rooms have a local-exhaust fan for ridding them of pollutants quickly, as well as an exhaust grille for removing pollutants slowly through a general ventilation system.

Fresh-air supply grille locations

With all ventilation systems, it is important to deliver fresh air to where the people are located. In office buildings, where the occupants are all adults who are seated most of the time, fresh air is often delivered to the headspace—about four feet off the floor. In houses, the headspace can vary from a few inches off the floor for a child, to four feet up for a seated person, to five or six feet high for a standing adult. An effective ventilation system should cause the incoming air to thoroughly mix and dilute the air in all Category A rooms so that it will reach people at all these heights. Good mixing is only important above a height of six feet if there are very tall people in a house, so this is the part of a room where air is often introduced and where the initial mixing (and tempering) takes place. Once the mixed air reaches the lower parts of the room, it should be adequately tempered and distributed. Mixing is enhanced by carefully selecting and locating the supply grilles in a room. The precise path and speed the air takes when it leaves a supply grille is determined by the design of the grille, the opening size, and the air pressure behind the grille.

The criteria for locating heating/cooling grilles is different than for ventilation grilles. For example, furnace supply registers are often located in the floor and directed to blow warm air up onto a cold window (condensation will be minimized on a warm window). There is no advantage to locating a ventilation grille near a window because the air should be introduced for freshness—not to warm the window. Ventilation supply grilles should always be located where they will throw fresh air into a room for best overall mixing.

Even after tempering in a cold climate, a ventilation system will often bring air into a room that is cool. When cold air is brought directly into a room, it is usually a good idea to introduce it higher rather than lower. This is because of cold air's natural tendency to fall. With central-exhaust ventilation systems, through-the-wall vents are often placed close to the ceiling. This gives cold incoming winter air a better chance to mix with the temperate room air. If cold ventilation air is introduced low into a room, say through a floor grille, it can hang close to the floor resulting in occupants complaining of cold drafts on their feet and legs. Cool air that enters a room low simply does not mix very well with the room air. It is a fine way to ventilate feet, but our heads are in greater need of fresh air, and we judge the air quality in room with our nose, not our toes!

In hot, humid climates, the incoming ventilation air is warmer than the air in the house, so supply grilles might be located lower on a wall. This allows the warm air's natural tendency to rise to enhance mixing with the air in the room. However, care should be taken when locating ventilation supply grilles low on walls in air-conditioning climates because the grilles can be easily blocked by furniture. Also, even warm drafts can be felt. Through-the-wall vents located low on a wall can minimize these drafts if they direct the incoming air upward, rather than out into the room. In reality, if supply grilles are designed to throw air far into a room for good mixing, they can often be located on or near the ceiling in all climates.

Stale-air exhaust grille locations

Bathrooms, laundry rooms, and kitchens are good rooms from which to remove stale air because of the high humidities and odors found there. If a house has a home office, it can also be a good idea to locate an exhaust grille near a copy machine or laser printer because both generate ozone. A hobby room or artist's studio would benefit from an exhaust grille located near a workbench or painting easel.

For a ventilation system to be truly effective, a stale-air grille should not only be located in the pollution-prone room, but it also should be as near the pollu-

Understanding Ventilation

tion source as possible. For example, a range hood above a stove will be more effective than an exhaust fan located on the opposite side of the room. A bathroom exhaust fan mounted above a shower will be more effective at removing moisture than if it were mounted above the toilet. But a fan above the toilet will be more effective at removing odors than one over a shower, so compromises must often be made. (In some recreational vehicles the toilet and shower are in the same compartment so a single fan can serve both quite well, but it is doubtful if this approach would be acceptable in houses.) The point is, by locating an exhaust grille near a pollution source, mixing of pollutants with the air in the room is minimized. When this is done, less ventilation capacity is necessary because pollutants are only pulled from a small area rather than an entire room.

In general, it is often a good idea to locate exhaust grilles near the ceiling because odors and moisture are often generated by heat, and they tend to rise. For example, cooking odors are warm and steam from

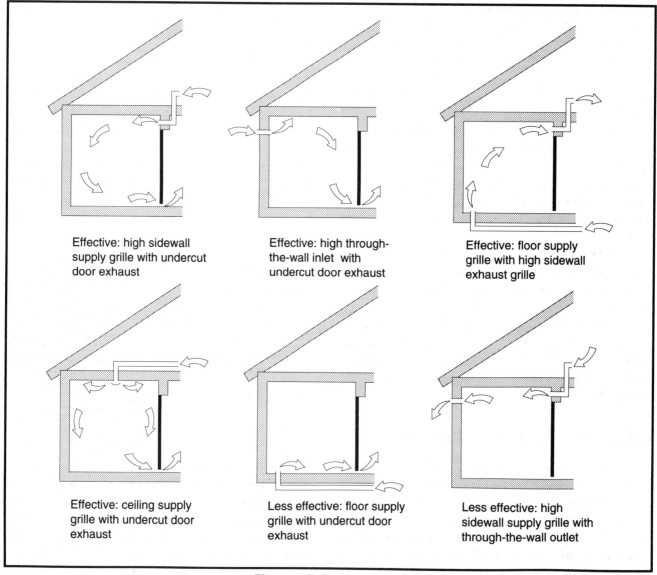

Figure 17–2. Examples of effective air distribution within a room.

a shower is warm. Body odors and exhaled breath are also warm—approximately 98.6°F. On the other hand, particulates such as house dust and skin flakes tend to settle toward the floor. Ventilation systems don't do a very good job of keeping particulates suspended in the air because they don't move that great a volume of air. Forced-air heating/cooling systems, on the other hand, move much more air than ventilation systems, so they can keep particulates suspended long enough to be removed by an air filter.

Short-circuiting of air within a room

When fresh air enters a room, good mixing prevents pockets of stale stagnant air. Without good mixing, the fresh air will take the shortest path through the room. This is often called *short-circuiting*. As an example, fresh air entering a room through a supply grille near the door may leave the room quickly through the doorway before it has a chance to mix with the room air. If this happens, the air in the rest of the room can become stale. The best mixing in a room can often be achieved by having the air move diagonally through the occupied space. Still, the type and size of grille, its location, the air pressure behind the grille all affect the distribution pattern in a room. Figure 17–2 shows some examples of air distribution.

If you are on a limited budget and can only afford to install a simplified ventilation duct system in a house, there are ways to maximize the system's effectiveness. For example, when there is only one supply grille and one exhaust grille in a house, they should be located at opposite ends of a house. In that way the ventilation air must travel a long distance, mixing along the way, as it moves from the supply grille to the exhaust grille.

Effective ventilation layouts

A ventilation system can be laid out in a wide variety of ways to move air through a house. In many cases, grille locations are selected to simplify the duct system, not for optimum effectiveness. For example, a complex floor plan may require a complicated distribution system, but a simple duct layout is chosen in order to save on costs. While the most effective layout might require a complex layout with several supply grilles in Category A rooms and several exhaust grilles in Category B rooms, simpler approaches are often possible.

There are instances where a single ventilation exhaust grille is desirable with a central ventilation system. Perhaps you have one particularly polluted room in a house, such as an artist's studio where oil paints are used. It might be a good idea to exhaust more air from that particular room (or even have the house's only exhaust grille in that room), and to have multiple supply grilles in other rooms. This layout provides fresh air to most rooms, and a flow pattern that moves air from less-polluted rooms, through the house, into the more-polluted room, then outdoors. This minimizes the chance of air from the contaminated room getting into the less contaminated rooms.

Another approach would be to place a central-ventilator's exhaust grilles in every room but have only one supply grille in the center of the house. This will result in a flow pattern that will deposit fresh air in one location, move air into the other rooms, then outdoors. This will prevent the rooms with the exhaust grilles from contaminating each other, and because the fresh air enters in one place, it can be easily filtered. A variation would be to place the exhaust grilles inside closets. After air enters the house through the supply grille, it will pass into the various rooms, then into the closets, then outdoors. This works well for sensitive people because it allows them to store potentially offensive products in the closets without having odors migrate out of them into the rest of the house. If this is done, be sure to allow a way for air to enter the closet when the door is shut (*i.e.* undercut the door or use a louvered door). If a closet contains relatively benign materials, such as clean linens or clothing, it is possible to keep the closet fresh by supplying it with air. This works well when a closet is in a Category A room. For example, if fresh air is introduced into a bedroom closet, it will then pass through the bedroom and eventually reach a Category B room having an exhaust grille.

There are two ways a house's ventilation system can be used to deal with a basement that is contaminated with pollutants such as radon or lawn chemicals. You could put extra exhaust grilles, but no supply grilles, in the basement so that fresh air enters the upper floor of the house first, then passes into the basement (per-

Understanding Ventilation

haps through a louvered basement door), then to the outdoors. The drawback to doing this is that the basement might become depressurized enough to cause additional soil gases to be pulled in. If you do just the opposite—place extra supply grilles in the basement and only have exhaust grilles upstairs—the basement may be pressurized enough to keep the pollutants out. Fresh air will flow into such a basement through the supply grilles, then it will move upstairs (picking up contaminants along the way), and finally return outdoors through the exhaust grilles. A better solution to dealing with soil gases is to either tighten the basement so they don't have a pathway to enter, or to install a separate fan specifically for radon removal. Actually, it is doubtful if a low-volume ventilation system can either pressurize or depressurize a house significantly, so if you plan to deliberately pressurize or depressurize all or part of the house, it is easier to do so with a higher-volume forced-air heating/cooling system.

When you combine a heating/cooling system with a ventilation system by introducing ventilation air into a forced-air furnace/air-conditioner's ducts, the ventilation air will take whatever path the larger system's ducts follow (as long as the furnace/air-conditioner fan is operating). With most forced-air heating/cooling systems, there is typically a supply grille in every room and one or more return grilles in the center of the house. Because these systems move 5–10 as much air as most ventilation systems, there is usually a sufficient volume of air moving through the house so that very good mixing and dilution takes place. Thus, this can be a very effective way of moving ventilation air through a house. The registers, in effect, have two functions: to supply both ventilation air and heated/cooled air.

There is no ideal ventilation-system layout for all houses. The design of a particular system will depend on the plan of the house, the occupants' sensitivities, the ventilation strategy chosen (central supply, central exhaust, or balanced), whether the system will be coupled with a forced-air heating/cooling system, and whether or not some rooms typically have more contamination than others.

Displacement ventilation

Most ventilation systems mix and dilute fresh outdoor air and stale indoor air during the ventilation process. Though used only in specialized situations, there is another, more effective way to move air through a structure—*displacement ventilation*. This approach is used when it is very important to remove as many air contaminants as possible, such as in surgical operating

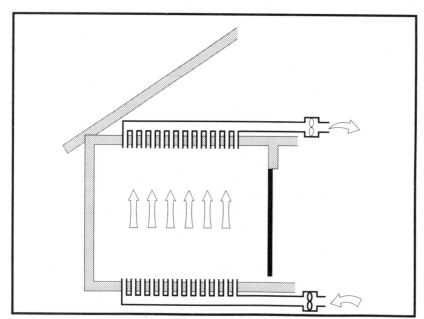

Figure 17–3. Displacement ventilation is very effective, but it isn't practical in residences.

Chapter 17 Distribution: moving air around a house

Figure 17–4. Round plastic supply and exhaust grilles are often used with residential ventilation systems.

rooms. With displacement ventilation, a great deal of fresh air (3–4 times as much as is common in houses) is introduced in such a way that is pushes all the stale air out of the way. All the stale air is then removed with equal effectiveness. One of the ways this is done is by having the ceiling and floor of a room perforated. Air enters the room through every square foot of the ceiling and exits through every square foot of floor (or vice versa). In effect, the entire ceiling is a giant supply grille and the entire floor a large exhaust grille (or vice versa). All the stale air in the room gets displaced by the incoming fresh air (Figure 17–3). Unfortunately, displacement ventilation is generally not suitable for use in residences because of the expense and technical design constraints involved.

Grilles

A grille is a decorative metal, plastic, or wooden grating, lattice, or screen through which air enters or leaves a room. In Sweden, grilles are called *valves* because they control the volume of air passing through them. Air enters a room through a supply grille and air leaves a room through an exhaust grille. When air is distributed through ducts, a grille is mounted on the wall or ceiling and the duct is connected to the back side of the grille. Through-the-wall vents have a built-in grille on the side facing the interior of the house. Many local-exhaust fans have built-in grilles. Grilles can have several purposes: They can improve the appearance of what would otherwise be an unattractive hole in the wall or ceiling, they can direct the air in a certain direction, they can cause better mixing of air, and they can restrict the flow of air.

Not all grilles are created equal. For example, some grilles are adjustable, some aren't. Furnace and air-conditioning grilles and registers work fine for high-volume airflows, but they don't perform nearly as well with low-volume ventilation systems. Many ventilation-equipment suppliers sell attractive round, adjustable plastic grilles that are designed specifically for lower airflows (Figure 17–4). Most suppliers offer one style of grille that can be used for different purposes. **Nutech Energy Systems, Inc.** calls its grilles Techgrilles; **Fantech, Inc.** calls its Hush Grilles.

Eneready Products Ltd. offers three different Whisper Grilles: two supply grilles and one exhaust grille. Figure 17–5 shows how a Whisper ceiling supply grille has a wide throw to move the air to all parts of a room and a Whisper wall supply grille has more of a straight throw to move air far into a room.

Understanding Ventilation

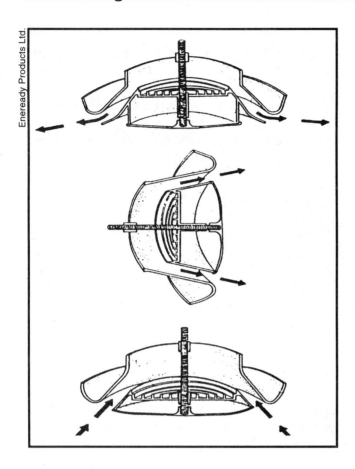

Figure 17–5. Some round plastic grilles are specifically designed for certain applications. This illustration shows a ceiling supply grille (top), a wall supply grille (center), and an exhaust grille (bottom).

Supply grilles are more correctly called diffusers because they tend to diffuse the air into the room air. This minimizes drafts and improves effectiveness by dispersing the fresh air into the headspace. In order for a supply grille to be more effective at dispersing air into a room, it is often sized slightly smaller than normal and coupled with a larger-than-normal duct. When this is done, there is less air resistance in the duct and more at the grille, so the total combined resistance is the same, but performance is improved. During operation, the air in the duct has a lower velocity, then it speeds up as it goes through the grille. This causes the air to be thrown further into the room for better mixing. With this approach, you should not blow the air into a room where it will be felt by the occupants. However, introducing high-velocity air near the ceiling, above the headspace, usually works very well.

Motorized grilles aren't commonly found in residential ventilation systems, but they can be very useful to periodically vary the amount of air flowing to or from a room. For example, the PoshTimer and Whisper POWERGrill from **Eneready Products, Ltd.** is a timer and motorized grille designed to work as a system (See also Chapter 14). These products were developed so a conventional central ventilation system could be used to temporarily provide extra capacity to one individual room. In other words, a central system provides general background ventilation throughout the house, and with activation of the PoshTimer, that same system will temporarily divert most of its capacity to the room served by the companion POWERGrill. This system was primarily designed to be used to boost the amount of exhaust air leaving a room, but it can also be used to boost the amount of fresh air entering a room. When you turn on a PoshTimer, it causes two things to happen: The POWERGrill in that room opens wide and the central ventilation fan is switched to high speed. Most of the capacity of the fan is now pulled through the POWERGrill because it is the path of least resistance. After 28 minutes, the timer shuts off, the central ventilator returns to low speed, and the POWERGrill closes down to its manually preset condition. The grille doesn't shut completely, so there is still some general ventilation for that room. By locating a PoshTimer and POWERGrill in each bathroom, you can eliminate conventional bathroom-exhaust fans. In effect, you have a single central ventilation system that can also function as one or more local-exhaust fans. PoshTimers and POWERGrills can also be used in laundry or hobby areas, or near a smoker's chair, to provide extra ventilation on a temporary basis.

Airex has an exhaust grille that opens and closes when their Opto X infrared motion sensor detects motion in a room. The extent to which it opens and closes

Figure 17–6. Manual dampers are used inside ducts so you can adjust the airflows when a ventilation system is first started up.

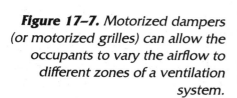

Figure 17–7. Motorized dampers (or motorized grilles) can allow the occupants to vary the airflow to different zones of a ventilation system.

Understanding Ventilation

can be adjusted to different occupancy levels. With this system, there is automatically more ventilation capacity in a room whenever the room is occupied. The Opto X was developed primarily for use in offices to increase the amount of ventilation when an office is occupied.

Trol-A-Temp also offers motorized grilles in 6–14" sizes and in round and square styles; however, these are designed primarily for higher-volume heating/cooling systems. They do not provide very good mixing with low-volume residential ventilation systems, but can be used for high-volume commercial ventilation systems.

Dampers

A damper is a flapper-like device that fits inside a duct to vary the airflow. Various styles are available from most ventilation-equipment suppliers. Manually operated dampers have a small lever outside the duct that repositions the damper itself to partially restrict the flow of air (Figure 17–6). These are generally adjusted once, when a ventilation system is first installed—then left alone.

Backdraft dampers are designed to automatically prevent airflow in one direction. Spring-loaded butterfly dampers are popular (see Figure 16–2). Because a certain amount of energy is needed to push open a spring-loaded damper, it is important to only use them when a fan is powerful enough to overcome the resistance of the spring. Simple backdraft dampers without springs are often used on ventilation fan outlets and clothes dryer outlets to prevent air from entering.

Several companies make motorized dampers that can be activated by a control of some type. For example, a dehumidistat could open a motorized damper in a duct serving a bathroom to move more ventilation air through that room when the humidity is high. When a control is used to activate a motorized damper (or a motorized grille) to vary the airflow, the damper and the control must operate at the same voltage—either high voltage (110 volts) or low voltage (24 volts). See Chapter 14 for the various control strategies.

Some motorized dampers have only two positions: open (or partially open) and closed. Other motorized dampers have three or more positions, such as fully open, partially open, and closed. Usually, the stopping positions of multi-position dampers are adjustable. There are also models, called modulating dampers, that can be opened a variable amount (Figure 17–7).

Suppliers of 4–6"-round motorized dampers include **Broan Mfg. Co., Inc.**, **Conservation Energy Systems, Inc.**, **Duro-Dyne Corp.**, **Honeywell Inc.**, **Tjernlund Products, Inc.**, and **Ztech**.

For larger motorized dampers, **Trol-A-Temp** offers round ones as large as 14", as well as various rectangular sizes. Their products are primarily used in heating/air-conditioning applications, but they can also be used for high-volume ventilation systems. They also offer special controls for multi-position dampers.

Ducts

Ducts can be a very important way to distribute air in a house, and most ventilation systems utilize them to some extent. Sometimes, extensive and complex duct layouts are used to distribute the air, but many systems require only a minimum number of ducts. A few installations use almost no ducts at all.

Ventilation ducts are made of the same materials as heating/air-conditioning ducts, although they are usually smaller in size. They may be rigid or flexible, insulated or not, metal or plastic. Occasionally, lightweight rigid-plastic sewer-drain pipe is used for a ventilation duct. This can be useful for a bathroom exhaust where excess moisture could rust a steel duct. As a cost-saving measure, building cavities are sometimes used to transmit ventilation air, but building cavities are usually very leaky and difficult to clean; therefore, this is now considered a poor construction practice and should be discouraged.

Rigid metal ducts made of galvanized steel are very inert from an outgassing standpoint. However, they sometimes have a thin oil film on them left over from the manufacturing process that is bothersome to sensitive individuals. The oil can be easily washed off with a TSP (trisodium phosphate) solution prior to installation. TSP is a heavy-duty cleaner available from paint and hardware stores.

Flexible ducts are popular because they are easy to route through a house. However, there is about twice as much resistance to airflow with flexible ducts when compared to rigid ducts. Flexible ducts are more difficult to clean than rigid ducts because of their irregular inner surface. For sensitive people who are bothered by the odor of flexible plastic ducts, flexible aluminum ducts are readily available.

Most insulated flexible ducts have a plastic liner. Building codes require that plastic liners in supply ducts be noncombustible. For this reason, low-cost flexible dryer ducts should not be used for supply ducts. (Besides, they are more odorous than ducts designed specifically for heating/cooling systems.) The plastic used to line insulated ducts is fairly innocuous, but it can sometimes be a problem for hypersensitive people, especially when used in a heating system because the warm air can cause a slight amount of outgassing. However, if the air in them is at or near room temperature, plastic-lined ducts are often tolerable for many sensitive people. If not, insulated flexible ducts are available with a more inert aluminum-foil liner at a slightly higher cost.

Unfortunately, insulated flexible ducts, no matter what they are lined with, can be easily damaged. This often accidentally occurs when a duct-cleaning service runs their equipment and hoses inside flexible ducts, and it won't be readily noticed. If they are routed through inaccessible locations, damaged ducts can be very difficult to repair or replace.

Conventional ducts can be avoided if building cavities are used as conduits for ventilation air. While this saves on the expense of the ducts, building cavities are notoriously leaky and can easily become contaminated. They are typically lined with a variety of materials such as wood framing lumber, insulation, plywood, drywall, the rear of bathtubs, galvanized metal, etc. Wood and drywall are porous so they can absorb moisture and harbor mold and other microorganisms. Extra care is necessary to seal building cavities to prevent excess air leakage, but this is often impractical unless the installer is especially conscientious. Unless properly sealed, it is relatively easy for air to move from garages, attics, and crawl spaces into the building cavities and contaminate the air within them. Because of the difficulty in sealing and the potential for contamination, it is highly recommended that ventilation air be moved through conventional ducts rather than through building cavities.

In general, it is important that ducts supplying a house with fresh air be as inert as possible so they don't contaminate the air within them. Ducts whose purpose is to exhaust stale or moisture-laden air from a house can be less inert because they are not transporting air that will be breathed by occupants.

Duct insulation

Ducts don't always need to be insulated, but there are two important reasons why insulation is sometimes necessary: 1) for increased energy efficiency and 2) to reduce the chance of internal or external sweating. R-4 is the minimum amount of recommended insulation, but in very cold climates, more may be necessary. Without insulation, it isn't unusual for 20% of the heat in ducts located in an attic or crawl space to be lost during the winter. Sweating occurs on cool ducts, such as those used for air conditioning, and it can lead to mold growth.

Insulating for energy efficiency

Insulating a duct for energy efficiency is only important when it is desirable to preserve the temperature of the air inside the duct. For example, consider a duct containing room-temperature air that is located in an unconditioned space, such as an attic. If the duct is uninsulated, in the winter it will lose some of its heat to the cold attic, and during the air-conditioning season it will absorb heat from the hot attic. If the room-temperature air within this duct is going to be blown back into the house, it would be wise to preserve its temperature, so it should be insulated.

On the other hand, if the air within a duct is going to be exhausted from the house, or if a duct contains outdoor air that is being brought indoors, then there is no need to preserve the temperature of the air, so insulation for energy efficiency isn't important. Insulation for energy efficiency is only needed when it is desirable to preserve the temperature of the air inside the duct. However, there is another important reason to insulate ducts: to prevent sweating.

Insulating to prevent sweating

It is not unusual for moisture from the air to condense, or sweat, on either the inside or outside of a cool duct. When ducts sweat, the moisture can lead to mold

Understanding Ventilation

growth, rot, or corrosion. Sweating can occur on the outside of ducts when they are run through spaces that are warmer and more humid than the air within them—for example, in the summer when cool house air is run through a duct located in a hot, humid attic. Sweating is a potential problem in the winter on the outside of a duct when cold outdoor air is drawn through a duct located within a conditioned part of a house. It is also possible for sweating to occur inside a duct. For example, when a duct is transporting moist air to the outdoors (as from a bathroom fan), and the duct is run through a cold attic or crawl space, moisture can condense inside the cold duct. In all these situations, the ducts should be insulated. (Sometimes a duct will be sloped so that any internal condensation will drain out, but this is prohibited by some building codes and the water may drain somewhere that results in damage.)

Where sweating is possible on the exterior of a duct, the outer surface of the insulation should be covered with a well-sealed vapor retarder. The insulation and vapor retarder will prevent the warm, humid air from reaching the duct and condensing on it. The joints between ducts and fittings should be sealed prior to insulating them (see below). The ducts running between an HRV and the outdoors are generally insulated to prevent both internal and external sweating.

Insulating materials

Ducts are often insulated with blankets or sleeves consisting of a 1–2"-thick batt of fiberglass covered with a vapor retarder. The sleeves are made in several sizes and designed to be slipped over round ducts. The blankets can be wrapped around large rectangular or irregularly-shaped ducts. If sweating on the outside of a duct is a possibility, a vapor retarder should always be used on the outside of the insulation, and all seams in the vapor retarder should be securely sealed.

Sometimes the insides of ducts are lined with fiberglass insulation that is exposed to the airstream. A widely used product is often called *ductboard*. Ductboard consists of a semirigid fiberglass board with an aluminum-foil facing material that functions as a vapor retarder. It can be cut into pieces and assembled with aluminum foil-faced tape into rectangular sections. The aluminum foil is on the outside to prevent sweating, and the raw fiberglass is on the inside. Besides preserving the temperature of the air in the duct and preventing external sweating, the exposed fiberglass also absorbs sounds, so it is a popular way to make a forced-air heating/cooling system quieter. However, there are a number of disadvantages. 1) Fiberglass insulation is porous, so the inside of ductboard can easily store dust.

Figure 17–8. *Ventilation-equipment suppliers offer specialized fittings, such as the Speedi Sleeve (left) and the Albo (right), to make an installation go easier.*

Figure 17–10. Easy-to-apply duct-sealing mastics are more effective at preventing leaks in a duct system than duct tape.

2) If it gets wet for any reason, it can become a perfect breeding ground for mold, mildew, and other microorganisms. 3) Because of its porosity, ductboard is virtually impossible to clean. 4) Because the fiberglass particles are held together with a formaldehyde-based resin, a small amount of formaldehyde will be released into the airstream. 5) Some of the fiberglass particles can break off so they too can enter the airstream. One manufacturer produces a ductboard that has a black coating on the inside to seal the fiberglass. It is less porous and it will minimize the shedding of fibers and the release of formaldehyde. Such a duct will be somewhat easier to clean, but it could still release odors bothersome to sensitive people.

Insulated flexible ducts are much more inert than ductboard. Most are made of a "sandwich" consisting of an aluminum-foil or plastic lining, an insulating layer of fiberglass, and an outer aluminum-foil or plastic vapor retarder. Because the fiberglass is not exposed to the airstream, it cannot contaminate the ventilation air.

However, this is only true if the lining is durable and never becomes damaged. (Duct cleaning services can sometimes accidentally—and unknowingly—damage the lining.) Some special noise-attenuating flexible ducts have fiberglass exposed to the airstream to absorb noise.

Duct locations

When ducts are run within conditioned spaces, rather than unconditioned spaces such as attics and crawl spaces, they only need to be insulated if they contain cold air (to prevent external sweating) or hot humid air (to prevent internal sweating if the house is air conditioned). If they contain air that is close to room temperature, they needn't be insulated. Such ducts can be run in out-of-the-way places such as within thick walls, above dropped ceilings, or between floors. By running ducts within the conditioned space, it is easier to build a house in an airtight manner because there are fewer penetrations into the attic or crawl space.

Understanding Ventilation

Figure 17–10. Diagnostic equipment is readily available to evaluate duct system leakage.

Ventilation ducts are often too large to fit into a 2x4 wall cavity. One solution is to build thicker walls in some locations in order to hide ducts inside them. **Eneready Products Ltd.** has a simple 90° plastic elbow that can be used to fit a 4"-round duct into a 2x4 wall cavity (which actually measures only $3^1/_2$"). It is called an Albo, after its inventor Al Koehli who is a past chairman of Canada's housing code. This is a 90° transition fitting with one side being a full 4" in diameter (to accept a round grille) and the other side is oval-shaped so a 4" round duct can be compressed slightly to fit into a $3^1/_2$" space. For ducts that pass from an attic or crawl space into the living space, Eneready offers a 4"-diameter Speedi Sleeve that can be easily caulked to the back of drywall for an airtight seal (Figure 17–8).

Duct sealing

A system of ducts is typically composed of straight lengths and a variety of elbows, boots, and other fittings. It is important to seal the joints between the sections securely to prevent pollutants from entering the ducts and to prevent ventilation air from escaping. As discussed earlier in Chapter 6, leaky ducts can cause major air-pressure imbalances in a house. Though such imbalances will be more severe with leaky heating/cooling ducts than with ventilation ducts, it is important to seal all duct systems for optimum efficiency and effectiveness. But before sealing a duct system, it is important that all the individual ducts and fittings be securely fastened together (usually with at least three sheet-metal screws). This will prevent excessive flexing of the joints, which could result in a broken seal.

In the past, duct sealing has been done with cloth or aluminum-foil duct tape, but the cloth tape often has a limited life span. Because the glue deteriorates with age, it isn't unusual to inspect an old system of ducts and find that all the cloth tape has come loose. Because ducts are often located in inaccessible locations, loose tape can be difficult to replace. With brand-new clean ducts, a high-quality aluminum-foil tape will generally be long-lasting if it is applied conscientiously. It is important that ducts be clean, because even the thin oil film often found on new metal ducts can prevent tape from adhering properly. With old ducts that are dirty and oxidized, even the best aluminum-foil tape will not adhere properly or last very long.

Duct-sealing mastics have been developed in recent years as a much more effective alternative to either cloth or aluminum-foil duct tape. They are available in caulking tubes or pails. About the consistency of mashed potatoes, duct-sealing mastics can be applied to joints and small gaps in ducts with a bare hand, putty knife, or stiff brush (Figure 17–9). Water-soluble mastics are easier to clean up and less noxious than solvent-based products. Manufacturers of water-based mastics include **Foster Products Corp.** (30–80 & 30–90), **Hardcast, Inc.** (Versa Grip), **Mon-Eco Industries, Inc.** (44–41 & 44–40), **RCD Corp.** (RCD #6), **RectorSeal Corp.** (Air-Lock) and **United McGill Corp.** (several products). Sealing a duct system with mastic is generally easier, faster, and more effective than using duct tape.

If ducts are to be insulated, the seams should be sealed first, then the ducts insulated. This prevents particles of insulation from getting through the seams into the airstream. Because duct sealing can be time-consuming, some contractors only seal some of the joints as a cost-saving measure. (A shady contractor will only seal the joints that are visible.) Ideally, all seams and joints should be sealed, but if there are monetary restrictions and a perfect job isn't feasible, the most important joints to seal are those between dissimilar materials, between separate sections of duct, and between fittings. Less important are the longitudinal seams along the length of rigid sheet metal duct and the individual seams of adjustable elbows. In other words, joints and seams that are fabricated in a factory are tighter than field-fabricated joints. Still, the best thing to do is seal all joints.

To evaluate an existing duct system and locate the hidden leaks, specialized diagnostic fans are available that can pressurize the entire duct system and measure leakage rates. They are available from **The Energy Conservatory** (Duct Blaster) and **Infiltec** (Duct Buster). These devices are often used by weatherization contractors (Figure 17–10). The blower doors discussed in Chapter 11 can also be used to evaluate duct leakage, but these diagnostic duct fans are designed especially to evaluate duct systems.

Using heating/air-conditioning ducts for ventilation air

Because a forced-air heating/air-conditioning system already has a system of ducts, it may seem obvious to use those same ducts to distribute the ventilation air. After all, a second set of ducts would mean additional expense. In fact, this is an important advantage; coupling a ventilation system to a forced-air heating/air-conditioning system saves some money during installation. However, there can be some disadvantages.

First of all, you can't simply blow ventilation air into heating/air-conditioning ducts and expect it to reach all the rooms. This is because heating/cooling ducts are designed to distribute air when a heating/air-conditioning fan is running. This fan generally moves 10–20 times more air than a ventilating fan, so it requires larger ducts. When a ventilating fan blows a relatively small amount of air into a heating/cooling duct system (when the larger fan is not running), the air falls out of the nearest grille. The ventilating fan simply can't generate enough pressure in the larger ducts to move the air to the farthest reaches of the duct system.

To use the heating/air-conditioning ducts to distribute ventilation air effectively, you must run both fans simultaneously. By introducing the ventilation air into the larger duct system and operating both fans at the same time, the small fan simply adds fresh air and mixes it with the air being moved by the larger fan. There are several methods of controlling such a system (see *Controlling a ventilation fan and a heating/cooling fan simultaneously* in Chapter 14). Operating two fans when only one is really necessary results in more electrical consumption. This disadvantage can be minimized by using a furnace/air conditioner that has a high-efficiency variable-speed motor such as an ECM motor (see *Fan efficiency* in Chapter 11).

Before a final decision is made to use one set of ducts for two systems, it is important to check with the heating/cooling fan manufacturer to see if its fan is rated

Understanding Ventilation

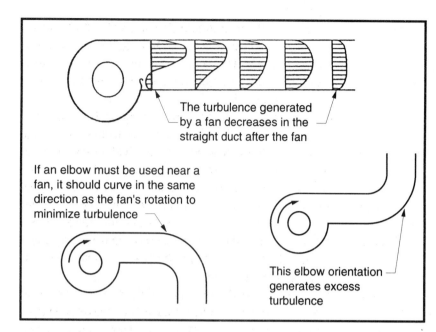

Figure 17–11. Turbulent airflow, which is generated by fans, elbows, and sudden changes in duct size, is inefficient and should be minimized. This can be done by using a length of straight duct after a fan, orienting elbows correctly, and using smooth transitions.

for extended operation; low-cost shaded-pole motors often aren't, but the more popular permanent-split-capacitor motors generally are. Running a cheap fan for longer-than-normal periods can sometimes result in premature failure and costly motor replacement. The cost of electricity can be minimized by using a multiple-speed (or variable-speed) furnace fan that operates at high speed when heating or cooling are called for, and at low speed when only ventilation is needed. By operating a fan at its lowest speed, less energy is used and the fan's bearings should last longer. At low speed, a fan will also be quieter and drafts will be minimized.

The ducts for a forced-air heating/air-conditioning system are usually laid out for optimal heating or cooling—not for optimal fresh-air distribution. For example, most heating systems deliver air to a room at floor level near windows. This is because warm air at floor level will rise and mix with room air, and most of the heat loss in a room is through the windows. If heat isn't called for, but ventilation air is, the furnace fan will distribute cooler air through its system; thus, cool air will exit through the floor registers. In the winter, cool air at floor level won't rise and will not keep windows warm. So, distributing ventilation air through a furnace system can work well when heat is called for, but it isn't as effective when the heat is off. Also, the amount of air delivered to a particular room by a furnace/air-conditioning system is based on the *heating or cooling* requirements of that room, not the room's *ventilating* requirements.

Despite the drawbacks, in reality, the high volume of air being moved by a forced-air furnace/air-conditioning fan generally tempers the incoming air and mixes it with house air very well. This approach also has the advantage of being easy to couple with a good-quality air filter. If an energy-efficient fan motor is used, this can be an effective, and economical, way to distribute ventilation air in a house. If the existing forced-air system contains a permanent-split-capacitor (or a shaded-pole) motor and you plan to run it continuously, you should see if it is possible to upgrade it to a more energy-efficient ECM motor.

Sizing and laying out a duct system

The size and layout of a duct system is determined by several interrelated factors: the amount of air flow, the resistance to airflow (friction) caused by the surface of the duct, the resistance to airflow caused by the fittings, the velocity of the air, the pressure created by the fan, etc. Designers often use rules of thumb to determine duct sizes for simple layouts, but some calculations are generally necessary for complex duct systems. Many ventilation-equipment suppliers offer a design service to help their customers create a workable duct system.

Forced-air heating/cooling ducts are almost always larger than ventilation ducts. These larger duct systems are often designed in such a way that the size of the ducts restricts the amount of air flowing to and from different rooms. In other words, you use a small duct where you want a smaller airflow and a larger duct where you want a larger airflow. The grilles or registers at the end of the ducts are then left wide open. With ventilation systems, it is often a good idea to use slightly larger-than-normal ducts throughout the system, then control the amount of air flowing into and out of the rooms by partially closing the grilles at the ends of the ducts. In this way, the air speed increases at the grille where the flow is restricted. For supply grilles, an increase in airspeed causes the air to be thrown into the room for better mixing.

One popular way of designing a forced-air heating/cooling duct system is called the *equivalent length method*. This technique recognizes that each length of duct and fitting has a certain amount of resistance to airflow. For example, an 8" 90° elbow has as much resistance as 10' of straight rigid duct, a 5" 90° elbow has as much resistance as 5' of straight rigid duct, a register elbow boot has the same resistance as 30' of straight rigid duct, and a flexible duct has twice as much resistance as a rigid duct.

The equivalent length of a fitting is the amount of straight rigid duct it would take to have the same resistance. In other words, the equivalent length of an 8" 90° elbow is 10'. To calculate the total resistance of a duct system, you first add up all the equivalent lengths of all the individual components, and you use total equivalent length to determine the system's airflow characteristics. For the least resistance, and the most airflow, a duct system should be designed with the least total equivalent length.

An in-depth discussion of duct design is beyond the scope of this book, because for relatively simple layouts where grilles are used to restrict the airflows, rules of thumb will generally suffice. For complicated layouts, forced-air heating/cooling duct design principles can be used. Manuals and computer programs are available to help the designer create more complex duct systems. The **Home Ventilating Institute** has an installation manual for HRVs, *Installation Manual for Heat-recovery Ventilators*,(Home 1990) that contains a good section on duct design. The Air Conditioning Con-

Figure 17–12. For most ventilation duct systems having short runs and simple layouts, the duct and grille sizes in this chart can be used as rules of thumb. With the sizes shown, the grille is adjusted to regulate the airflow.

SUGGESTED DUCT AND GRILLE SIZES				
	up to 40 cfm	40 to 90 cfm	90 to 120 cfm	120 to 180 cfm
Rigid round duct	4"	5"	6"	7"
Rigid rectangular duct	2¼" x 6"	2¼" x 10"	3¼" x 10"	3¼" x 14"
Flexible round duct	5"	6"	7"	8"
Round grilles	3–4"	5–6"	6–7"	7–8"

tractors of America has a *Manual D: Duct Design for Residential Winter and Summer Air Conditioning and Equipment Selection*(Air Conditioning) that is very complete, and it also has a duct-design computer program available. **Nutone, Inc.** has a helpful *Duct Design Chart* that works like a slide rule to aid in the selection of duct sizes as small as 3" in diameter.

A number of general recommendations apply to all well-designed duct systems. For example, you should keep the duct runs as short and straight as possible to minimize resistance to airflow. A larger diameter will have less resistance than a smaller diameter (but it will take up more physical space). All joints should be sealed to reduce duct leakage. Use as few bends and elbows as

Figure 17–13. A Magnahelic gauge can be used in conjunction with an airflow grid to determine the amount of air flowing through a duct.

Figure 17–14. This handy device allows you to both measure and adjust the airflow in a duct.

Figure 17–15. A flow-measurement hood allows you to quickly determine the amount of air entering or leaving a grille.

possible to minimize the turbulence that causes resistance and noise. It is especially important to avoid bends and elbows near a fan outlet or near inlet and outlet grilles (Figure 17–11). Keeping room-temperature ducts within the conditioned space will eliminate the need for duct insulation. When necessary, use insulation on the outside of ducts so it won't contaminate the airstream, and use a well-sealed vapor retarder over the insulation to minimize sweating. You should use rigid ducts because of their reduced resistance, but if flex-duct must be used, allow for the increased resistance to airflow by using the next-larger size.

Some additional suggestions: If you use larger-than-normal ducts, the air will be moving at a slower velocity; thus, it will be quieter. Use an adjustable damper in each duct run so you can fine-tune the amount of air flowing through it (this isn't necessary if adjustable grilles are used). Keep in mind that reducing airflows with adjustable dampers and grilles can result is a certain amount of air noise. This is often less noticeable with dampers because they are inside the ducts—some distance from the living space—whereas a grille is in the living space. (The round plastic grilles often used in ventilation systems are especially designed to be quiet.) If duct runs are fairly long, or if several elbows are used, it is wise to select the next-larger size to minimize resistance to airflow. For example, bath fans often have a connection for a 3" duct, but the fan will function much more effectively if it is connected to a 4" duct (unless there are no elbows at all and the duct is extremely short). If you plan to install a motorized grille or damper in a duct to vary the airflow, the duct must be sized for the largest flow desired, then the grille or damper is used to restrict the flow to a lower capacity.

The rules of thumb in Figure 17–12 for duct and grille sizes are recommended for simple ventilation duct systems where adjustable grilles are used to restrict the airflow. These recommendations work well if duct runs are short, if there aren't too many elbows, and if the joints and seams have been sealed. When in doubt, se-

Understanding Ventilation

lect the next-larger size. Rectangular grilles are commonly used in forced-air heating/cooling systems, but they are not listed in the table because they cannot be used effectively to control small airflows and throw supply air into a room.

Adjusting the airflows

Once a ventilation system has been designed and installed, the airflows will probably need to be adjusted to ensure that the correct amount of air is going where you want it to. With a balanced ventilation system (with or without heat recovery), the amount of air entering the house should equal the amount leaving the house. This will ensure that the house will be maintained under a neutral pressure. Because central-supply and central-exhaust ventilation systems inherently cause a house to be pressurized or depressurized, overall system balancing is unnecessary with them.

Adjusting airflows to individual rooms is also necessary so one room isn't overventilated at the expense of another. In most instances you will only adjust the airflows in a house once, when the ventilation system is installed, and then leave the settings alone. If motorized dampers or grilles are used to vary the airflow in a system, you should check the airflows with the damper open as well as closed.

Balancing a ventilation system

Most balanced ventilation systems use a system of ducts to supply air to and exhaust air from a house. The length and layout of the supply ducts are often different from the length and layout of the exhaust ducts. (In most balanced ventilator installations, the exhaust ducts have less total resistance than the supply ducts, but this will vary from one installation to another.) This means that the incoming and outgoing airflows will have different resistances and different flow characteristics. This can result in the house being either pressurized or depressurized. The result can be wasted energy, potential condensation (or freezing) problems, or cool drafts in the winter. For a house to be under a neutral pressure, the airflows must be balanced.

From a practical standpoint, perfect balancing is not necessary. In fact, it is unlikely that a system will ever be perfectly balanced at all times because of all the natural and accidental pressures that can effect house pressures. Balancing the volume of incoming air to within 10% (±) of the volume of the outgoing air is generally close enough. Before actually attempting to

Static pressure (in. w.g.)	VLR-80 3" wall supply grille — Number of turns grille is opened					Static pressure (in. w.g.)	VLK-150 6" exhaust grille — Number of turns grille is opened			
	0	3	6	9	12		4	6	10	20
0.02	.5	2	2	2.5	3	0.02	4	5	8	18
0.04	.5	3	4	5	6	0.04	6.5	9	15.5	32
0.06	1	5	6	7	8	0.06	10	12.5	22	45
0.08	1.5	7	8	9	10	0.08	14	16	28	56
0.10	2	8	9.5	11	12.5	0.10	16	19.5	31	65
0.12	2.5	9	11	12	14	0.12	17.5	23	35	72
0.14	3.5	10	12.5	13	15	0.14	20	25	40	78
0.16	5	11	13	14	16	0.16	22	27	42	83
0.18	6	12	13.5	15.5	18	0.18	24	29	46	89
0.20	6.5	12.5	14.5	17	19	0.20	25	31	50	94
0.25	7.5	13	15.5	18	21	0.25	28	35	55	105
0.30	8	13.5	16	19	23	0.30	29	37	58	120

Eneready Products Ltd.

Figure 17–16. Some round plastic grilles are supplied with a chart that lists the airflow when there is a certain pressure difference across the grille.

balance a ventilation system, make sure that all local-exhaust ventilation fans are turned off, and the forced-air furnace/air-conditioning system is off (unless the furnace/air-conditioning ducts are being used to distribute ventilation air). Check the ducts (and the HRV core if there is one) for blockages, and turn the ventilator on at high speed. Balancing will be affected by wind and stack effect to a certain extent, so a system balanced on a calm, warm day might be out of balance on a cold windy day. This is why ±10% is close enough.

There are several different ways of measuring airflows in a ventilation system. One of the most common methods utilizes *airflow grids* that are available from many ventilation-equipment suppliers. Airflow grids are mounted inside a duct (Figure 17–13). Two grids are installed in a balanced system—one in a main supply duct and one in a main exhaust duct. (Airflow grids are often permanently installed, but some installers remove them once the system has been balanced.) An air-pressure sensing device, such as a Magnahelic gauge, is attached to the connections on the grid to measure the air pressure within the duct. The pressure reading can be converted to an airflow rate (usually in cfm) by using a chart supplied with the grid. The accuracy of an airflow grid is affected by turbulence, so they should not be located too near elbows or fans—something that is covered in installation instructions.

Eneready Products Ltd. has a handy PRA Flow Grid that contains both a flow-measuring grid and an adjustable damper in one unit (Figure 17–14). (PRA Flow Grids must be left in place because removing them would also remove the damper.) Pressure-measuring instruments can be purchased from many ventilation-equipment suppliers such as **Broan Mfg. Co., Inc.**, **Conservation Energy Systems, Inc.**, and **Nutech Energy Systems, Inc.** They can also often be rented from local ventilation-equipment dealers and distributors.

To actually balance a ventilation system, the pressures in the supply and exhaust ducts are measured at the airflow grids and the numbers are converted to flow rates, then the higher airflow is restricted with an adjustable damper until it matches the lower flow. This is done by partially closing the damper in the duct having the higher flow or by adjusting a grille at the end of the duct. When the airflows are within 10% of each other, the system is considered balanced. In houses built to the Canadian government's R-2000 energy-efficiency standards, building inspectors check the airflow grids to see if the system is balanced.

Adjusting the airflows to individual rooms

Proper distribution also involves adjusting grilles and dampers so the correct amount of air reaches specific rooms. This is generally done one time only—just after installation. Many grilles can be adjusted manually to vary the amount of air passing through them. Motorized grilles and dampers can be opened and closed by a remotely mounted electrical control of some type. In either case, it is important to measure the airflows at the grilles to determine that the correct amount of air is getting into and out of each room.

Measuring airflows at grilles can be done in several different ways. For example, an airflow grid can be used for each separate duct run, but this can be expensive in a complicated ventilation system having a number of grilles. Some ventilation equipment installers use a portable flow-measurement hood, such as those made by **Shortridge Instruments, Inc.** and **Pacific Science & Technology Co.** These hoods have a built-in flow measurement device and can be moved from grille to grille quickly (Figure 17–15). Unfortunately, many can't accurately measure airflows below 15–25 cfm. As a result, these flow-measurement hoods are used more often in commercial installations, where higher airflows are common, than by residential contractors.

Another way to determine the airflow at a grille involves using one of the round plastic adjustable grilles on the market today that come with cfm flow tables, such as the Whisper grilles from **Eneready Products Ltd.** (Figure 17–16). To determine the airflow, you need a gauge capable of sensing air pressures (such as a Magnahelic gauge) attached to a probe connected to a length of tubing. When the probe is pushed through the grill, it will sense a difference in air pressure behind the grill when the ventilation fan is running. By comparing this pressure to the table supplied with the grille, and counting the number of turns the grille has been opened from the fully closed position, you can determine the airflow in cfm. Because the grilles are adjustable, the tables will list several different flow rates, depending on how much the grille is opened.

It is possible to make an inexpensive flow-measurement hood yourself that can be used to measure the smaller airflows typical of residential ventilation systems. All you need is a cardboard box with one open side, a length of gasket material, and a pressure-measuring gauge (Figure 17–17).(Nelson, Gary 1994a) The box should be large enough for the open side to fit over the grille whose airflow you want to measure. You will need to cut two holes in the side of the box opposite the open end. The larger hole should have an area (in square inches) that is about half the flow (in cfm) you want to measure. For example, if you want to measure a 40 cfm airflow, cut a 20 sq. in. hole (say, 4" x 5"). The box should be large enough so the side you cut the hole in is at least twice the length and width of the hole (*e.g.* if the hole is 4" x 5", the box should be at least 8" x 10"). The second hole is much smaller—it should be poked in same side as the large hole with a pencil, and should be located near a corner. Now, insert a piece of tubing into the small hole so one end sticks inside the box about $1/2$" and connect the other end to your pressure-measuring gauge. To use the box to measure an airflow at a grille, you first need to place it in a stable position *next to* the grille and adjust the pressure gauge to zero. (Some gauges have auto-zeroing capability.) Once the gauge has been zeroed, place it over the grille and read the pressure difference on the gauge. (This will be the difference between the pressure inside the box and that outside the box.) If the pressure is less than 4 Pa. (or less than 1.5 Pa. if you are using a digital gauge capable of reading 0.1 Pa. increments), then you need to make the hole a little smaller in order to get the best accuracy. (With a Magnahelic gauge, you will have the best accuracy if the pressure reading is between 4–7 Pa. With a more accurate digital gauge, you will have the best accuracy if the pressure reading is between 1.5–4 Pa.) To calculate the flow, multiply 1.07 times the area of the hole (in sq. in.) times the square root of the pressure reading (in Pa.). Obviously, calculator will make this a little easier. The formula is:

$$\text{Airflow} = 1.07 \times (\text{Area of hole}) \times (\sqrt{\text{Pressure}})$$

While the above description may seem somewhat complicated, using a cardboard box can be an accurate and useful way to measure airflows. If you use a Magnahelic pressure gauge you can determine an airflow within 10% accuracy. If you use a digital manometer capable of measuring in 0.1 Pa. increments, you can get within 3%. Cardboard boxes being cheap, you may want to make two or three boxes to measure different

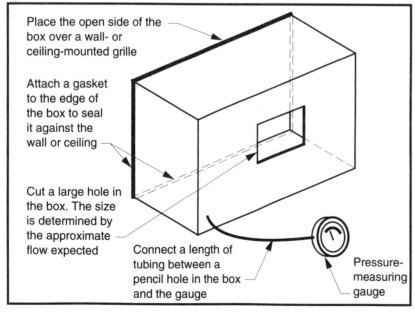

Figure 17–17. A do-it-yourself flow-measurement hood can be made from a cardboard box and a pressure-measuring gauge. See text for explanation.

flows. Simply cut different-sized holes in each box and mark the size and flow-measurement range on the box.

Airflows can sometimes be determined very crudely by using a length of string. Though not nearly as accurate as the above methods, this can occasionally yield satisfactory results. With this method, a short length of light-weight string, or thread, is suspended next to a grille. The amount the string is deflected away from vertical by the moving air is indicative of the airflow. A larger deflection means more airflow than a smaller deflection. While this method won't give you actual cfm readings, it can tell you if the flows are roughly comparable at two or more different grilles. However, if some grilles are to have different airflows, this method will not give you the percentage of air that is going to each grille.

Duct cleaning

It is a good idea to have both central-heating/cooling ducts and ventilation ducts cleaned periodically. Sometimes an annual cleaning is called for, but in some houses, several years may pass before cleaning is required. Because much less air travels through ventilation ducts than heating/cooling ducts, longer intervals between cleaning are generally possible.

Over time, the interior of all ducts become dusty. This is because air is rarely 100% pristine. It often contains small amounts of contaminants that eventually build up on the inside surface of the ducts. Commercial duct-cleaning services exist in many cities that can remove this dust. They generally utilize a large truck-mounted vacuum with a long suction hose that is run into the house to remove the buildup of dust from the ducts. Sometimes a mechanical device is inserted inside the ducts to vibrate or loosen the dust so the vacuum can remove it easier.

Some duct-cleaning contractors spray or fog a chemical inside the ducts after cleaning to "seal" or kill any remaining microorganisms. This is generally not necessary if the ducts have been cleaned well, and because the chemical itself should be considered a contaminant, it is not recommended.

When planning a system of ducts, it is important to consider how they will be cleaned in the future. For example, there will need to be access points to insert the long vacuum hose into the ducts. If the ducts twist and turn and have very few straight sections, there may need to be numerous openings cut in them at cleaning time. On the other hand, when duct runs are relatively straight, there may only need to be a single access point at one end, perhaps by removing a grille. In commercial duct systems, operable doors are often used, but in residential ventilation systems the openings are often cut with tin snips, then after cleaning, sealed shut with tape or mastic. When designing a system of ducts, it is relatively easy to plan how it will be cleaned. Without such pre-planning, some ducts may be inaccessible without cutting holes in walls, ceilings, or floors.

Chapter 18

Miscellaneous considerations

This chapter discusses of several important miscellaneous factors to consider when selecting and designing a ventilation system for a house. These apply to all the different strategies to some degree.

Transparency

In an ideal installation, no one in a house should be aware that a ventilation system is operating. The ventilation system should be invisible, or transparent, to all of your senses. Except for some inconspicuous grilles and control switches, you shouldn't be able to see any indications that a house even has a ventilation system, or that it is operating. You shouldn't be able to feel any air movement. Nor should you be able to hear the system running. And a ventilation system should remove all odors from a house and not add any odors to the air. While it is unlikely that a ventilation system would introduce any unusual tastes into a house, it certainly shouldn't leave a bad taste in your mouth due to a high installation cost or high energy bills. The idea of transparency simply means that you shouldn't be able to perceive when a ventilation system is running with any of your five senses.

Out of sight

A window fan is fine for temporary ventilation, but it is generally not suitable for year-round fresh air. Because most people don't care for the appearance of a window-mounted air conditioner, they wouldn't be much happier with ventilator that sits in a window. That is why there are not many window-mounted HRVs on the market. Of course, there are applications where a window ventilator is the best option—a rental, for example, where the tenant can't modify the building. A window unit may also be a viable option where there

Understanding Ventilation

isn't enough money available to install a central system. But in general, people don't want a ventilation system that they can see.

Central ventilation equipment can usually be located in a basement, garage, or utility room where it will be easy to maintain, yet be no more visible that a furnace or water heater.

Controls and grilles are the only components that should be visible in the living space. Grilles should be located so air leaving them won't cause objects in the room to move—billowing curtains or papers blowing off a table or desk. Grilles can't be made invisible, so they are sometimes placed in inconspicuous, out-of-the-way places, such as behind doors or in corners. This is a big mistake. Ventilation is a very important component of a house and grilles should always be located where they will be the most effective. Otherwise, optimum system performance and occupant satisfaction is not possible.

A good way to make grilles inconspicuous is to paint them the same color as the wall or ceiling. The popular adjustable round plastic grilles are generally white in color, but they can easily be spray-painted. The use of a low-tox paint is especially important for supply grilles because all the air you breathe will pass through them.

Can't be felt

A person's sense of touch can easily detect air movement as well as air that is colder or warmer than room temperature. Therefore, cold or warm drafts should be avoided. This can often be done by proper grille selection and location. In general, air leaving ceiling grilles and those located high on a wall will not be felt by occupants unless they reach up and place their hand directly in the airstream. Grille placement is covered in more detail in Chapter 17.

Sometimes a ventilation system will vibrate slightly from an out-of-balance fan. If this vibration is transferred to the structure of the house, the occupants may feel it. It may not be a significant, obvious sensation, but it can be irritating. For this reason, ventilation fans and HRVs are often hung from flexible straps or springs or are mounted on rubber pads or cushions to minimize vibrations. Sometimes fans are rigidly mounted to a concrete basement wall. The concrete has enough mass that it absorbs any vibrations without transferring them to the rest of the house. In addition to a sound-absorbing mounting, it is also a good idea to use a resilient duct connector, such as a short length of flexible duct, between the fan cabinet and any rigid duct (Figure 18–1). If this isn't done, the vibrations will be

Figure 18–1. *A short length of flexible duct, or a special resilient fitting, can be used to prevent vibrations from being transferred from one part of a duct system to another.*

transferred from the fan to the ducts, and to the structure of the house.

A few manufacturers use a dynamic computer-balancing procedure on their fans that is similar to how automobile wheels are balanced. This will yield a fan that is less susceptible to vibration, has longer lasting bearings, and is quieter. Some motors are mounted in a special vibration-isolation frame.

Quiet

Noise is undoubtedly the number one complaint people have about ventilation systems, and it isn't unusual for an occupant to disconnect and totally disable a ventilation system because it is too loud. One study found that kitchen-range exhaust hoods were used only 40% of the time when people were preparing dinner.[Nagda] Sixty percent of the time nothing was done to rid the house of cooking odors—primarily because the fan was too noisy.

Some fans are inherently quieter than others. For example, although there are certainly exceptions, axial fans are often noisier than centrifugal fans because their blades chop through the air. There are several methods of expressing the loudness of a fan: the decibel (dB) is measured on a logarithmic scale (3 dB is 10 times as loud as 2 dB). The more commonly used sone is linear (4 sones is twice as loud as 2 sones). The Phon, used in Japan, is rarely used in North America. To understand what these units mean, a running refrigerator in a quiet kitchen will produce approximately 1 sone and a very quiet room is 30–35 dB. If a fan is going to be running continuously, it should be as quiet as possible—usually less than 1.5–2 sones. Most people will complain if a ventilation system is significantly louder than their refrigerator. At least one manufacturer—**Panasonic**—produces an incredibly quiet, $1/2$-sone fan. When a kitchen range hood is as loud as 5 sones, it will probably only be used rarely—it is simply too noisy.

Noise can be created in a variety of ways in a ventilation system. For example, vibrations can cause noise, but this can be minimized by the methods mentioned above. Noise can also be generated by poor-quality equipment or sloppy installation: loose screws, cheap fans, or poor overall design. Low-quality fans may get louder with age as they wear out. These problems can be minimized by selecting high-quality equipment and a conscientious installer. The sound of a ventilation fan is always more noticeable at night when there is less background noise from traffic, appliances, televisions, people talking, etc.

Air noise occasionally results from air moving through a fan or duct. A whistling or whooshing sound, air noise can be minimized during the design phase by keeping the air speed (velocity) slower than 500 feet per minute (fpm). The velocity can be slowed down by increasing the duct size. Air speed can also be reduced by using a smaller-capacity fan to move less air through a duct, or by slowing down the fan with a speed controller. A duct layout with many elbows and fittings can generate a great deal of turbulence in the air, resulting in excessive noise.

Air noise can also be reduced by using an *air silencer* (more correctly called a *sound attenuator*). These devices are similar to automotive mufflers and can deaden the sound of air rushing through a ventilation system. Unfortunately, many of them are made for commercial applications and are too large for residential use. They are also often lined with raw fiberglass, or a similar material, to absorb the sound. Therefore, these devices have the potential to contaminate the airstream slightly with small fibers or from minor outgassing of resins that hold the fibers together. Sound attenuators can be a good medium for mold to grow in if they ever get damp. **American Aldes Corp.** has a 3"-diameter noise-absorbing duct and a 3"-diameter sound attenuator. Residential-sized sound attenuators are also available from **Kanaflakt, Inc.** in 4–6" diameters.

Fan noise in a bathroom or kitchen can be reduced by using an exhaust fan that is located away from the room it serves. Instead of mounting an exhaust fan in a bathroom ceiling, for example, the fan could be mounted in the attic or basement with a length of duct connecting the fan to the bathroom. Kitchen range hoods are readily available that can be ducted to a fan mounted outside the house on an exterior wall or roof. A remote-mounted fan might be fairly loud if you are standing near it, but if it is mounted away from where people sleep or sit, it is often not very noticeable. Basements are good places to mount fans—but not directly under the bedroom if the fan will run at night when people are trying to sleep.

Odor-free

Occasionally, disagreeable odors are associated with ventilation systems. This can come about if the system has been contaminated with something—mold growth, a spilled cleaning product, wood smoke, etc.—or because of outgassing from plastic- or insulation-lined ducts. Contamination can generally be minimized by routine maintenance.

Odors can also be released by a warm electric fan motor. In the vast majority of installations, this is a very minor consideration, but hypersensitive occupants can sometimes be bothered by bearing oil, insulation, or lacquers as a motor gets warm. To minimize contaminating the airstream with a motor, sensitive people should look for a ventilator that does not have its motor in the fresh airstream. Some HRVs use a single motor to power both the fresh-air fan and the stale-air fan. If the motor is mounted in the stale airstream, it should not be able to contaminate the incoming fresh air. Similarly, if the fan and motor are mounted completely outside the HRV cabinet, they can't contaminate the fresh airstream. (Of course, such a motor will still generally be within the living space. But if the ventilator is located in a room that is not regularly occupied, such as a basement, it probably won't have the same effect it would if it were in the incoming airstream.) In reality, most HRV's have motors located within one or the other airstreams so that any heat given off by the motor can be utilized in the heat-recovery process. This can be advantageous in the winter when a house needs heat, but it is a disadvantage in the summer because will result in slightly warmer incoming air and a bit more work for the air conditioner.

Many ventilation-system cabinets are lined with insulation to prevent sweating and excess heat loss. Because some insulations have a distinctive odor, they should be covered with aluminum foil to minimize outgassing into the airstream. Unfaced fiberglass insulation has the advantage of absorbing unwanted sound, but foil-faced insulation is preferred because it is less likely to introduce glass fibers and odors into the air. It is also less likely to become contaminated with microbial growth if it gets wet.

Many ventilation cabinets use flexible rubber or plastic seals and gaskets to prevent air leakage. These seals can occasionally yield an odor that is offensive to a sensitive person, but usually a disagreeable odor from a gasket can be eliminated by covering it with aluminum-foil tape. If done carefully, the gasket should still be able to be compressed to seal the cabinet.

Taking responsibility

The people who design and build houses, as well as the people who live in them, need to take more responsibility for indoor air quality. We should all quit leaving something so vital to our health up to chance. It is foolish to think that we can continue to pollute our indoor environment, then expect Mother Nature to keep it healthy for us. Experience has shown that natural ventilation rarely keeps the air in a house clean—especially when a house is filled with polluting materials. In fact, many mechanical ventilation systems often can't keep the air in a house clean when the house is filled with polluting materials.

Because we live in a highly technological age, it is certainly possible to design a complex, high-powered ventilation system that will keep the air in any house pristine, under all circumstances, no matter what we fill it with. But that solution would be unbelievably expensive to purchase as well as to operate. If builders and designers are careful about what materials they use in constructing houses, and if occupants are careful about what they use to furnish and maintain their houses, then a simple ventilation system will work very well—and a simple ventilation system is much easier to install and far more affordable.

Actually, anyone who lives in a house has a tremendous amount of control over how polluted the air becomes in that house—just by being careful about what they bring indoors. By using low-tox cleaning supplies and untreated fabrics, by banning smoking, and by selecting remodeling materials carefully, the indoor air will be far healthier than it will be in a similar house where the occupants aren't as careful—even if there is no ventilation system! With a ventilation system, the air will be even better. So, if we are truly concerned about indoor-air quality, we must all take personal responsibility for the materials we use to build, and what we bring indoors. All experts agree that the most effective way to deal with a polluted house is source control. In other words, control the source of the pollution first, then ventilate using either local-exhaust ventilation, a central ventilation system, or both.

Providing adequate ventilation for indoor tobacco smoke can be very difficult. If a family member is a smoker, the best solution (for many reasons) would be for them to quit smoking. The next best solution (to protect nonsmoking family members) would be for them to only smoke outdoors. If neither of those options is feasible, the smoker should confine his or her smoking to a single room, and that room should be provided with extra local ventilation, rather than installing an extra-powerful general ventilation system to remove smoke from the entire house. A smoking room should be ventilated at a rate of at least 8–10 ACH. Bathrooms are sometimes suitable as makeshift smoking rooms.

The KISS principle

In our technological society, we have gotten used to operating extremely complicated pieces of equipment on a daily basis. Automobiles, answering machines, computers, and televisions are vastly more complex than the buggies, butter churns, and quill pens of our forebears. As a result, we often tend to think that something must be complicated to be better. A major drawback to this thinking is the fact that not only can complex equipment be expensive, when something goes wrong, repairs can also be costly. A repair bill for a fax machine or an automatic camera can easily be half the cost of new equipment, that is if you can locate someone properly trained in making such repairs. If a central ventilation system is overly complex, it will be difficult to 1) understand, 2) install, 3) operate, and 4) service. So, what is the solution?

In an article titled *Indoor Air: Is it Dangerous to Your Health* that originally appeared in the October/November 1986 *Solplan Review*, editor Richard Kadulski suggested three points to keep in mind in selecting a ventilation system: "Firstly, above all, the system should follow the KISS principle: 'Keep It Simple, Stupid.' Secondly, Murphy's law should be accepted as a given: 'If something can go wrong it will.' A simple design with a minimum of complex parts that can break down should be used. Thirdly, when selecting equipment, look for the best, most durable, easily serviceable equipment. If it worthwhile to use a complex piece of equipment, look for its track record, warranty provisions, and availability of service and parts." Even though Kadulski's advice is several years old, it is still sound. Simple is often better than complex.

There are some very real advantages to small-capacity, continuously running central ventilation systems. They are simple, inexpensive to install and operate, easy to understand, and they can work very well. However, we often still tend to make things more complicated than they need be. Complex ventilation systems are often promoted by salesmen who profit by selling those same more complicated and expensive systems. Keep it simple!

Accessibility and maintenance

Ventilation equipment should always be located where it is relatively easy to get to. This is so the occupants will be able to take care of the periodic servicing that all mechanical equipment requires. No one likes routine maintenance, but ventilation equipment will last longer, operate more efficiently, and provide cleaner air if some simple routine maintenance operations are performed regularly.

Because a ventilating fan might be running for extended periods, day after day, it must be taken care of if it is to last. A year contains 8,760 hours—a long time for a continuously operating central ventilation system to operate unattended. The life span of a fan is typically determined by the life of the bearings, so periodic lubrication is important. (Some bearings are permanently lubricated.) In addition, a ventilator will operate more efficiently if any buildup of dust is cleaned off the fan blades and the inside of the cabinet occasionally. Inlets, outlets, grilles, and through-the-wall vents should also be cleaned periodically. The frequency of cleaning depends on how dirty the air is that is passing through the system. With very clean air, only light occasional cleaning is required, but it is a good idea to do so at least once a year.

Some ventilation units incorporate filters of various types (most are low-efficiency filters) that need ei-

Understanding Ventilation

ther periodic cleaning or replacement. A clogged filter can mean extra resistance to airflow and reduced ventilation capacity. It is possible for a filter to become so dirty that the airflow is only a fraction of what it should be, and a clogged filter can cause the airflows in a balanced ventilation system to be out of balance. HRVs have a core that should be cleaned at least once a year. Insect screen at a fresh-air inlet and blades on fans can get especially dirty. Detailed maintenance information can generally be found in the equipment's operating manual. If the manual is missing, contact the manufacturer for a new one.

Ventilation equipment takes up room in a house, just like a furnace, air conditioner, water heater, and any other mechanical device. Because every square foot of a house is valuable, mechanical equipment is often installed in out-of-the-way places such as crawl spaces and attics. This is a big mistake. Ventilation equipment needs to be accessible if you expect it to ever be maintained. When ventilation equipment is hidden away somewhere, and you must crawl 30' on your belly just to find it, it will probably be forgotten until it breaks—and even then it may be ignored because it is too hard to get to. It is essential that a ventilation system be maintained and in good working order for the occupants to derive any benefit from it.

In selecting a location for ventilation equipment, don't choose unconditioned or cramped spaces (*e.g.* attics and crawl spaces) if it can be helped. Cubbyholes under stairwells are often too small to physically fit the equipment into, or to run ducts into and out of. Better, more accessible locations include the basement, laundry room, or utility room. A location above a dropped ceiling or above a washing machine and clothes dryer can be suitable if the equipment can be reached easily by standing on a stepladder.

Occupant education

Builders and designers need to be careful not to create a complex ventilation system that the occupants won't be able to understand or maintain. They also need to make sure that people know where their ventilation system is located and precisely what it does. If the occupants have little understanding of their system, they won't know when it is malfunctioning, and they may inadvertently sabotage it. Ecotope, Inc., of Seattle, WA, studied 66 households in 7 multifamily buildings that had a variety of ventilation systems installed in them.(Ventilation Systems) Among other things, they found the following:

- Over 25% of the occupants were not even aware that they had a ventilation system.
- Only 30% of the occupants could describe how their system worked.
- 65% of the occupants had complaints about noise, drafts, stuffiness, odors.*
- One occupant taped shut all the exhaust grilles because of the mistaken perception that the system was "sucking all the warm air out of the apartment."
- One occupant found a way to shut off a system that was designed to run continuously.
- At one site, 25% of the heat-recovery ventilators were completely disabled.

* These systems obviously weren't transparent.

In an earlier study that examined local-exhaust fan use, 16 households out of 112 *never* used the fan.(Yuill) Many respondents found the fans too noisy. According to one expert, "Since environmental effects from indoor air pollutants are not well-understood, it is likely that many people will not perceive a need to operate these devices very frequently. This means that homeowner education regarding the need to operate ventilation systems...is crucial."(Lubliner)

Based on these findings, it is obvious that better occupant education is essential. Even the best ventilation systems won't work if they are turned off, disabled, or malfunctioning. Every day, people operate very complex pieces of equipment—automobiles, televisions, computers, etc., but at the same time, most of us have difficulty programming our VCRs. Ventilation systems are quite simple in comparison, and most are easy to learn how to operate, but basic occupant education is still necessary.

Whenever someone moves into a house having a mechanical ventilation system, they should be given a simple explanation of how the system operates, or they should be provided with an easy-to-follow operating manual. This is especially important when a house is tightly constructed and the ventilation system is the primary source of fresh air. (Tight houses have minimal natural ventilation.) There should be, at least, a well-labeled switch or control in a conspicuous location. The responsibility of occupant education originally rests with the architect or designer, but he passes that responsibility on to the builder, who in turn may pass it on to a Realtor or a landlord. In the end, the person who hands the front door key to a new homeowner or tenant has the responsibility of telling them how to operate the ventilation system. Unfortunately, that person often has very little knowledge of anything about the house.

Future occupancy

Houses are bought and sold every day. As a result, sooner or later, most houses will end up having quite a few different occupants. In order that future occupants understand how to operate their ventilation system, it is a good idea to place a small engraved plate next to the controls, where it will be readily seen by the occupants. The actual cover plate of the switch itself can be labeled. A central hallway or a kitchen are good locations. The plate should contain basic operating and maintenance instructions. Following are three examples:

> This ventilation system is capable of changing the air in the house once every 3 hours at low speed, a rate that is suitable for 4 occupants under normal conditions. If extra people or odorous activities create additional indoor air pollution, the speed control should be set on high. For annual maintenance, see the label on the ventilation-equipment cabinet in the basement.

> This timer controls your energy-efficient heat-recovery ventilation system. The system should be operated for at least 2 hours per day per occupant— more if polluting activities or hobbies are performed indoors. Air is filtered through an electronic air filter that needs to be cleaned at least once a month. Other maintenance is required yearly. For more information, see the owner's manual.

> This switch controls your central-exhaust ventilation system. This system is designed to provide fresh air and remove stale air continuously. It should only be shut off when the house is unoccupied, or for short periods when the outdoor air quality is especially poor. Do not leave the ventilation system turned off for extended periods. The operating manual is hanging next to the furnace.

***Figure 18–2.** Typical instruction plates.*

Part 5

Mechanical ventilation systems

Chapter 19 Central-exhaust ventilation systems
Chapter 20 Local-exhaust ventilation systems
Chapter 21 Central-supply ventilation systems
Chapter 22 Balanced ventilation systems
Chapter 23 Balanced ventilation with heat recovery
Chapter 24 Passive ventilation systems
Chapter 25 A nine step design and installation process
Chapter 26 Typical central ventilation systems

Chapter 19

Central-exhaust ventilation systems

This chapter summarizes the advantages and disadvantages of central-exhaust ventilation systems and discusses how they can be used in homes. It also describes various pieces of equipment currently available.

Exhaust ventilation systems (both central and local systems) depressurize buildings by blowing air out of them. This can be accomplished with something as simple as a portable window fan, although this isn't a year-round solution to ventilation in many climates. (Actually window fans can be used to either blow air outdoors (depressurizing a house) or blow air indoors (pressurizing a house).) Some high-capacity window fans can move so much air—from several hundred cfm up to about two thousand cfm—that it isn't possible for sufficient make-up air to infiltrate through the random holes in a house. For a high-capacity window fan to operate efficiently, there should be at least one other open window somewhere else in the house so make-up air can easily enter.

Central-exhaust ventilation systems are often called *exhaust-only* ventilation systems. However, no ventilation system can only exhaust air from a house because an equal volume of make-up air will always enter. Therefore exhaust-only probably isn't the best terminology to use.

With a very simple central-exhaust ventilation system, a fan is used to blow the air out of the house from a central location. Fresh make-up air from the outdoors enters either through random holes in the structure or through deliberate inlets to try to relieve the depressurization caused by the fan. Prior to the widespread use of air conditioning, many houses were cooled in the summer with a central-exhaust fan. Often located in the ceiling of the uppermost story, these whole-house fans blow a great deal of air outdoors (the largest are capable of moving 10,000 cfm). An equal amount of make-up air then enters through open windows on the lower level. Large fans used for cooling are generally

Understanding Ventilation

too powerful for ventilation in the winter, but they can work well in warm weather. Manufacturers include **Fasco Consumer Products, Inc.**, **Grainger, Inc.**, **Lomanco, Inc.**, and **Penn Ventilator, Inc.** A 1,000-cfm exhaust fan is available from **Tamarack Technologies, Inc.** that comes with an insulated, motorized cover that automatically closes to prevent unwanted infiltration or exfiltration when the fan is not in use.

High-capacity exhaust fans are used primarily as a low-cost way of cooling a house in the summer—ventilation is a side benefit. These fans should only be considered ventilating fans if they can be used continuously in all seasons of the year. Window fans are best used for temporary ventilation because they leave a window accessible to burglars. When an exhaust fan is designed specifically for ventilation air, it doesn't need much capacity—usually less than 100–200 cfm is sufficient—and it doesn't provide an entry route for thieves to enter a house.

An exhaust fan used to provide a low level of ventilation throughout an entire house is called a central-exhaust ventilation fan. If an exhaust fan is used primarily to remove a buildup of pollutants or excess moisture from one part of a house, it is called a local-exhaust ventilation fan. Local-exhaust fans will be covered in Chapter 20.

Central-exhaust ventilation strategies

There are several ways central-exhaust fans can be used to ventilate houses. The simplest method is to pull air from a single location in the middle of the house, often a hallway, and blow it directly outdoors (Figure 19–1). A variation is to use a bathroom exhaust fan to ventilate the whole house. In either case, you must use a fan that is quiet, designed to run for long periods at a time, and sized to ventilate the whole house. Using a bath fan as a central-exhaust fan is sometimes called an upgraded bath fan. You will need to either undercut the bathroom door ($^1/_2$" is usually sufficient for low-capacity fans) or install a pass-through (for high-capacity fans) to allow make-up air to enter the bathroom when the door is closed.

Another approach is to connect a central-exhaust fan to a system of ducts and fittings so air is pulled from two or more rooms simultaneously—usually rooms with excess moisture such as the kitchen and the bathrooms—then the air is exhausted to the outdoors through a single duct (Figure 19–2). When there are several pickup points located near pollution sources, stale air will be blown outdoors before it has a chance

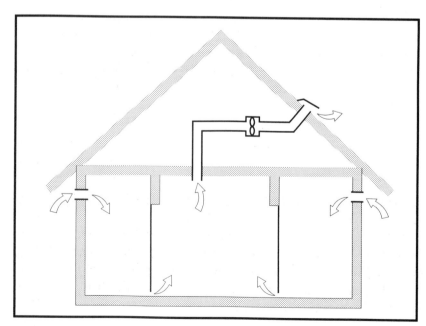

Figure 19–1. A popular central-exhaust ventilation strategy involves expelling air from a central location in a house and allowing make-up air to enter by way of through-the-wall vents.

Chapter 19 Central-exhaust ventilation systems

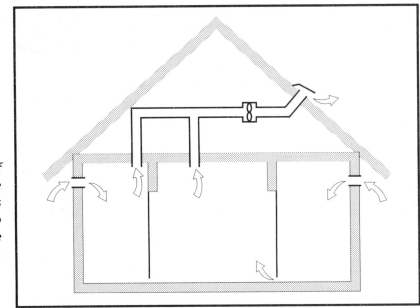

Figure 19–2. The distribution of air in a house will be more effective if a central-exhaust fan is connected to a series of ducts to pull stale air from more than one room at a time,

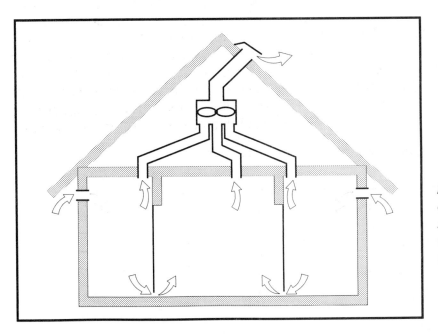

Figure 19–3. "Multi-port" central-exhaust ventilation systems have several connecting points on their cabinet to which stale-air exhaust ducts can be attached.

to mix with air in other rooms. Some central-exhaust-fan cabinets have connections for several ducts. These are called *multi-port* fans (Figure 19–3). They are designed so ducts from several rooms can be connected directly to the fan cabinet to ensure good airflow patterns. Even with multiple suction points, it is important to remember that you often still need a range hood in the kitchen because many exhaust fans aren't designed to handle grease-laden air.

Central-exhaust ventilation strategies are often installed with through-the-wall vents so make-up air can easily enter the house. (Through-the-wall vents are most suitable in small, tight houses or apartments.) It is also possible to have the make-up air enter the house through

279

Understanding Ventilation

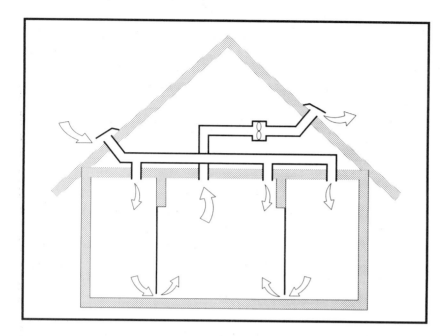

Figure 19–4. Though it is seldom done, it is possible to have make-up air enter a house through a system of ducts rather than through-the-wall vents

a single duct with branches to several rooms (Figure 19–4), but this isn't done often. Or, the make-up air can enter through a duct connected to the heating/air-conditioning system, then the heating/air-conditioning ducts are used to distribute both conditioned air and ventilation air. (Actually, when properly installed, this setup will *not* depressurize a house—it will result in a neutral pressure, so some of the advantages and disadvantages listed below aren't applicable.) When cold air enters through a fresh-air duct in the winter, and the duct is within the heated conditioned space, it must be insulated, and the insulation must be covered with a vapor retarder to prevent condensation from forming.

Advantages of central-exhaust ventilation

There are three basic advantages to using a central-exhaust ventilation system to provide a house with fresh air. From an economic point of view, these systems are often very reasonably priced. They are also fairly simple and easy to maintain. From a technical standpoint, they can prevent hidden moisture-related problems in cold climates.

Low installation cost

Central-exhaust ventilation systems are often modestly priced because many installations require only a minimal amount of equipment: an exhaust fan, a few ducts, perhaps a few through-the-wall vents, and a simple control switch. Note: Here we are only talking about a low installation cost. In a harsh climate, a central-exhaust ventilation system can have a high *operating* cost associated with tempering of the incoming air.

Easy to maintain

Because they are often quite simple installations and because they don't require much in the way of complicated equipment, central-exhaust ventilation systems can be easy for occupants to understand and to maintain. This is very appealing to people whose lives are already too complicated.

Minimizes hidden moisture problems in heating climates

Exhaust systems depressurize houses. As a result, the potential for moisture condensation and eventual mold or rot inside building cavities is reduced in the winter. This is because a certain amount of cold, dry make-up air will be infiltrating passively through the random holes in the structure whenever the exhaust fan is running. This prevents warm moist indoor air from getting into those cavities and condensing on cold surfaces such as the siding or sheathing. See Chapter 8 for a discussion of hidden moisture problems.

Disadvantages of central-exhaust ventilation

Four conceivable disadvantages are associated with central-exhaust ventilation systems. As was already stated, in harsh climates, operating costs can be high. Second, excessive depressurization can bring unwanted pollutants into the living space. Third, extended periods of depressurization can cause hidden moisture problems in hot, humid climates. And fourth, too much depressurization can cause backdrafting and spillage.

High operating cost

Any ventilation system that does not incorporate heat recovery can have a high operating cost in very cold or very hot climates because tempering the incoming air places an extra burden on the furnace or air conditioner. Most central-exhaust ventilators aren't capable of recovering heat; however, there is one specialized device—a central-exhaust heat pump that is coupled to a water heater—that can recover heat that would otherwise be wasted. It is described later in this chapter. Before deciding that this is a major disadvantage, it is important to also consider the cost of running the exhaust fan itself because central-exhaust fans typically use less energy than heat-recovery ventilator fans.

The tempering-cost disadvantage is primarily applicable to cold climates with high ventilation rates. (If a ventilator is especially powerful, it can have a high operating cost in any climate, so it is important to size a system correctly.) In mild climates, where outdoor temperatures are not extreme, there won't be much energy needed to heat or cool the incoming air, so there is not a great deal of energy to recover. Even in many warmer areas of the U.S., there isn't enough cost savings to make heat recovery a cost-effective option. Therefore, a high operating cost is generally only considered a disadvantage in very cold climates. Chapter 13 covers costs.

Unwanted pollutant entry

When a house is depressurized, make-up air will be drawn from the outdoors to the indoors through any opening it can find. Typical random holes in houses include the gaps around window and door frames, around electrical outlets, the gap between a floor and a wall, and the dozens of holes in the framing and drywall that were cut by electricians, plumbers, and heating/air-conditioning contractors. When make-up air infiltrates through random holes in the structure of a house, it can bring unwanted pollutants indoors such as small particles or odors from the insulation, soil gases, lawn chemicals, or termite chemicals. This disadvantage can be minimized by tightening the house or by reducing the degree of depressurization (*e.g.* installing more make-up air entry points). See also Chapter 9.

Actually, even though the potential exists for pollutants to be sucked indoors by a ventilation system that depressurizes a house, in many cases there is more fresh air being brought in than polluted air. The net result is cleaner indoor air than if no ventilation system were operating. Depressurization-induced pollution entry is often more of a problem with leaky ducts in forced-air heating/cooling systems and high-capacity local-exhaust fans than with modestly sized central-exhaust ventilation systems.

Cause of hidden moisture problems in cooling climates

In an air-conditioned house in a hot, humid climate, depressurization can cause hot, humid outdoor air to be pulled passively indoors through the random holes in the house, resulting in hidden moisture condensation inside the structure of the house. As noted above, depressurization prevents hidden condensation problems in a cold climate; now we see that it can cause them in a hot, humid climate. The same laws of physics apply in both cases. When hot, humid outdoor air infil-

Understanding Ventilation

trates through the structure of a house, it will get cooler as it reaches the air-conditioned interior. Because cool air can't hold as much moisture as warm air, the RH of the infiltrating air rises as the temperature falls. In an air-conditioned house, the drywall (or plaster or paneling) will be cool. When the infiltrating air hits the back of that cool surface, water can condense on it. This can result in mold growth or rot—hidden inside the structure. This will not occur if the interior of the house isn't air conditioned because there must be a cool surface on which the moisture can condense. As was discussed in Chapter 8, this disadvantage is limited primarily to the hot and humid southern U.S.

Cause of backdrafting or spillage

Because a chimney is simply a hole in a house, make-up air often enters a depressurized house by passively flowing down the chimney. This can result in two potentially serious problems associated with depressurization—backdrafting or spillage—that can result in death. When warm combustion by-products flow up a chimney, they generate a certain amount of upward pressure called a *draft*. If the depressurization pressure in a house is greater than the upward pressure of the draft in the chimney, then it can be difficult for combustion by-products to rise up the chimney. The problem is potentially so serious that central-exhaust ventilation systems are usually not recommended in any house having a combustion device that depends on a natural draft in a chimney to function correctly. This includes the vast majority of natural gas and oil furnaces in the country, as well as combustion-fired water heaters, wood stoves, and fireplaces. See Chapter 10 for a more in-depth discussion. The bottom line: If you use a central-exhaust ventilation system, don't use any natural-draft combustion appliances in the house.

Minimizing the disadvantages of depressurization

Pressure-related disadvantages are less serious in tight houses where negative pressures are not excessive. Tight houses have fewer random holes through which air can infiltrate, and small negative pressures won't pull much air through the holes that do exist. So, tight construction is advantageous. In such a house, the inlets (*e.g.* through-the-wall vents) that are deliberately placed in the walls do two things: 1) They provide a dedicated pathway for air to enter and 2) they help minimize the depressurization.

Figure 19–5. *In-line fans work well as central exhaust fans because they are long-lasting and available in a wide variety of capacities.*

Chapter 19 Central-exhaust ventilation systems

Figure 19–6. A fan mounted in a box insulated for sound control is a popular central-exhaust fan option.

If a loosely built house has deliberate air inlets such as through-the-wall vents, the make-up air will enter through both the inlets and the random holes in the structure. When you have both types of openings as pathways for air movement, you can't specify which openings you want the air to move through. (It isn't unusual for air to enter through-the-wall vents easier than it enters through a random openings as defined as an ELA. See Chapter 16 for a discussion.) Thus, it can be difficult to predict exactly where and how much make-up air will enter when there are a variety of openings in a house. On the other hand, in a tight house, most of the make-up air will enter the inlets because there are few other entry points. Therefore, the disadvantages of depressurization causing unwanted pollutants or moisture to travel through the structure can be reduced to a minimum by constructing an airtight house.

When a house is depressurized, backdrafting and spillage can be prevented by using combustion appliances equipped with an induced draft fan, a totally sealed combustion chamber, or by locating them in a sealed mechanical room that is not affected by depressurization somewhere else in the house. (In practice, this is not always very easy to do.) Or, you can use non-combustion forms of heating such as solar or electric. For a discussion of backdrafting and spillage, see Chapter 10.

Central-exhaust ventilation equipment

The following section covers the actual equipment that can be used in central-exhaust ventilation systems. It is possible to purchase individual components (*e.g.* fans, through-the-wall vents, controls, etc.) or pre-engineered packaged systems. When you select individual components, you will need to do some calculations to determine fan capacities, duct sizes, etc., and you will need to decide exactly which features are needed. With a packaged system, most of the guesswork is eliminated. All you have to do is follow the installation instructions. If you run into difficulties, the manufacturer or distributor will usually be available to answer your questions. Some suppliers offer complete ventilation-system packages that include the fan, ducts, grilles, control, etc.

To minimize noise inside the living space, central-exhaust fans are often mounted in unheated attics and crawl spaces. If they are not operated continuously, they can occasionally freeze up if they are shut off during cold weather. This is most likely when a fan is in a cold attic because stack effect often causes humid house air to escape through the fan. (This can happen at a low

283

Understanding Ventilation

flow rate even when the fan is off when there is no damper to prevent such air movement.) This humid air can then condense and freeze inside the cold fan. If a fan is mounted in a cold crawl space, freezing is less likely because stack effect will tend to cause cold, dry, outside air to enter through the fan.

Individual fans

A number of equipment choices are available if you would like to use a central-exhaust fan with a single centrally located exhaust grille. For example, several manufacturers offer fans specifically for this purpose. It is also possible to upgrade either the bath fan or kitchen range hood to a piece of equipment that is designed for continuous operation. However, a range hood is often too powerful and too loud—unless it has a solid-state variable-speed control to slow it down. If a range hood runs slower, the air noise is substantially reduced, making extended operation tolerable, but the speed control may cause an annoying hum.

An individual exhaust fan that can be left running for extended periods is one of the simplest ventilation strategies. Besides a rating for continuous operation, look for high-quality bearings, permanent lubrication, a low energy requirement, and quietness. When combined with some type of control, and perhaps some through-the-wall vents, this can be one of the lowest-cost ventilation systems currently in use.

Some of the best individual fans to use are the *in-line* fans used by radon contractors (Figure 19–5). They are often quieter and longer-lasting than ceiling- or wall-mounted fans. Ducts can be easily hooked to each end of these fans with the suction side connected to a grille in a central location in the house and the exhaust connected to the outdoors. Companies that sell in-line fans include **Fantech, Inc.**, **Kanaflakt, Inc.**, **Rosenberg, Inc.** and **Vent-Axia, Inc.**

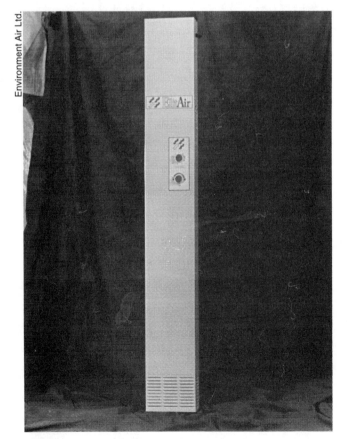

Figure 19–7. This unique exhaust fan pulls air from near a basement floor and blows it outdoors.

Chapter 19 Central-exhaust ventilation systems

Figure 19–8. Reversible through-the-wall fans are available that can work as either a supply fan or an exhaust fan.

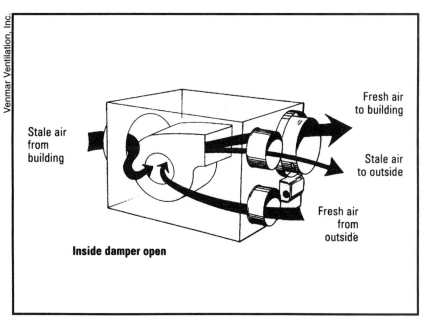

Figure 19–9. This central ventilating system tempers the incoming air by mixing it with house air. More stale air is exhausted than fresh air brought indoors, so this system results in a slight depressurization.

285

Understanding Ventilation

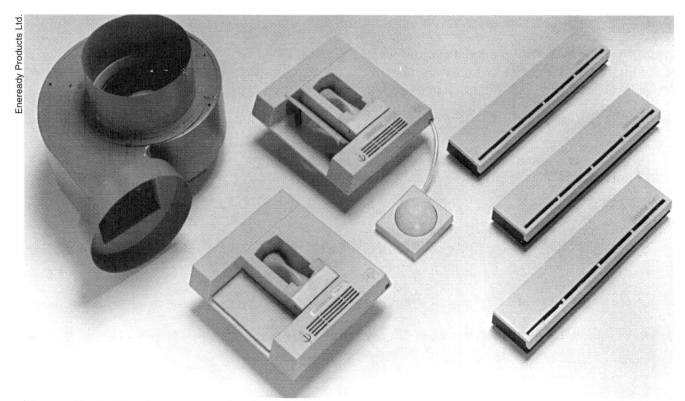

Figure 19–10. The Aereco central-exhaust ventilation system is a well-thought-out approach to exchanging air in a house.

American Aldes Corp., **Broan Mfg. Co., Inc.**, and **Kanaflakt, Inc.** have *single-point exhaust ventilators* that are basically an exhaust fan mounted in a metal box that can be used either as remote-mounted exhaust fans with a single pickup, or connected to a branch system of ducts for multiple pickup points (Figure 19–6). Any of the local-exhaust fans mentioned in the next chapter, especially the quiet ones, are also suitable as central-exhaust fans. In fact, a wide-variety of general-purpose fans can be used for central-exhaust ventilation, available from companies such as **Broan Mfg. Co. Inc.**, **Fasco Consumer Products, Inc.**, **Grainger, Inc.**, **Penn Ventilator Co., Inc.**, and **Reversomatic Htg. & Mfg. Ltd.** (**Panasonic** fans are especially good because they are extremely quiet and use very-little energy. They have 4 models between 50–110 cfm rated at 13–20.7 watts and $1/2$–1 sones.) In addition, **Greenheck** manufactures a number of industrial-quality fans appropriate for use in residential applications. In general, fans used for central-exhaust ventilation should not have cheap shaded-pole motors. Permanent-split-capacitor motors and ECM motors are more energy-efficient, although ECM motors are not yet widely used.

Environment Air Ltd. has a specialized 40 watt, 145-cfm basement ventilating fan (Figure 19–7). It pulls air from the basement along the floor, directs it upward through a $4^{1}/_{2}$" x $10^{1}/_{2}$" white cabinet, then blows the air outdoors through the above-ground portion of the wall. This unit comes complete with a dehumidistat, speed control, and power cord.

Vent-Axia, Inc. has several British-made fans rated from 150 to over 1,000 cfm (55–130 watts) that are reversible (Figure 19–8). This means they can be used as either supply or exhaust fans at different times of the year. Their wall-mounted controller has three selector switches: on/off, high/medium/low fan speed, and intake/exhaust direction. There are models that can be installed through a wall or roof, or connected to a duct. Some of these fans can be fitted with a multi-port adapter plate. Vent-Axia also has a unique fan that mounts directly on a window pane. To install it, you need to cut a round hole in the glass through which the fan can blow

Chapter 19 Central-exhaust ventilation systems

air. The size of the hole depends on the fan capacity. These fans can even be mounted in insulated windows. One of the advantages of a reversible fan is that it can be used as a supply fan during cold weather and an exhaust fan during warm weather to avoid hidden moisture problems. However, this point might be confusing to the occupants unless there is a very clear instruction plate next to the control.

Venmar Ventilation, Inc. has a unique 200 watt central-exhaust ventilator that pulls a certain amount of air from the outdoors and mixes it with stale air that is drawn from the house (Figure 19–9). This tempers the incoming air so it isn't uncomfortable. The fan then blows most of the mixture indoors and part of the mixture outdoors. It pulls in about 30 cfm of fresh air from the outdoors and exhausts about 80 cfm of the fresh/stale-air mixture, so it results in a slight depressurization. About 280 cfm of air is recirculated throughout the house for good mixing. When the control panel is set to the recirculation mode, a damper closes off the outdoor air duct. This gives the option of circulating air through the house without any stale air being exhausted or any fresh air being brought in. Because this system has both a dedicated exhaust and a dedicated inlet, it doesn't need through-the-wall inlets. Of course, when it depressurizes the house, some air—50 cfm (80 – 30) to be precise—will enter through random gaps and holes in the structure.

Multi-port systems

Several companies make central-exhaust ventilation fans that are already set up for a multi-port installation. They usually consist of a cabinet with several connecting points for stale-air ducts that are run from various rooms in the house. A single exhaust duct is run to the outdoors. These manufacturers generally offer through-the-wall vents to create paths for make-up

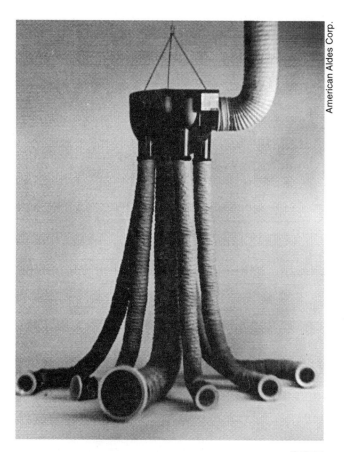

Figure 19–11. The well-engineered American Aldes central-exhaust system comes complete with everything needed for installation.

air to enter the house. These systems are often meant to run continuously, but they certainly can be turned on or off with some type of control.

Airex has a well-engineered Aereco central-exhaust ventilation system, complete with controls that allow you to vary the amount of air being exhausted from different rooms (Figure 19–10). Technically, this is a single-port fan, but it is almost always connected to a duct system to pull air from more than one room at a time. This system has been used very successfully in France and North America for a number of years. The ventilation fan itself is designed to run continuously and only consumes 38 watts. The fan has two 6" connections—one is run to the outdoors and the other is branched to several rooms in the house, typically the kitchen and bathrooms. The suction lines are fitted with Aereco air exhausters—special grilles that automatically vary the amount of air leaving a room based on the relative humidity. There are three sizes of exhausters. When a room is dry, they allow 3, 9, or 12 cfm to pass through them, respectively. If the air becomes more humid in one room, the exhauster located there automatically opens to allow up to 30–50 cfm to be pulled from that particular room. The exhausters also have a manual override (a unique pneumatic control) that will boost the room output up to 100 cfm (depending on the model).

If, for example, someone is taking a shower, only a minimum amount of air will be exhausted from the bathroom until the relative humidity starts rising, then the exhaust rate will increase automatically until the humidity decreases. If offensive odors are created in the bathroom, a manual override button can be pushed to increase the exhaust rate even more. Airex also distributes humidity-controlled through-the-wall vents (see below) that are typically located in each bedroom and the living room.

American Aldes Corp. also produces well-engineered multi-port central-exhaust ventilation equipment (Figure 19–11). They have well-integrated systems in capacities of 160, 190, 230, 330, and 400 cfm (60–145 watts), complete with ducts, grilles, and through-the-wall vents. They also offer noise-absorbing ducts, controls, backdraft dampers, and special duct fittings. A typical installation might have 3" ducts running from the bathrooms and laundry room to the exhaust fan, a 6" or 8" duct running from the kitchen to the fan, and a 6" duct running from the fan to the outdoors. Their residential models have connection points for up to six rooms, but larger units are available that are suitable for ventilating several apartments with a single ventilator. The Aldes system is usually supplied with Constant Airflow Regulators on some ducts. These special self-

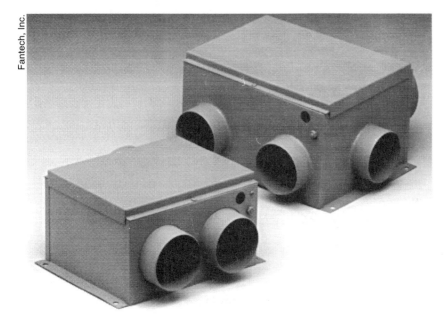

Figure 19–12. Several manufacturers offer simple-to-install multi-port ventilators that are basically "a fan in a box."

Chapter 19 Central-exhaust ventilation systems

Figure 19–13. Therma-Stor's central-exhaust heat-pump water heater is an efficient way to obtain both ventilation air and water heating throughout the year.

adjusting dampers contain a special pressure-sensitive "bladder" that automatically maintains a certain airflow no matter the fan speed or static pressure in the ducts. A popular approach is to use Constant Airflow Regulators on the 3", lines but not on the 6" kitchen line. On low speed, a small amount of air will be pulled through all the ducts, but on high speed, the constant airflow regulators keep the flow uniform in the smaller ducts, and the kitchen gets all the extra capacity. This allows you to use the same ventilation system for both general ventilation and local kitchen ventilation. Most central ventilation systems aren't designed to handle kitchen grease, but this one can be connected to a range hood if some minor modifications are made. The Constant Airflow Regulators are made in 3–8" sizes for airflow ranges of 9–235 cfm.

Aston Industries, Inc. has a Model 2000 central-exhaust ventilator with connections for three 5" stale-air ducts. It is rated at 80 cfm and 109 watts and designed to run continuously.

Broan Mfg. Co., Inc. has four fairly straightforward multi-port central-exhaust ventilators with capacities of 110, 150, 210, and 290 cfm (50–210 watts). These units have a single 6" outlet that is run to the outdoors and up to four 4" suction lines that are connected to different rooms in the house. This is a fairly simple system basically consisting of "a fan in a box." Broan also has a through-the-wall vent, the SV1 Fresh Air Inlet,y of controls.

Fan America, Inc. offers three sizes (75, 100, or 140 cfm ranging from 48–100 watts) of their Engineered Ventilation System. These systems consists of an in-line tube fan connected to a separate collector box. The collector box has 3 or 4 suction ports, depending on the fan size. Each suction line is fitted with a Constant Airflow Regulator to maintain a uniform exhaust capacity at each room. A roof scupper, mounting bracket, grilles, and accessories are included.

Fantech, Inc. has four CVS Series multi-port central-exhaust fans in capacities of 190, 275, 330, and

289

Understanding Ventilation

400 cfm (89–158 watts) (Figure 19–12). A speed controller can be used to obtain lower airflows. The two smaller models have a pair of 4" suction ports and a single 5" exhaust. The larger models have four suction ports and either a 6" or $7^1/_2$" exhaust. Backdraft dampers and grilles are available. Fantech also markets in-line fans as multi-port exhaust fans by connecting them to one or more Y fittings.

Reversomatic Htg. & Mfg. Ltd. has several multi-port ventilators designed to pull air simultaneously from 2–4 rooms (usually bathrooms and the kitchen). Their capacity range is 200–325 cfm at 100–345 watts. Some are designed to handle the exhaust from a clothes dryer without a lint buildup, as well as humid or odorous air from the bathroom.

Therma-Stor Products offers a two-speed (165/295 cfm, 50/95 watt) Quiet-Vent multi-port central-exhaust ventilator, complete with a programmable 7-day timer and a simple metal filter. With the timer control, this unit can be easily set up to be an intermittently running system. The cabinet is insulated and has a single 6" outlet and one 6" and three 4" suction ports. Therma-Stor also offers duct fittings, grilles, through-the-wall vents, and other controls. They often recommend that the high speed of the fan be activated by a dehumidistat in the bathroom.

Central-exhaust heat-pump water heater

The only central-exhaust devices capable of heat recovery are exhaust-vent heat-pump water heaters (Figure 19–13). These high-quality devices are manufactured by **Therma-Stor Products**. The Therma-Vent and the Envirovent operate like any other multi-port central-exhaust system in that they depressurize a house by pulling air from several rooms and blowing it outdoors. Make-up air enters the house either through random holes in the structure or by way of through-the-wall vents. Heat-pump water heaters are quite popular in Scandinavia—more than half of all new Swedish houses have similar devices.

What makes these systems different is the fact that they contain a small heat pump that extracts heat from the air that is being exhausted. During operation, the stale air from the house (72°F ±) passes through the heat pump where heat is extracted before the air is blown outdoors. By the time the stale air passes through the heat pump, it is reduced in temperature to about 35°F. Therefore, about 37° of heat (72 – 35) is reclaimed from the stale air, no matter what the outdoor temperature.

This 37° of heat is transferred either into a hot-water storage tank or as warm air back into the house.

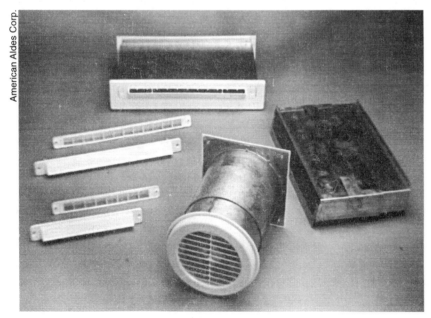

Figure 19–14. American Aldes offers several different types of through-the-wall vents.

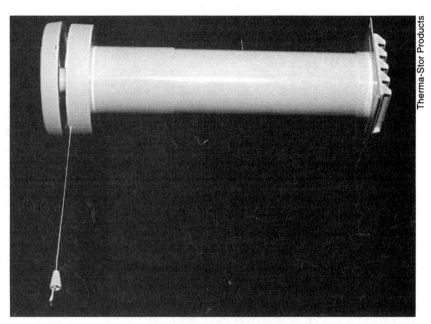

Figure 19–15. Fresh 80s are popular through-the-wall vents.

When the heat is transferred into the water tank, it is available as hot water for residential use. If the air in the house needs to be warmed up, the heat pump sends the heat to the house rather than the water tank. In the summer, the heat pump provides the house with a certain amount of air conditioning. To make hot water always available, the water tank has a supplemental heater that maintains the water temperature when the ventilation system is shut off, or if most of the output from the heat pump is being used to heat the house.

The ventilation fan can be programmed to run at specific times of the day, and it will also turn on automatically whenever there is a demand for hot water. So, ventilation is actually controlled by two things: occupant demand and hot water usage.

Because this device will recover heat from outgoing air no matter what the indoor or outdoor temperature, it can be somewhat more energy-efficient than a conventional balanced HRV (see Chapter 23), especially in a mild climate. In fact, it is the most energy-efficient water heater on the market. And it is very quiet compared to some other ventilation systems. The primary drawback is the fact that it has a higher up front cost than other ventilation strategies. The life of the storage tank itself should be similar to other hot-water tanks: 10 years or so.

Through-the-wall vents

There are several companies currently producing through-the-wall vents specifically for residential ventilation systems. Custom-made vents can be constructed in a variety of different ways[McLeister], but if vents are to be custom-made, it is important that they be lined and sealed with something that will prevent insulation from being drawn indoors. Ready-made vents are often easier to install than making vents from scratch, they are specifically designed for the purpose and are reasonably priced ($25–35). Because a through-the-wall vent is basically just a hole in the wall, air can move through it in either direction when the exhaust fan is turned off; thus, even though a through-the-wall vent is installed as an inlet, it can occasionally become an outlet. For a discussion of guidelines on locating inlets and outlets, see Chapter 16.

When specifying through-the-wall vents, you must know how much air can be expelled from the house when all the exhaust devices are operating simultaneously. It will then be possible to determine the locations, quantity, and sizes of the vents. Several small through-the-wall vents are more effective than a few large ones, because they will allow fresh air to be introduced into more parts of the house. One of their pri-

Figure 19-16. Thinking Vents manufactures through-the-wall vents designed to be mounted in a window sash.

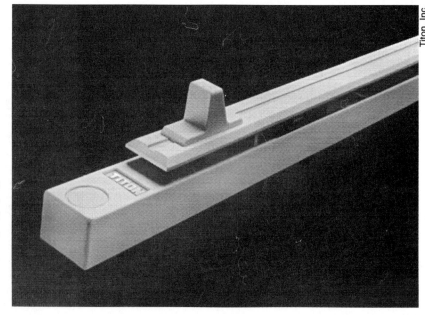

Figure 19-17. Titon offers a complete line of low-profile through-the-wall vents, called Trickle Ventilators, that can be mounted in a window sash.

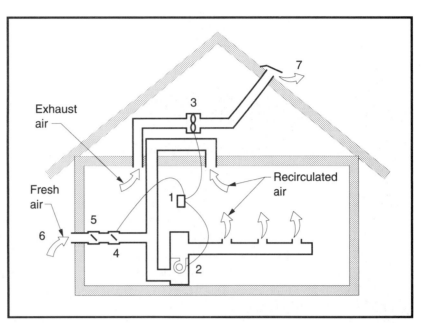

Figure 19–18. This illustration shows the components necessary to couple a central-exhaust ventilation system with a forced-air heating/cooling system. A ventilation control (1) activates the forced-air furnace/air-conditioner fan (2), the central-exhaust fan (3), and also opens a motorized damper (4). The adjustable damper (5) is adjusted so that the amount of fresh air (6) entering the house is equal to the amount of stale air leaving (7).

mary functions is to be a fresh-air distribution system without ducts. When installed correctly, the incoming air mixes quickly with room air near the ceiling and drafts are not felt in the room. When specifying through-the-wall vents, keep in mind the fact that they are most effective in small, tight houses. In larger houses or in loosely-built houses there are generally enough random holes that most of the incoming air infiltrates through them rather than the deliberate openings.

Through-the-wall inlets are not very effective in large houses or in loosely built houses because there are so many random holes through which air can enter, so little air enters by way of through-the-wall vents. However, they are quite effective in tightly built houses, small houses (less than 1,500 sq. ft.), and apartments, because there are few random holes. Through-the-wall inlets function at optimum capacity when a central-exhaust system is designed to maintain a negative 10 Pa. in a house because this amount of depressurization is sufficient to overcome wind and stack-effect pressures in most parts of North America.

Airex produces through-the-wall vents that sense indoor relative-humidity levels. They have two operating ranges that can be easily changed (23–58% RH or 35–70% RH). When a room is dry, the airflow passing into the house through a vent will be low, say 3 cfm, but if a room is occupied, moisture from any source, such as exhaled breath, will cause the RH to rise and the vent in that room will open automatically to admit approximately 18 cfm. The actual airflows will depend on the precise degree of depressurization created by the exhaust fan.

American Aldes Corp. has several different through-the-wall vents called Airlets. These have built-in Constant Aiflow Regulators so they are designed to allow only a certain amount of air to pass through them even if the depressurization in the house varies. They have models rated at either 10 or 20 cfm that can be mounted in a wall (up to 15" thick) or in a window sash. Two of Aldes' Airlets are designed to open or close, based on the relative humidity; one operates in a range of 25–50% RH (for dry climates) and the other at 40–75% (for humid climates). These open up to allow more airflow when the RH is high. For example, when the indoor RH rises as a result of "people activities" (bathing, laundry usage, washing dishes, exhaled breath, etc.), they open automatically. Airlets can be ordered with a flat exterior flange or a beveled flange for use on clapboards (Figure 19–14).

(The humidity-controlled inlets offered by **Airex** and **American Aldes Corp.** are not designed to operate rapidly. In fact, they operate very slowly, taking as long as a few hours to open or close when the humidity fluctuates. Thus, when a room is occupied by a number of people, the RH in the room due to exhaled breath rises slowly, and the vents open slowly. Then, when the people leave the room, the RH falls slowly, and the vents close slowly.)

Broan Mfg. Co., Inc. has a narrow through-the-wall SV1 Fresh Air Inlet with a washable filter and an insect screen that is used with its central-exhaust ventilators. The filter is removable from the indoors. During installation, the inlets can be telescoped to accommodate wall thicknesses up to 11".

Therma-Stor Products' imports a very popular Fresh 80 through-the-wall vent from Sweden (Figure 19–15). It consists of an outdoor grille, a $3^1/4$"-diameter tube, a coarse filter to keep out insects, and an indoor diffuser. These can be mounted in a wall that is up to 13" thick. Fresh 80s have a convenient pull cord that can be used to close the vent. The amount of air entering a single vent will vary, depending on how much a house is depressurized, but it is usually in the 20 cfm range. They also have a larger model, the Fresh 100, that is 4" in diameter, and a $3^1/4$" Fresh 99 that opens and closes, based on the relative humidity indoors.

Thinking Vents, Inc. has a simple vent that can be mounted in a window sash (Figure 19–16). It is 7" long and will allow approximately 10 cfm of air to enter, depending on the amount of house depressurization. A ceiling-mounted make-up air inlet is also available. These products were primarily developed to use with a passive ventilation system (see Chapter 24) but can also be used with central-exhaust ventilators.

Titon, Inc. distributes a complete line of Trickle Ventilators that mount in a window sash. Made of aluminum and plastic, these units have several different styles. Most are adjustable and can be opened or closed manually, but a few are designed to always remain open. The amount of net free area depends on the style and the length, but it ranges from about 3 sq. in. to 30 sq. in. Lengths run from approximately 11" to 59". Some models contain thermal breaks for energy-efficient construction. Because cold air passes through them in the winter, window vents can cause a window to be slightly colder than normal; however, condensation on the glass is unlikely because the air entering is dry (Figure 19–17).

Vent-Axia, Inc. has a simple Air Replacement Grille Set that can be mounted in doors and panels to allow make-up air to pass through.

Coupling a fresh-air duct with a forced-air furnace/air conditioner

In many central-exhaust ventilation systems, air enters the house by way of random holes in the structure and/or several through-the wall vents. But if a house has a central heating or air-conditioning system, it is possible to have the air enter through a single inlet connected to the furnace/air-conditioning ducts. Because furnace/air-conditioning ducts are already running to each room, this can be an effective way to distribute the fresh incoming air throughout the house.

This approach is shown in Figure 19–18. It has been well thought-out and is approved in the Pacific Northwest by **Bonneville Power Administration's** Super-Good Cents Program. The specific details of this system have been featured in several recent magazine articles.(Jackson)(Using)(Sullivan 1993b) Outdoor air enters the house through a *fresh-air duct* connected to the return-air side of the furnace/air-conditioner's air handler. Outdoor air is actively drawn into the return-air duct because that duct is under negative pressure whenever the air-handler fan is running. The fresh-air duct should contain a motorized damper that opens and closes when ventilation air is called for (this damper is not needed if the system is designed to bring in ventilation air continuously), and an adjustable damper that should be positioned (usually one time only) to regulate the amount of air that enters the house when the system is running. When the outside air passes into the return-air duct, it mixes with the house air inside the duct, then the air handler blows the mixture through the supply ducts into the rooms of the house. At the same time, a central-exhaust fan expels stale air from the house. Fresh air enters the house in one location, mixes with house air, the air handler distributes it throughout the house, and the exhaust fan blows an equal amount of air outdoors.

For this system to work effectively, all the joints in the heating/air-conditioning ducts and the ventilation ducts must be sealed. If they aren't, the volume of air infiltrating and exfiltrating through the leaky ducts (especially the heating/air-conditioning ducts) could easily be several times the volume of the air coming in the fresh-air duct.

The air entering the fresh-air duct is pulled indoors *actively* by the furnace/air-conditioner fan, and the air leaving the house is *actively* exhausted by the central-exhaust fan. Once the system has been installed, the air leaving the house should be measured (perhaps by using a flow-measuring grid or a flow-measuring hood), then the incoming airflow should be adjusted so it is equal to the amount of air being exhausted. When this is done, the house will not be depressurized or pressurized—it will experience a neutral pressure. (If the system isn't adjusted properly—and many installations aren't—the incoming and outgoing airflows may not be equal, and the house will be either pressurized or depressurized.) Because of this neutral pressure (if properly adjusted), this is actually a balanced ventilation system. However, it is usually classified as a central-exhaust system, so it will be discussed in this chapter.

It is important to note that this type of system might be might be balanced correctly when it is installed, but that doesn't mean that it will always be adjusted properly. Occupants can change the adjustable-damper setting, either on purpose without realizing the possible consequences, or by accidentally bumping into it. Dirty filters or closed heating/cooling registers can also affect the pressures in the system, throwing the incoming and exhaust airflows out of balance.

If for some reason the central-exhaust fan is running, and the motorized damper is open, but the furnace/air-conditioner fan is not running, the house will be depressurized and most of the make-up air will enter the house *passively* through the random holes in the structure. (Some air will probably also enter passively through the fresh-air duct, but most of the incoming air will likely pass through random holes unless the house is very tightly constructed.) This will result in less effective distribution of fresh air to all the rooms in the house. A fresh-air duct will always be more effective at mixing and distributing fresh air to all rooms only if the furnace/air-conditioner fan runs whenever the central-exhaust fan runs.

Coupling a central-exhaust fan with a forced-air furnace/air conditioner requires several mechanical components that must be selected, installed, and adjusted properly in order to work together. Unfortunately, this makes the system rather complicated, and if any one component is malfunctioning, it can affect operation the entire system.

Figure 19–19. The TMAC is an automatic timer that is often used as a controller when a central-exhaust ventilation system is coupled to a forced-air heating/cooling system.

Fresh-air duct

To ensure good airflow, a fresh-air duct should be smooth on the inside and as short as possible. Avoid using an abundance of elbows, keep the length to less than 10 feet, and insulate the duct to prevent sweating. Six-inch diameter insulated flexible duct is often used for short runs, and 7" is often used for fresh-air ducts longer than 10 feet. Sometimes a larger size is needed, depending on how much air it will need to carry. The fresh-air duct should be attached to the main return-air duct of the furnace/air conditioner. It is usually recommended that the fresh-air duct be attached to the return-air duct as close to the air handler as possible—say, within 2 feet.

In very cold climates, when excessive outdoor air is brought in through a fresh-air duct, the shock of cold air rushing into a hot furnace can crack the furnace's heat exchanger or result in internal corrosion. This type of damage isn't likely when moderate amounts of fresh air are brought in, inasmuch as a small amount of cold incoming air will be tempered quickly as it mixes with a larger amount of warmer recirculating air. In most cases, the air mixture reaching the furnace will be warmer than 13°C (55°F), so there will be no problems.

Air-King Ltd. has a simple thermostatically controlled fresh-air damper system (Fresh Air 970) that allows a small amount of heated air from the furnace supply duct to mix with incoming cold air from the outdoors. This raises the temperature of the incoming air so it won't damage the furnace. **Skuttle Mfg. Co.** has a similar product called a Make-Up Air Control.

Adjustable damper

A simple manually adjustable damper is familiar to most heating/air-conditioning installers. It is intended to let you adjust the amount of air entering the house and balance it to match the amount leaving through the central-exhaust fan. It is generally installed in conjunction with a flow-measurement grid so that you can accurately adjust the damper while reading a gauge. This damper should be adjusted once, after the system has been installed, then left alone. **Eneready Products Ltd.** has a high-quality damper with a built-in flow-measurement grid called a PRA Flow Grid.

American Aldes Co. produces a fresh-air supply kit containing a Constant Airflow Regulator that can be used instead of an adjustable damper. It contains an internal bladder that automatically adjusts itself to only allow a given amount of air to pass through it. These are available in several sizes, ranging from 15 to 235 cfm. Constant Airflow Regulators will automatically allow the rated amount of air to enter the house, but only when the heating/air-conditioning fan is running and the electrical damper is open. They work well when a furnace/air-conditioner fan may be operating at different speeds. (It isn't unusual for such a fan to operate at low speed for ventilation, medium speed for heating, and high speed for air conditioning.) At different speeds, there will be different pressures in the duct system. Constant Airflow Regulators automatically compensate for the different pressures and maintain a relatively uniform rate of incoming fresh air. One drawback to these devices is the fact that there must be at least 0.2" w.g. of negative pressure in the return-air duct for them to work properly, and many return-air ducts don't have that much negative pressure in them.

In reality, the typical differences in return-air-duct pressures, resulting from different heating/cooling-fan speeds, generally don't account for significantly different airflows—usually the airflow will be within 10%. However, the change in airflow due to a dirty medium-efficiency filter, or closed heating/cooling registers, can be more significant.

Motorized damper

If the central-exhaust fan is to operate continuously, you don't need a motorized damper. However, if the central-exhaust fan only operates intermittently, then there may be times when the furnace/air-conditioner fan is running and you don't want fresh air to enter the house. In that case, this damper should be closed. Because it is motorized, the electrical connection can be wired to open only when the central-exhaust ventilation fan is running. A typical damper for this application would be kept closed by a spring, and would open only when it receives power from the control.

The Energy Vent (by **Furnace Tender**) is an electrically operated damper that can be used for this purpose. It contains either low-voltage or 110-volt

Chapter 19 Central-exhaust ventilation systems

connections for the exhaust fan, the air handler, and a control. The Energy Vent is available in sizes of 4–12". When the ventilation control (*e.g.* a timer, see below) calls for fresh air, the central-exhaust fan turns on, the furnace/air-conditioner fan turns on, and the motorized damper opens.

Broan Mfg. Co., Inc., **Conservation Energy Systems, Inc.**, **Duro-Dyne Corp.**, **Honeywell Inc.**, **Tjernlund Products, Inc.**, and **Ztech** also offer similar electrically operated motorized dampers.

Central-exhaust fan

The central-exhaust fan itself can be any of those listed earlier in this chapter. It can have a single exhaust grille or it can be a multi-port fan with several suction lines connected to pollution- or moisture-prone rooms. It should be quiet, long-lasting, and sized appropriately. Continuously running central-exhaust fans often require less capacity than many local exhaust fans. For reliability and energy efficiency, the fan should have a low-wattage, permanent-split-capacitor motor (or an ECM motor, if available) rather than a cheaper, less efficient shaded-pole motor.

System control

For this type of system to operate most effectively, the furnace/air-conditioner fan must run whenever the central-exhaust fan is running. Therefore, the control that turns on the central-exhaust fan must also open the electrical damper and turn on the furnace/air-conditioner fan. If the control is wired to turn on the ventilation fan and open the motorized damper—but not turn on the furnace/air-conditioning fan— the house will be depressurized and most of the incoming air will enter through random holes. Fresh air will, no doubt, find its way into the living space, but it will not be distributed very effectively.

The control itself can be anything from a simple on/off switch to a device that senses humidity or carbon dioxide. See Chapter 14 for a more complete discussion of controls and how to operate a furnace/air-conditioner fan and a ventilation fan simultaneously. Following are four popular timers that are often used to interconnect the three items being controlled—the damper, the central-exhaust fan, and the furnace/air-conditioning fan.

Duro-Dyne Corp. manufactures a complete control package called the Air Quality Control Center. It includes an electrically operated motorized damper, a programmable-timer, a 110-volt connection for the central-exhaust fan, and a connection for the furnace fan.

Tjernlund Products, Inc. also has a complete FAI-6 system that consists of an electronic timer, an adjustable damper, a motorized damper, and an outdoor air intake hood that connects to the fresh-air duct.

Trol-A-Temp produces a Timed Make-Up Air Control (TMAC) that is basically a clock-timer control with low-voltage (24-volt) electrical connections for the central-exhaust fan, the air handler, and the electrical damper. The timer can be set to activate up to 96 intervals per day for at least 15 minutes at a time. A 24-volt motorized damper is included. The TMAC control is also marked by **Honeywell, Inc.** (Figure 19–19).

Ztech has a Smart Vent system that includes a motorized damper and a controller. The automatic controller opens the damper and turns on the furnace/air-conditioner fan based on the indoor and outdoor temperatures. This gives you fresh air based on temperature—not necessarily when the occupants need it. As an alternative, the company also has a timed controller that can be set to turn the fans on and off and open the damper as often as every 15 minutes, so it can be set to match the times when the house is occupied.

Air-handler fan

With this type of installation, the forced-air furnace/air-conditioner fan will be running continuously, so it will be consuming electricity all the time. For reliability and energy-efficiency, it should have a low-wattage motor, such as an ECM motor. A high-wattage fan motor will contribute significantly to the electric bill.

Summary

As long as they are installed conscientiously, central-exhaust ventilation systems can work quite well at supplying occupants with fresh air. They have been quite popular in Europe for over 20 years. In fact, the Aldes

Understanding Ventilation

and Aereco systems that are distributed in the U.S. and Canada were both originally developed in France. These systems are relatively simple to understand and can often be installed at a reasonable cost. Though they are more costly to install, using a central-exhaust heat-pump water heater is one of the most popular approaches to ventilation in Sweden.

Central-exhaust ventilation systems that utilize deliberate through-the-wall inlets work most effectively in tight buildings such as apartments and small houses (less than 1,500 sq. ft.). When through-the-wall inlets are used in larger, looser houses, there are enough random holes so that a larger percentage of make-up air enters through the random holes rather than deliberate ones. When make-up air enters through random holes in a loose house, distribution is less effective.

The most serious disadvantage of a ventilation system that depressurizes a house is the fact that it can cause backdrafting or spillage, but this will only occur if the depressurization is strong enough to overcome the natural draft in a chimney. Because combustion gases can kill, safe chimney operation is always an important consideration. However, for many houses (*e.g.* those that are electrically heated, or those containing appliances with sealed combustion chambers) this is not a concern because there is no chimney to backdraft.

When evaluating the pros and cons of a particular central-exhaust system, keep in mind the fact that, when properly installed and adjusted, a central-exhaust fan, coupled with a fresh-air duct on a forced-air furnace/air conditioner, will result in a neutral pressure, so it is actually a balanced system.

Chapter 20

Local-exhaust ventilation systems

All houses have occasional periods of excess pollution or moisture that is confined to specific rooms, so it is a good idea to equip such rooms with local-exhaust fans. General background pollution and moisture, from people or from building materials, occurs throughout the house, so it is better handled with a central ventilation system. Most houses can benefit from one or more local-exhaust fans as well as a central ventilation system. While the majority of today's houses contain one or more local-exhaust fans, they were relatively uncommon before the 1940s.

Bath fans and kitchen range hoods are two examples of local-exhaust devices, but other fans also fall into this category—for example, an exhaust fan in a workshop, hobby area, home office, or darkroom. Actually, many exhaust fans can be used as either central-exhaust fans or local-exhaust fans; it just depends on where and how they are used. After all, a fan is just a fan—it doesn't know what its function is! Of course, it is extremely important to choose a fan that is designed for its intended purpose. For example, avoid running a fan continuously that is only designed to operate for a few hours a week.

A local-exhaust fan will generally be the most effective when it is located near the pollution or moisture source, as well as above it. For example, a bath fan whose primary purpose is to remove excess moisture will be more effective if it is located above the shower rather than on a sidewall on the other side of the bathroom. A kitchen range hood located directly over the range will be more effective than an exhaust fan located several feet away, or a downdraft range exhaust.

Because local-exhaust fans depressurize houses the same as central-exhaust ventilation systems, make-up air will enter either through random holes in the structure or through deliberate inlets to relieve the depressurization caused by the fan. Local-exhaust fans are often more powerful than central-exhaust fans, and in some cases they are considerably more powerful. So, they often depressurize a house more easily than a central-exhaust fan. However, local-exhaust fans usually don't operate for extended periods—perhaps less than an hour at a time—and they only depressurize a house when they are running. Central-exhaust fans, on the other hand, may run continuously, resulting in chronic depressurization.

Advantages of local-exhaust ventilation

Local-exhaust ventilation has the same theoretical advantages as central-exhaust ventilation—namely low installation cost and preventing hidden moisture problems in cold climates. In addition, a local-exhaust fan will remove pollutants and moisture from a particular room faster than a central ventilation system. After all, this is their primary purpose, and it is why local-exhaust fans are often recommended in addition to a central ventilation system.

Low installation cost

Local-exhaust fans often require little maintenance and they are generally the simplest, least expensive ventilation devices on the market. Unfortunately, the lowest-priced models rarely provide much ventilation capacity. For example, cheap bath fans are often rated at 50 cfm in free air (0.0" of static pressure). This means they will deliver 50 cfm only when they aren't hooked up to any ducts. Once installed, their capacity may drop to 25 cfm because of the air resistance of the duct, damper, and outlet. Low-cost fans also usually have a short life span.

Local-exhaust fans are important pieces of equipment and should be selected with care. Choose a fan that will have a long life and will expel enough air to be effective. The inexpensive fans on the market rarely meet these criteria, but there are many high-quality fans available that are only modestly more expensive ($100±). Low-quality fans typically have shaded-pole motors and axial fans, so for better quality, look for a permanent-split-capacitor motor and a centrifugal fan.

Minimizes hidden moisture problems in heating climates

Any ventilating fan that depressurizes a house increases the infiltration of air through the random gaps and cracks in the structure. If cold outdoor air is infiltrating in the winter, the humid indoor air will be prevented from entering the wall cavities. This keeps the cavities dry and prevents mold growth or decay hidden within the structure of the house. Because local-exhaust fans don't often run for extended periods, this is rarely a significant advantage.

Eliminates pollutants quickly

This is the main reason local-exhaust fans are installed: to rid a house of pollutants at their source rapidly (usually from Category B rooms). If this isn't done, the pollutants will disperse throughout the house and overtax the central ventilation system. This increases total ventilating costs because it requires the central system to have more capacity.

Because there are episodes in all houses when pollution levels rise suddenly, all houses need some method of flushing those pollutants out quickly. The easiest way to do this is to use local-exhaust fans in the rooms that are most susceptible to being contaminated. Traditionally, the bathroom and kitchen are thought of most usually in need of local-exhaust fans, so building codes often require exhaust fans in these rooms. But any room that has periodically high levels of pollution or moisture can also benefit from a local-exhaust fan.

Disadvantages of local-exhaust ventilation

Four disadvantages were listed in the last chapter for central-exhaust ventilation systems. Theoretically, local-exhaust fans have similar disadvantages, such as those related to creating negative pressures indoors. However, in many cases the drawbacks are really not very serious, simply because a local-exhaust fan operates for a shorter period of time than a central-exhaust ventilation system.

High operating cost

A ventilating fan that doesn't recover heat from the air being exhausted will have a higher operating cost than a similar fan that does incorporate heat recovery. And a powerful fan that runs continuously will have a higher operating cost than a low-capacity fan only intermittently operated. However, in general, most local-exhaust fans simply aren't operated long enough for operating costs to be a concern. For example, a typical bath fan in Chicago or Toronto only costs about $3.00 to operate for an entire year. The cost of ventilation is covered in Chapter 13.

Brings in unwanted pollutants

It doesn't matter whether depressurization is caused by a local-exhaust fan or a central-exhaust ventilation system, depressurization can cause pollutants such as radon, termiticides, insulation, etc. to be sucked indoors. But this only occurs when the fan is running. So, since a local-exhaust fan doesn't run for very long, this may not be as much of a concern as with a central-exhaust fan that runs continuously. For a discussion, see Chapter 9.

Because a local-exhaust fan is designed to remove pollutants quickly, most of the time the fan will exhaust more pollutants than are being sucked in by depressurization. Even though a local-exhaust fan might pull in something undesirable, the net result is often cleaner indoor air than if the fan weren't running. However, each house must be evaluated on an individual basis.

Cause of hidden moisture problems in cooling climates

One of the primary reasons for installing local-exhaust fans is to remove moisture. Bath fans remove the moisture generated while bathing; kitchen exhaust fans remove moisture released from dishwashers and cooking; laundry fans exhaust moisture released from washing machines. While this reduces the chance of mold growth within these particular rooms, depressurization can still cause moisture from the outdoors to be pulled into building cavities. However, this is only a concern in hot, humid climates when a house is air conditioned. Like the other disadvantages, this is less of a concern than with central-exhaust systems because local-exhaust systems don't operate for extended periods of time. This gives the wall cavities an opportunity to dry out at times when the fan is off. Most of the time, local-exhaust fans reduce more moisture problems than they cause. Hidden moisture problems are discussed in detain in Chapter 8.

Cause of backdrafting or spillage

Both local- and central-exhaust systems result in depressurization, which can cause make-up air to enter a house by coming down a chimney. This will make it difficult for combustion gases to escape upward through the chimney and out of the house. This is always a major concern in houses because backdrafting and spillage can be deadly. Carbon monoxide is deadly in the smallest amounts, so even a short-term episode can be serious. Any type of exhaust fan, central or local, has the capacity to depressurize a house enough for there to be a potential problem. High-capacity kitchen range exhausts are particularly likely to cause backdrafting or spillage because they move so much air out of a house. When a local-exhaust fan has a capacity of 300 cfm or more, you should make sure that all chimneys function correctly when the fan is operating.

This disadvantage is potentially very serious, but it only applies if a house has a combustion appliance connected to a chimney that operates with a natural draft. As was pointed out in Chapter 10, houses containing electric furnaces, heat pumps, and electric water heaters, as well as ones that are solar-heated are immune from backdrafting and spillage. Many high-efficiency gas furnaces are also immune. Chapter 10 contains a more complete discussion of backdrafting and spillage, as well as a discussion of furnaces with induced draft fans or sealed combustion chambers.

Bear in mind that natural gas ranges are combustion devices capable of releasing combustion by-products into the living space. As such, they always need a range hood capable of expelling those by-products to the outdoors. But the hood's exhaust fan can cause enough depressurization to result in backdrafting or spillage somewhere else in the house.

Avoiding depressurization when using a powerful local-exhaust fan

The seriousness of depressurization caused by local-exhaust fans is not always a problem, simply because local fans are operated for much shorter time than central ventilation systems. Some houses are capable of withstanding a degree of depressurization without suffering ill effects, such as a house having no natural-draft combustion devices. A tightly constructed house with a very powerful local-exhaust fan can experience a great deal of depressurization if there aren't enough deliberate openings for make-up air to enter. For example, downdraft kitchen exhaust fans are not as effective at removing pollutants as overhead exhaust hoods, so they must be considerably more powerful at getting their job done. The result is that the fan with the less effective design has greater potential to cause problems. If a house is very loosely constructed, there will be plenty of random holes through which make-up air can enter easily, so there will be less depressurization.

When depressurization is a concern, there are several ways to minimize its impact. First, extra through-the-wall vents can be installed. Second, a single large through-the-wall vent could be located near the exhaust device. Third, the occupants could open a window whenever they are operating a powerful local-exhaust fan. Fourth, you could install a supply fan. These solutions all allow an existing exhaust fan to function more efficiently by relieving the depressurization.

Extra through-the-wall vents

When a local-exhaust fan depressurizes a house, through-the-wall vents can be used to relieve the depressurization and provide paths for make-up air to enter. If a house has a central-exhaust ventilation system, several through-the-wall vents may have already been installed in conjunction with it. However, if a local-exhaust fan is more powerful than a central-exhaust fan, and the inlets have been sized for the central-exhaust fan, there may not be enough inlets for the larger local-exhaust fan. One solution (but not a very good one) is to size the through-the-wall vents for the largest exhaust fan in the house. In other words, simply install some extra through-the-wall vents. Because the extra vents are only needed for the short periods of time when the powerful local-exhaust fan is operating, this means there are more openings in the house through which more air can enter when the fan is off. Thus, the house will end up having more openings than it needs under normal circumstances. Not a very good idea.

One large through-the-wall vent

Sometimes a single extra-large through-the-wall vent, or a special inlet duct, is installed near a high-capacity local-exhaust fan such as a kitchen range hood. Unfortunately, a large vent is not much different than a large open window. It can become a pathway for a great deal of air when the ventilation system is off. Also, because this would probably be the largest hole in the house, all of the make-up air for the entire house would probably enter here rather than through any other inlets. In other words, if there are smaller through-the-wall vents located elsewhere in the house, they would become less effective.

A good way to make this a viable option is to install a motorized damper on the large vent. Under normal conditions, the damper would be closed most of the time. A single switch could be wired to activate the local-exhaust device, and the electric damper simultaneously. Thus, whenever the powerful exhaust fan is running, the damper would be open. When the fan is shut off, the damper is closed. This type of vent must be strategically located to avoid drafts, and at the same time remove locally generated pollutants quickly. For example, a large vent near a kitchen range should be positioned to allow air to move across the stove and up into the range hood without disturbing the air elsewhere in the kitchen. Motorized dampers are available in a variety of sizes from **Broan Mfg. Co., Inc., Conservation Energy Systems, Inc., Duro-Dyne Corp., Furnace Tender, Honeywell Inc., Tjernlund Products, Inc., Trol-A-Temp**, and **Ztech**.

It is also possible to have a motorized damper open automatically whenever negative pressures are sensed indoors. This can be done by using one of the differential pressure transducers discussed in Chapter

Chapter 20 Local-exhaust ventilation systems

14 to open the damper whenever a local exhaust fan generates a significant negative pressure indoors. However, this generally requires a certain amount of specialized engineering expertise that is not common in residential work. Actually, the amount of pressure that can backdraft a chimney is often too small to be detectable by these controls.

Open a window

The option of opening a window offers the least complicated way of relieving depressurization. With only one very powerful local-exhaust fan in the house—usually the kitchen range hood—it can be a simple matter to crack open a kitchen window whenever operating the range hood. Opening a window when operating a clothes dryer will prevent the area around the dryer from becoming depressurized and will allow the dryer to operate more efficiently. The primary drawback to this approach is the fact that the window might not be in a conveniently accessible location. For example, it may be difficult to reach across a kitchen counter, or it may be on the opposite side of the room from the exhaust device. The further the window is located from the exhaust fan, the more likely there will be uncomfortable drafts. Still, this can be a low-cost, low-tech solution to depressurization when it is the result of one particular local-exhaust fan.

Add a supply fan

With the three aforementioned approaches, you are exhausting air actively with a fan, but allowing make-up air to enter passively. For air to move passively through an opening, the opening must be somewhat larger than the fan's opening. In other words, if you exhaust air through a 6"-diameter duct with a fan, you may need a 12"-diameter passive inlet to adequately relieve the depressurization. (The actual size of a passive opening depends on several factors, but it is typically 2–3 times the cross-sectional area of the fan's duct.) So, a powerful exhaust fan might need an excessive number of through-the-wall vents, an inordinately large single through-the-wall vent, or a wide-open window to avoid depressurization-related problems.

A supply fan is simply a fan that blows the same amount of air indoors that the exhaust fan is blowing outdoors. When you use a supply fan, both fans should be controlled by the same switch. The supply fan can blow air anywhere into the house, but it will be most effective if it blows air indoors near the exhaust fan. Be sure that wherever it blows air, drafts are minimized. A

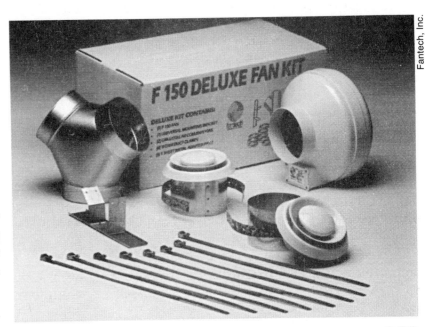

Figure 20–1. To minimize the number of local-exhaust fans in a house, a single fan can be used to serve two rooms.

diffuser located high on a wall will tend to mix and temper the incoming air with the room air before it can be felt by occupants. In effect, you are creating a balanced, local ventilation system with two fans of equal capacity—one blowing air in, one blowing air out.

Sometimes a supply fan will be controlled by a combustion appliance. For example, if a furnace is susceptible to backdrafting or spillage, you might use the thermostat to turn on a supply fan whenever it turns on the furnace. This can be a good option when there is more than one exhaust device in a house capable of causing excessive depressurization (*e.g.* a clothes dryer, a range hood, and a powerful bath fan). Another option would be to use a differential-pressure transducer (see Chapter 14) to activate a supply fan when the draft in a chimney is too weak.

Minimizing the number of exhaust fans in a house

Because a local-exhaust fan and a central-exhaust fan both blow stale air outdoors, it is sometimes possible to use the same fan for both purposes. What is needed is a way to temporarily boost the amount of air being exhausted from specific rooms when those rooms contain a high level of pollution or moisture.

Consider a house that has a central-exhaust ventilation system that usually operates at a rate of 60 cfm. The ducts are arranged so that 20 cfm is exhausted from each of the two bathrooms and from the kitchen. When someone is using the bathroom, it would be desirable to exhaust at least 50 cfm from that room. This can be done is a variety of ways with a single fan.

First of all, the central ventilation fan could have a two-speed motor that operates on low speed most of the time. The high speed would only be used when the indoor air quality declines substantially, as when someone is in the bathroom. The duct system would have to be sized for the high-speed airflows. There could be a simple switch in the bathroom that would activate the high speed of the fan. The **American Aldes Corp.** multi-port ventilation system described in Chapter 19 allows for extra air to be exhausted from one room—usually the kitchen—when operated at high speed.

Another way to minimize the number of local-exhaust fans would be to connect a single local-exhaust fan to several different rooms (Figure 20–1). An improved version of this approach would be to install motorized grilles or dampers in the ducts that are normally closed. When a switch is activated in one of the rooms, it would turn on the fan and open the grille or damper in that particular room. In this way, a single exhaust fan can be used to serve multiple rooms. A similar setup can be used to provide both local exhaust and central exhaust with the same fan. During normal operation, all the dampers or grilles would be partially open and the fan would pull a small amount of air from all rooms. When a switch in one room opens its damper or grille all the way, more air will be pulled from that particular room. This is basically how the PoshTimer and POWERGrill from **Eneready Products Ltd.** (see Chapter 14) and the **Airex** Aereco system (see Chapter 19) operate.

It is also possible to combine a two-speed system with adjustable dampers or grilles in pollution-prone rooms. All grilles would be partially open during normal operation. Then, when more ventilation air is needed in a particular room, the switch controlling the high fan speed could also cause the exhaust grille in that room to open completely to allow more air to flow through it. This is the approach used by **Eneready Products Ltd.** with its PoshTimer and POWERGrill (see Chapter 14). Eneready's system was specially developed to allow you to use a central ventilator for both whole-house ventilation and local ventilation.

Local-exhaust ventilation fans

This section covers the basic requirements of local-exhaust ventilation fans as well as a description of some of the equipment currently available. Most of the suppliers mentioned sell their products through local distributors, but some also sell direct to consumers and builders. Some pieces of equipment have built-in controls, and many of the fans can be easily used in rooms other than a bath or kitchen—rooms such as a laundry, hobby room, artist's studio, or home office. In general, because a local-exhaust fan blows air outdoors like a central-exhaust fan, many of the fans mentioned in

Chapter 19, *Central-exhaust ventilation systems,* can also be used for local-exhaust ventilation.

Kitchen-range exhausts

Besides pleasant aromas, cooking releases a variety of pollutants in the form of moisture, smoke, and grease into a kitchen. Gas ranges emit by-products of combustion such as water vapor, carbon dioxide, nitrogen dioxide, and carbon monoxide. All these contaminants should be exhausted from the house.

There are several options for removing pollutants from a kitchen. In the past, through-the-wall local-exhaust fans were common, but they were soon replaced with exhaust fans built into hoods that mounted above the range. Range hoods can be very effective in removing pollutants, but they don't actually suck air up off the top of a range. Instead, cooking odors and moisture, which are warm, rise up due to their natural buoyancy into the hood. Then the fan carries the warm, polluted, moisture-laden air away.

Noise is a common complaint for all types of kitchen exhaust fans. Unfortunately, range hoods aren't available in quiet low-sone models like some other fans. Most are in the 4–5 sone range, but some are as loud as 7 sones at high speed. Noise can be minimized either by selecting a model that has a remote-mounted fan, or by using a solid-state infinitely-variable speed control to run a high-capacity range hood at a slower and quieter speed. However, these speed controls can result in an annoying hum.

Recirculating range hoods

Some range hoods are said to be *recirculating* or *ductless*. They capture the warm air rising from a stove, pull it through a thin filter that is supposed to remove grease and odors, then blow the air back into the room. The filters in these hoods generally contain some activated carbon, but they are extremely inefficient, so they do very little to remove contaminants—and they do nothing to remove moisture. They can give an occupant a false sense of security by not removing noxious gases from the kitchen. To be effective at removing pollutants, a range hood *must* be ducted to the outdoors. All ducted hoods also have a thin aluminum-mesh filter to prevent excessive amounts of grease from building up in the duct. These filters are only marginally effective and must be cleaned periodically. In general, ductless range hoods should be avoided.

Overhead range hoods

Contaminants from tall pots on a cooktop will be more likely to rise up into an overhead hood than those from short pots simply because they are released closer to the hood. For an overhead hood to be more effective at capturing the rising warm air from all heights of pots, it shouldn't be any more than 24" above the cooking surface, and the canopy should be a few inches wider than the cooktop on each side. For example, use a hood with a 42" wide hood and center it over a 36" cooktop. As warm air rises, it expands, so a wider hood will be able to capture the air more easily. For the best capturing ability, look for a model that extends outward 20" from the back of the stove; it will be more effective than the typical units that only extend out 17".

Because there can be more crosscurrents in the center of a room than near a wall, a hood used over an island or peninsula will need to be wider or have a more powerful fan than one mounted along a wall. When a hood is mounted over an island or peninsula, it is often raised up as much as 27" above the cooking surface to maintain a good line of sight. Because a hood is less effective at this height, it will need to be more powerful than one mounted lower. Fans up to 600 cfm are often used in these applications.

A rule of thumb for selecting a capacity for an overhead range hood (from the **Home Ventilating Institute**) is that there should be at least 40 cfm of capacity per linear foot of hood. (*i.e.* a 36" hood should exhaust at least 120 cfm.) People sensitive to cooking odors often require a fan with twice this capacity. (Commercial range hoods often exhaust as much as 100 cfm per foot.) When selecting a hood, be sure to consider the fact that the longer the duct is to the outdoors (and the more elbows it has), the greater the resistance to airflow and the less the operating capacity. It isn't unusual for a 200-cfm hood to only remove 100 cfm once installed, especially if it has an axial (propeller) fan.

It is often a good idea to oversize a range hood so it will deliver the correct amount of air even if the filters are slightly clogged. Because many range hoods are fairly noisy, builders often install high-capacity range

Understanding Ventilation

hoods with the assumption that when they are operated at slower and quieter speeds, they will still have adequate exhausting capacity.

Low-profile and microwave hoods

Hoods that are built into the bottom of microwave ovens usually only project out 13–15" from the wall, and they don't have a conventional canopy. As a result, they don't capture contaminants from the front burners very well. They generally do a reasonable job on the rear burners, but most people find the front burners more convenient to use.

Low profile, pullout *silhouette* hoods are sleek-looking. They have a flat shelf that slides back into the shallow cabinet when not in use. Because they project out further from the wall than microwave hoods, they work better at removing contaminants from the front burners, but without a canopy, they aren't as effective as conventional hoods.

Surface-mounted downdraft exhausts

When a range or cooktop is located on an island or peninsula, the duct from an overhead ventilator can be difficult to rout to the outdoors. It can also be unattractive, and block the line of sight through the kitchen. For these applications a downdraft exhaust can be an option. Downdraft fans are built into some cooktops—some are center mounted, between the burners, and some are mounted to the side.

Inasmuch as downdraft exhaust fans are flush with the cooking surface, they must be much more powerful than an overhead hood because they must overcome the natural tendency of warm air to rise. As a result, most are rated at over 500 cfm—ten times the capacity of a typical bath fan. Unfortunately, even with that much capacity, they often aren't as effective as an overhead hood. In fact, it is difficult for them to capture contaminants from a tall pot—say, one over 3" high. When used in conjunction with an island or peninsula, there can be significant crosscurrents across a stove, which renders them even more ineffective.

Because they are so powerful, surface-mounted downdraft exhausts can significantly depressurize a house, resulting in the disadvantages listed earlier. When installing one in a house having a combustion device that is connected to a natural-draft chimney, it is imperative that extra care be taken to ensure that backdrafting or spillage won't occur.

Pop-up downdraft exhausts

Pop-up downdraft exhausts were introduced several years ago as an alternative to overhead hoods and

Figure 20–2. Remote mounted kitchen range fans are much quieter indoors, but they can be noisy outdoors. Roof-mounted models are also available.

conventional downdraft exhausts. They are usually located behind the cooktop, with the fan mounted in the base cabinet. When in use, they stick up about 8" above the surface of the range, pull the warm air into them, then blow it outdoors. When not in use, the intake retracts into the cabinet, resulting in a very uncluttered appearance. Because the intake is raised up off the surface during operation, these units don't have to work as hard as conventional downdraft exhausts to overcome the natural tendency of warm air to rise. But they still usually require over 500 cfm—enough capacity to seriously depressurize a house.

Pop-up exhausts work best with shallow pots (3" or less in height) on the rear burners. Some contaminants from a tall pot on a rear burner, or a short pot on a front burner, will escape into the room and will not be captured or exhausted. And most of the contaminants escaping from tall pots on the front burners will simply dissipate into the kitchen.

Remote-mounted kitchen fans

If a fan and motor is located outside the living space, in an attic, crawl space, or basement, or on the roof or exterior wall, it will be significantly quieter than a fan in the kitchen itself. Many overhead-range-hood manufacturers produce remote-mounted fans designed to be connected to a hood in the kitchen by a length of duct. They include **Bowers/Thermador**, **Broan Mfg. Co., Inc.**, **Fantech, Inc.**, **Nutone, Inc.**, **Reversomatic Htg. & Mfg. Ltd.**, and **Viking Range Corp.**

Remote-mounted kitchen fans are often powerful enough to pull air through long duct runs. They are also usually very noisy, sometimes as loud as 10–20 sones when you are standing next to them. However, when they are mounted far from the kitchen—up on a roof, or on the back of the house for example—they can be fairly quiet indoors. Because the actual amount of noise that is detectable in the kitchen will vary—depending on the particular installation—these fans are not sound rated. Outdoors, they can be distracting, so don't locate one near a window that will likely be open, or near a deck or patio (Figure 20–2).

Most central ventilation systems aren't designed to handle the grease found in kitchen air, but one of the multi-port central-exhaust ventilation system produced by **American Aldes Corp.** can (see Chapter 19). It has a two-speed fan that, when operated on high speed, will boost the amount of air leaving the kitchen. This remote-mounted fan can actually be connected to a range hood, so you can use one fan for two purposes: It can function as a general ventilation system on low speed (40 cfm) and a kitchen range local-exhaust on high speed

Figure 20–3. Quieter-than-normal local-exhaust fans are now available from most manufacturers. A quiet fan will be more likely to be used than a noisy fan.

(120 cfm). For safety reasons, any exhaust fan that pulls air from near a range in a kitchen must be connected to metal ducts to reduce fire risks. (This is actually required by most building codes.) So, if an Aldes unit is connected to a range hood, the plastic duct between the hood and the fan, and the duct between the fan and the outdoors, must be replaced with metal.

Bath fans

Moisture removal is the primary function of a bath fan, while odor removal, though also very important, is secondary. To adequately remove moisture, a fan must usually run for 20–30 minutes after a shower. According to the **Home Ventilating Institute**, a bathroom fan should be capable of exhausting air at a rate of at least 8 ACH. A small bathroom (5' x 9' with an 8' ceiling) contains 360 cubic feet. To provide 8 ACH in this size bathroom, a fan would need a capacity of 48 cfm (360 x 8 ÷ 60). Thus, according to the Institute, a 50-cfm fan is sufficient for a small bathroom.

In practice, many of the 50-cfm fans on the market won't remove moisture fast enough to prevent mirror fogging. One of the reasons has to do with the fact that bath fans are often rated at 50 cfm at a resistance 0.1" w.g. There can be that amount of resistance in about 15 feet of straight smooth duct and two 90° elbows. When an installation has more resistance, it won't deliver as much air. For example, bath fans are often connected to the outdoors with long, small-diameter flexible ducts, something that creates a great deal of resistance to airflow. As a result, it isn't unusual for bath fans to have to work harder to overcome the resistance—and not move very much air. Where flexible duct, or long duct runs, are necessary, a fan should deliver at least 50 cfm at perhaps 0.4" w.g.

In most situations, a 100-cfm capacity fan is probably a better choice in an average-sized bathroom be-

Figure 20–4. *This remote-mounted bath fan has a motorized cover that automatically shuts when the fan is turned off to prevent unwanted infiltration.*

Chapter 20 Local-exhaust ventilation systems

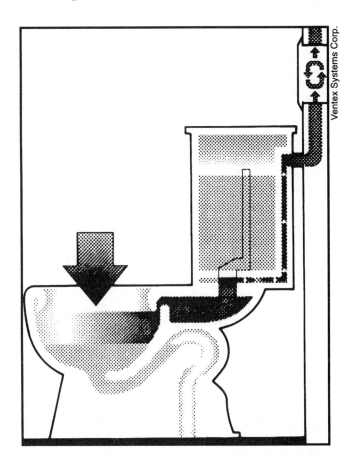

Figure 20–5. This unique toilet-fan system pulls stale, odorous air directly from the toilet bowl and exhausts it outdoors.

cause it can remove both odors and moisture quicker. If a bathroom contains a hot tub, a great deal of moisture can evaporate so the fan should be rated at 200–300 cfm. If a hot tub has a tight-fitting cover, the cover will prevent evaporation when the tub is not in use.

One of the biggest complaints about bath fans is noise, and noisy fans rarely get turned on. The easiest way to minimize noise from a local-exhaust fan is to either select a fan that is inherently quieter, or to mount the fan some distance from the room it is meant to serve and run a connecting duct between the fan and the room. On the other hand, a noisy fan does have one advantage: overpowering any embarrassing noises generated when occupants use the toilet.

Quiet fans

Because so many people complain about noisy fans, it is important to select equipment that is as quiet as possible. Many manufacturers produce at least one line of well-made, quiet fans (usually at a slightly higher cost) that can be mounted in either the wall or ceiling of a room (Figure 20–3). They make excellent local-exhaust fans for bathrooms and other rooms, but they should not be used as kitchen range fans because they are not designed to handle grease. Most are made for extended operation, so they are suitable for use both as local-exhaust fans and as central-exhaust fans.

Panasonic has a series of Super-Quiet fans ranging up to 340 cfm. Their 50- and 70-cfm models are rated at only $1/2$ sone, making them virtually impossible to hear. They are also very energy-efficient, consuming a mere 13 and 14 watts respectively. Following is a listing of several other manufacturers of quieter-than-normal fans that are currently on the market. (Most consume 2.5–3 times as many watts as comparably-sized **Panasonic** fans.) LoSone Ventilators are available from **Broan Mfg. Co., Inc.** in ranges of 100–1,000 cfm. **Fan America, Inc.** offers a SoLo series and **Fantech, Inc.** has a line of Luxury Centrifugal Fans, the smallest of which is rated at 100 cfm and less than one sone. **Fasco Consumer Products, Inc.** has Super Q exhaust ventilators in sizes from 95 to 825 cfm. **Grainger Inc.** offers several sizes of Quiet Exhaust Ventilators and Extra-Quiet Exhaust Ventilators. (Grainger's fans are sold under the brand name Dayton.) **Greenheck** produces a quiet SP and CSP series, and **Kanaflakt, Inc.** has several fans in its Quiet Pack series with ranges of 90–1,000 cfm. **Nutone** has a QT Series, **Penn Ventilator, Inc.** offers a Zephyr Series ranging from 60 to over 3,000 cfm, and **Reversomatic Htg. & Mfg. Ltd.** has several styles of quiet fans sized at 50–1,500 cfm.

Keep in mind the fact that all fans must be installed properly in order to deliver the correct airflow at a specific noise rating.

Remote-mounted local-exhaust fans

A fan mounted directly in the ceiling or wall of a room will be noisier than a fan mounted some distance from the room such as the aforementioned remote-mounted range fans. For rooms other than the kitchen, there are two possibilities for remote mounted local-exhaust fans. First, a fan can be separated from a room with a length of duct, then connected to the outdoors with another length of duct. Such a fan should be located where it can be maintained occasionally. An inaccessible attic or crawl space are not as preferable as a garage or basement. In-line fans work well and are designed to be long-lasting. Four through 12"-diameter in-line fans are available from **Fantech, Inc.**, **Kanaflakt, Inc.**, **Rosenberg, Inc.**, and **Vent-Axia, Inc.**

For a fan that mounts outside the house, **Environment Air Ltd.** has a 165-cfm exhaust fan that mounts on the wall outside the house with a round duct extending into the interior. **Fantech, Inc.** has a line of wall-mounted exhaust fans that can either be used as through-the-wall fans or be connected to the indoors with a duct. They also have a line of roof-mounted fans that can be separated from the room they serve with a length of duct. **Tamarack Technologies, Inc.** produces a Preventilator bath fan that mounts outside the house on a wall with a duct running into the bathroom. It has an insulated, gasketed exterior cover that is motorized so it will close tightly when the fan is off to prevent uncomfortable drafts (Figure 20–4).

It should be pointed out that remote-mounted exhaust fans can be subject to freezing when they are mounted in unheated attics and crawl spaces. If they are not operated continuously, and local-exhaust fans aren't, they can sometimes freeze up if they are off during cold weather. This is most likely when a fan is in a cold attic because stack effect often causes humid house air to escape through the fan. (This can happen at a low flow rate even when the fan is off when there is no damper to prevent such air movement.) This humid air can condense and freeze inside the cold fan. If a fan is mounted in a cold crawl space, freezing is less likely because stack effect will tend to cause dry, outside air to enter through the fan.

Miscellaneous local-exhaust fans

A number of miscellaneous fans can be used for local exhaust. For a temporary situation, a portable window fan can be used to air out a room quickly. This can be useful when repainting the walls or refinishing a floor. In such cases, local exhaust is needed, but it wouldn't make sense to install a permanently mounted fan.

Broan Mfg. Co., Inc., **Fasco Consumer Products, Inc.**, and **Grainger, Inc.** all produce several different types of general-purpose exhaust fans in a variety of sizes. They also offer bath fans with built-in lights or infrared heaters. In addition, **Greenheck**, **Grainger, Inc.**, and **Penn Ventilator, Inc.** carry a good selection of industrial-quality fans that can be adapted to a variety of residential uses.

Broan Mfg. Co., Inc. also offers a Sensaire bath exhaust fan with a built-in humidity sensor that turns the fan on when there is a rise in the relative humidity. (See Chapter 14 for a discussion.) Broan also has a more sophisticated Sensaire bath exhaust fan that contains both a humidity control and a motion sensor.

Ventex Systems Corp. produces a unique low-flow toilet/ventilation-fan combination (Figure 20–5). To remove odors from the toilet quickly and efficiently, this system pulls 10 cfm of air directly into the toilet bowl, then exhausts the stale air outdoors before it can dissipate into the rest of the room. Ventex offers both a complete toilet-fan system as well as a retrofit package that can be adapted to existing toilets. This approach works very well in removing odors with a minimum amount of air from a small room such as a powder room that doesn't have a bathtub or shower. But 10 cfm isn't enough air exchange to remove the moisture generated while bathing. For both moisture and odor control, the company offers a dual-exhaust fan that pulls 10 cfm through the toilet for odor control and 95 cfm from another point in the bathroom for moisture removal.

Most local-exhaust fans are suitable for removing moisture and odors, but occasionally a fan is needed for a very specific purpose. For example, fans for photographic darkrooms require special baffles to prevent light from passing through, and explosion-proof motors are used for exhaust fans in hazardous locations. They minimize the possibility of a spark from the motor igniting something in the air passing through the fan. While we don't often think of a house as a hazardous location, hobby materials such as paints and adhesives can be quite flammable, so some home workshops or garages might benefit from an explosion-proof fan motor. If the air contains corrosive chemicals, such as chlorine from a swimming pool, a corrosion-resistant exhaust fan is usually in order.

Summary

Most houses can benefit from one or more local-exhaust fans in order to remove stale air from pollution- or moisture-prone rooms quickly. In fact, local-exhaust fans are commonly found in kitchens and bathrooms of most houses today.

A local-exhaust fan should not be considered a substitute for a central-exhaust system designed to exchange the air in the entire house; both types of systems are needed in most houses. However, with proper design and selection of controls, it is possible to have a single system that will perform the functions of both local-exhaust and central ventilation.

Chapter 21

Central-supply ventilation systems

This chapter covers the advantages and disadvantages of central-supply ventilation systems and it discusses how this type of equipment can be used to exchange air in houses. Though this approach is less common than other ventilation strategies (*i.e.* central-exhaust and balanced ventilation), central-supply ventilation has some definite advantages.

Central-supply ventilation systems are often called *supply-only* systems. However, no ventilation system can only supply air to a house because an equal volume of air will always escape somewhere. Therefore supply-only probably isn't the best terminology to use. Central-supply ventilation systems pressurize houses by blowing air into them. As mentioned earlier, this can be done with an ordinary window fan. But window fans are generally only used for temporary ventilation (usually in mild weather) because they are not designed to run continuously and a window is not very secure against burglars with a portable fan in it. Whenever a house is pressurized, air leaves either through random gaps and holes in the structure or through deliberate openings specifically planned to be outlets. The air exits passively in order to relieve the pressurization caused by the fan.

Central-supply ventilation strategies

There are three ways a central-supply ventilation system can be used in a house. The first is relatively straightforward: A fan is used to blow air directly into the living space (Figure 21–1). A variation would be using a system of ducts to direct the incoming air into two or more rooms (Figure 21–2). Although these two approaches offer a simple way to bring air indoors, they are, in fact, seldom used.

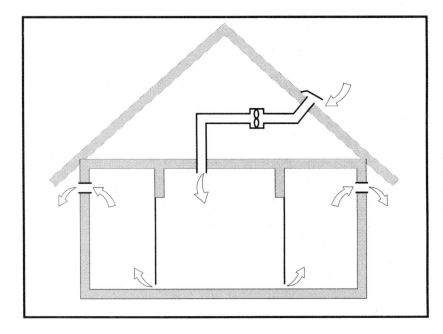

Figure 21–1. A simple central-supply ventilation system could consist of a fan that blows fresh air into a central location, and through-the-wall vents through which stale air can escape. This type of installation will pressurize a house.

The third approach involves attaching a *fresh-air duct* to the return-air duct of a forced-air furnace/air conditioner. With this setup, a certain amount of fresh air is pulled in from the outdoors through the fresh-air duct because of the negative pressure in the return-air duct. Occasionally, a fan is used in the fresh-air duct, but not often. The incoming fresh air mixes with the stale house air inside the return-air duct, then the furnace/air-conditioning fan blows the fresh/stale-air mixture throughout the house.

With all three of these approaches, air leaves the house either through the random holes in the structure or through-the-wall vents. The through-the-wall vents do two things: 1) they act as pressure-relief valves, preventing excessive pressurization from developing indoors, and 2) by being located carefully, they contribute to the distribution of air through the house.

Advantages of central-supply ventilation

All ventilation systems have advantages as well as disadvantages. With central-supply ventilation, the costs, both of installation and operation, and maintenance requirements are usually comparable to central-exhaust systems with similar features installed in similar climates. The other advantages and disadvantages are all pressure-related. Because pressurization and depressurization have opposite positive and negative effects, the pressure-related advantages of exhaust systems are disadvantages with supply systems, and vice versa.

Low installation cost

A central-supply ventilation system can consist of a fan, some ducts, a few through-the-wall vents (if the house is tightly constructed or isn't very large), and a control. These are not overly expensive items, so these systems can usually be installed for a minimal amount of money. When coupled with a forced-air furnace or air conditioner, you usually don't need a separate ventilation fan, so the capital cost can be even less. However, there can still be a high operating cost, especially in a harsh climate (see *Disadvantages of central-supply ventilation*, below).

Easy to maintain

Because many central-supply systems are relatively simple installations and because they don't re-

Chapter 21 Central-supply ventilation systems

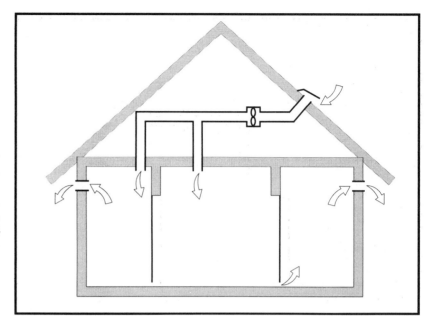

Figure 21–2. If a central-supply fan is connected to a series of ducts to blow air into more than one room at a time, drafts will be less noticeable. In this example, air leaves the house by way of through-the-wall vents.

quire much in the way of complicated equipment, they are often easy for occupants to understand and maintain. Simplicity is very appealing to people whose lives are already too complicated.

Minimizes hidden moisture problems in cooling climates

Pressurization can prevent hidden moisture problems when it is hot and humid outdoors and air conditioned indoors. It does so by causing cool indoor air to continually exfiltrate through the random gaps and holes in the structure (as well as any deliberate holes such as through-the-wall vents).

As the cool, air-conditioned indoor air moves through the walls, it becomes warmer as it nears the outdoors, so the RH of this exfiltrating air goes down, thus eliminating the potential for hidden condensation within building cavities. Because cool air is always exfiltrating, pressurization prevents the hot, humid outdoor air from getting inside building cavities where it could condense on the back of the cool drywall or interior wall paneling and result in mold growth, mildew, or decay. Hidden moisture problems are discussed in depth in Chapter 8.

Keeps out unwanted pollutants

Pressurization will keep a number of unwanted pollutants from entering the living space. When air is exfiltrating through random holes in the structure, outdoor air can't enter through those same holes. Therefore, pressurization prevents lawn chemicals, radon, and other soil gases from entering a house through holes or cracks in the foundation. Pressurization also prevents particles or gases from the insulation from getting into the living space from wall cavities and attics. This is covered more completely in Chapter 9.

Minimizes backdrafting or spillage

A big advantage of pressurizing houses is that it can eliminate a backdrafting or spillage problem. By blowing air into a house for ventilation, the building will experience a positive pressure, and air will usually escape up the chimney, as well as through other random and deliberate holes in the house. Pressurization actually helps the natural draft in a chimney to work more efficiently by giving the combustion gases a boost on their way out of the house. See also Chapter 10.

315

Disadvantages of central-supply ventilation

Central-supply ventilation has two primary disadvantages: the operating expenses can be high in harsh climates when airflows are excessive, and in cold climates it has the potential to cause hidden moisture problems. A minor disadvantage to using a central-supply ventilation system is the possibility that the fan motor could be located in the fresh airstream. As motors get warm, they sometimes give off odors from the insulation on the wires or the lacquer on the windings. In the vast majority of installations, this usually isn't a consideration. However, for hypersensitive occupants, the slight odor may be bothersome.

High operating cost

The operating cost of any ventilation system depends on the cost of tempering the incoming air, the cost of running a fan, and maintenance expenses. Operating costs are of most concern in harsh climates when ventilation rates are excessive. For more on the cost of ventilation, see Chapter 13.

More energy will be consumed by a ventilation system that does not incorporate heat recovery when compared to a similar-sized system that does have such capability. In the winter, a central-supply system will blow cold air directly into the house, and that cold air will need to be warmed up. This places a burden on the heating system. In the summer, the incoming outdoor air will be humid and warm, so the air conditioner will need to work harder to maintain comfort. In a mild climate, where outdoor temperatures aren't extreme, and heating and air-conditioning bills aren't normally high, the increase in utility bills attributable to tempering the incoming fresh air probably won't be significant. But in a harsh climate with expensive fuel (say, in a cold climate with electric-resistance heating), tempering costs can be high.

The cost of electricity to run a supply ventilation fan should also be considered. When a forced-air furnace/air-conditioner fan is used to pull outdoor air into the return-air duct, there is no separate supply fan. However, the cost of running the furnace/air-conditioner fan during periods when heating or cooling aren't needed should be considered. For the most effective ventilation, this fan should operate continuously. If it is not an energy-efficient fan, this can lead to a high operating cost. The use of a high-efficiency ECM furnace/air-conditioner fan motor (rather than a less-efficient shaded-pole or multispeed, permanent-split-capacitor motor) can often significantly minimize the cost of operating the fan continuously.

Cause of hidden moisture problems in cold climates

Pressurization has the potential to cause hidden moisture problems in cold climates. During the heating season, pressurization can cause warm, humid indoor air to exfiltrate through the random holes in the structure toward the outdoors. As the air reaches the sheathing or siding (which are cold), it can condense on those cold surfaces. Whenever water vapor condenses, there is a danger of mold, mildew, or decay—inside the wall where you can't detect it immediately.

Although the potential exists, not all pressurized houses in cold climates experience moisture problems. For moisture inside a wall cavity to lead to mold growth or rot, several factors should be considered, but the indoor relative humidity is the most significant variable. If the indoor RH is kept enough low enough in the winter, say 25%, it is doubtful if pressurization will cause any hidden moisture problems.

As it turns out, houses are pressurized in cold climates all of the time without experiencing any hidden problems. For example, in the winter, stack effect causes the upper part of a house to become pressurized throughout the winter. And leaky ducts or closed doors between rooms can routinely cause individual rooms of a house to become pressurized. Yet, it has been estimated that only 10% of houses in cold climates experience hidden moisture problems. When a house has a continuously operating central ventilation system (of any type), it will often be dry enough indoors so that there isn't enough moisture to cause a hidden moisture problem in the winter due to pressurization. The potential for such hidden moisture problems is discussed more fully in Chapter 8.

Central-supply ventilation equipment

The science of ventilation-system design in North America had its roots in the coldest parts of the U.S. and Canada during the energy crunch of the 1970s when houses were made more energy-efficient and tightened up. Those tight houses could no longer rely on uncontrolled pressures (accidental or natural) to supply them with sufficient fresh air. Rather quickly, people began experimenting with a variety of mechanical ventilation systems. It soon became apparent that supply ventilation systems had the potential to create hidden moisture problems in cold climates. Occasionally, horror stories circulated describing wet insulation and rotting framing inside wall cavities. As a result, builders and designers in cold climates began to rely primarily on central-exhaust ventilation or balanced ventilation (usually with heat recovery).

In the past, the need to conserve energy in moderate and warm climates hasn't been as critical as in very cold climates, so many houses in the southern U.S. haven't been tightened up as much as their northern counterparts. Therefore, mechanical ventilation hasn't been as important in milder climates. As a result, mechanical-ventilation design in warm climates has lagged behind what has been done in cold climates. In fact, only a few manufacturers are producing products specifically designed for warm-climate applications, such as stand-alone central-supply ventilation systems.

Individual fans

The in-line fans often used by radon contractors can also be used for central-supply ventilation systems. Long-lasting and quiet, they are available from **Fantech, Inc.**, **Kanaflakt, Inc.**, **Rosenberg, Inc.**, and **Vent-Axia, Inc.** Wattages vary, but the smallest-capacity in-line fans typically consume 40–50 watts. Unfortunately, most in-line fans are not particularly quite.

Companies such as **Grainger, Inc.**, **Greenheck**, and **Penn Ventilator, Inc.** have a variety of residential and commercial fans of different types that can be adapted to a residential supply ventilation system. These fans can be installed to blow air into a single room, or when attached to a series of fittings, to blow air into several rooms simultaneously. The latter approach can be coupled with motorized dampers or grilles to vary the airflow to different rooms.

Though they are not built to be mounted to blow air indoors, it is possible to adapt one of the low-watt-

Figure 21–3. The Fresh Air In-Forcer is designed to blow air into a basement, slightly pressurizing it.

age, quiet, exhaust fans described in Chapter 20 so it can be used as a supply fan.

Therma-Stor Products' Filter-Vent is a central filtration unit that can be used as a central-supply ventilation system to pressurize a house and filter the incoming air at the same time. It contains an activated-carbon prefilter and a high-efficiency particulate filter. With the standard model, 77 cfm of the unit's 300-cfm capacity is fresh outdoor air and the remaining 223 cfm is recirculated throughout the house (see also Chapter 15). The Filter-Vent has been used successfully in a variety of climates.

Tjernlund Products, Inc. manufacturers a Fresh Air In-Forcer system consisting of a small cabinet with a supply fan and timer control. The Fresh Air In-Forcer is used to slightly pressurize a house by blowing air indoors (usually into the basement) either at certain times of the day (*e.g.* when a house is occupied) or continuously (Figure 21–3).

Through-the-wall vents

Many of the through-the-wall vents on the market were originally designed to be used as inlets with central-exhaust ventilation systems. However, some models will allow air to flow in either direction, so they can also be used with central-supply ventilation systems as outlets. Through-the-wall vents that are specifically designed to work as inlets will generally have different airflow characteristics when used as outlets. Thus, before using an inlet as an outlet, be sure to check with the manufacturer to ensure you will have the desired airflows. (Some through-the-wall vents have a small flap that acts as a backdraft damper to allow airflow in only one direction.) Through-the-wall vents are discussed in Chapter 19. Manufacturers include **Airex**, **American Aldes Corp.**, **Broan Mfg., Co., Inc.**, **Therma-Stor Products**, **Thinking Vents, Inc., Titon, Inc.**, and **Vent-Axia, Inc.**

In a pressurized house, the through-the-wall vents will act as outlets, so they should be located where the indoor air is the poorest—in bathrooms, kitchen, etc. (Category B rooms). For the most effective distribution of air in the house, the fresh air should be deposited in the high-occupancy rooms such as bedrooms and living areas (Category A rooms).

Keep in mind the fact that air can often pass through a through-the-wall vent in either direction when a central-supply ventilation system is shut off, depending on which natural or accidental pressures are at work. Therefore, through-the-wall vents should be located where there will be no pollution problems if they function as either inlets or outlets (see also Chapter 16).

When a house is pressurized, air will escape through any opening it can find. It will leave through both random holes and deliberate openings. Therefore, only a portion of the air will escape through the through-the-wall vents. As was discussed in Chapter 16, the amount of air flowing through deliberate openings compared to the amount flowing through random holes is often proportional to the areas of their respective openings. In other words, if a house is loosely constructed, more air will escape through the random holes than through deliberate openings, such as through-the-wall vents. Thus, if you want an effective distribution system whereby stale air leaves the house primarily by way of deliberate openings (*e.g.* through-the-wall vents), the house should be tightly constructed. For this reason, through-the-wall vents work better as outlets in small houses (less than 1,500 sq. ft.) and apartments rather than in large houses (unless the large houses are very tightly constructed).

Coupling with a forced-air furnace/air conditioner

A central-supply ventilation system can be coupled with a forced-air heating/cooling system by connecting a length of duct (a *fresh-air duct*) between the outdoors and the main return-air duct of the heating/cooling system's air handler. A fresh-air duct is not the same as a combustion-air duct. A fresh-air duct is used to supply occupants with fresh air, while a combustion-air duct is used to supply air to a gas, oil, or wood furnace (or boiler, or water heater) for combustion. Combustion air and ventilation air should always be considered two separate issues and ideally are brought indoors independently of each other.

In any system that moves air, the ducts downwind of the fan are under positive pressure, while the ducts upwind of the fan are under negative pressure. When a fresh-air duct is connected to the negative-pressure side of a forced-air heating/cooling system, a certain amount of outdoor air will be automatically drawn into the house whenever the fan is running. The precise amount of air entering depends on the diameter and length of the fresh-air duct as well as the amount of negative pressure in the return-air duct at the point where the fresh-air duct is connected. (Return-air duct pressures can vary considerably, depending on the design of the system and the leakiness of the ducts.) This approach has been done for many years to supply occupants with fresh air, and in most of these installations air leaves the house through the random gaps and holes in the structure.

To understand how a fresh-air duct connected between the outdoors and the return-air side of a furnace/air conditioner pressurizes a house, you need to analyze the airflows. Suppose you have a furnace/air-conditioner fan rated at 1,000 cfm and it pulls 50 cfm from the outdoors through the fresh-air duct and 950 cfm from the living space of the house. At the same time it will blow 1,000 cfm into the living space. So, it is pulling 950 cfm from the house and replacing it with 1,000 cfm. This results in a slight pressurization, and 50 cfm will escape from the house through the random gaps and holes in the structure. This may cause no problems when a small amount of air is brought indoors, but if too much cold outside air enters a furnace in the winter, it can result in condensation, corrosion, or cracking of the furnace's heat exchanger.

The old way

Adding a fresh-air duct to a forced-air heating system has been done extensively in a variety of climates over the years, and it almost never causes any serious hidden moisture problems. One of the reasons this is so has to do with the fact that fresh-air ducts have been installed on far more forced-air gas and oil furnaces than used with heat pumps or air conditioners. The reason for this is because gas and oil heat have generally been less expensive than electrical heating methods, and contractors, trying to keep heating bills down, have been reluctant to bring additional outside air into a less-efficient heating system. Actually, these ducts have been installed primarily to supply houses with *combustion air*, but they have provided occupants with a certain amount of *fresh air* as well.

As a result, fresh air ducts have been widely used with combustion furnaces, most of which are connected to a natural-draft chimney. (The high-efficiency furnaces

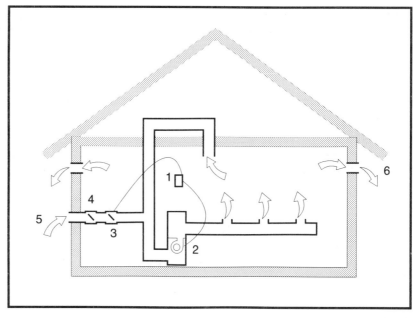

Figure 21–4. When used in conjunction with a forced-air heating/cooling system, a well-designed central-ventilation system requires several components. A ventilation control (1) activates both the forced-air heating/cooling fan (2) and a motorized damper (3), allowing outdoor air to enter through the fresh-air duct (5). The system also needs an adjustable damper (4) to restrict how much air enters the house through the system.

that do not need a chimney are a fairly recent development.) Generally, whenever such a system is running, combustion gases are going up the chimney. In effect, this is a crude balanced ventilation system—air is sucked indoors actively by the furnace fan through the fresh-air duct, and simultaneously air is being exhausted up the chimney. This means that there won't be very high positive pressures generated in the house, so there won't be much moisture being pushed into building cavities, and there won't be many hidden moisture problems. Because there is enough air exchange in the house to dilute any moisture generated indoors, the RH remains relatively low.

When there is a natural draft in the chimney, there is a negative pressure inside the chimney (the draft) due to the rising, hot, combustion gases and it can be fairly strong. As a result, the chimney acts like an exhaust fan as it sucks air out of the house. (A 100,000-Btu gas furnace will exhaust 50–80 cfm up the chimney.) This relieves some of the pressurization created by using a fresh-air duct. When the thermostat isn't calling for heat (furnaces don't run 24 hours a day except when it is very cold outdoors), but the fan is running anyway for ventilation, the chimney will still have some warm air rising in it due to stack effect during the winter (if it is not located on an exterior wall). Many of these systems have a combustion-fired water heater connected to the same chimney as the furnace, which also acts like an exhaust fan—year-round. So, there are many periods when the furnace itself isn't contributing any exhausting capability, but air is flowing up the chimney nevertheless. In these situations, the exhausting capability is just a little weaker. In any case, the air entering the house through the fresh-air duct is performing two functions: It is supplying outdoor air for the benefit of occupants, and supplying outdoor air to the house that will be used for combustion. The duct is actually a fresh-air duct *and* a combustion-air duct.

If a furnace fan is running during mild weather (when you want ventilation air but not heat), the chimney acts like a large through-the-wall vent. At these times, there is no natural draft in the chimney because there are no warm combustion by-products (unless there is a combustion-fired water heater connected to the same chimney), and there is little or no stack effect due to indoor/outdoor temperature differences. This means that the pressurized house will push air out the chimney passively. When the chimney acts like a through-the-wall vent rather than an exhaust fan, the house will feel slightly higher pressures. But this doesn't result in hidden moisture problems in mild weather because there are no cold surfaces inside building cavities. In any case, in mild weather the windows are often open.

An electrically heated house (heat pump or resistance heating) has no chimney, so it is inherently tighter than a house that does have a chimney. (Remember, a chimney is basically just a big hole in the house.) In these houses, air will still be pulled into the fresh air duct, but it will have no chimney through which it can easily escape. Therefore, more air will pass through the structure into building cavities. This can lead to hidden moisture problems, especially if the house is underventilated (underventilated houses often have higher indoor relative humidities). Actually, any house in a cold climate, loose or tight, that doesn't have a chimney, has the *potential* for hidden moisture problems when it is pressurized. (Of course, the high moisture levels could be reduced significantly by using a local-exhaust fan regularly.)

In reality, many of the fresh-air ducts that have been installed in the past have been in leaky houses with leaky duct systems. As a result, the air movement between the indoors and the outdoors was probably influenced by several different natural, accidental, and mechanical pressures. Air was, no doubt, entering through both the fresh-air duct and a variety of other random locations, and it was escaping through the chimney as well as random holes in the structure. In short, this isn't a very well-thought-out or predictable way to ventilate houses.

A better way

A combustion-fired furnace or water heater needs combustion air to operate. People need fresh air to breathe. The systems that provide these dissimilar needs have different functions and requirements and they are best kept separate. In other words, you need a combustion-air system for the furnace or water heater that operates whenever combustion by-products are expelled up the chimney, and you need a separate fresh-air system that operates whenever people are indoors. See

Chapter 10 for a discussion of combustion air and chimney requirements.

You can use a fresh-air duct to supply occupants with outdoor air in a predictable manner. Instead of using a chimney (or random openings) to exhaust air from the house, you can provide through-the-wall vents for the pressurized air to escape through. (As was discussed in Chapter 16, through-the-wall vents are only effective in a small (less than 1,500 sq. ft.), tight house or in an apartment. In a larger, looser house there are plenty of random holes for air to escape through, so less air passes through the deliberate openings.) The fresh-air duct should be insulated if sweating is a possibility, and it should contain an adjustable damper and a flow-measurement grid so you can accurately adjust the amount of air entering the house. There also needs to be a motorized damper that shuts off the fresh air when ventilation isn't needed, and a control that will open the motorized damper and turn on the furnace/air conditioner fan simultaneously whenever ventilation is called for. Actually, the controls, dampers, and installation guidelines for this type of central-supply ventilation system are identical to the central-exhaust system connected to a forced-air furnace/air conditioner described in Chapter 19, with one exception: Through-the-wall vents (or random holes) are used instead of an exhaust fan (Figure 21-4).

This system will only provide a house with fresh air when the furnace/air conditioner fan is running. In many houses, the fan only runs about 25% of the time, primarily during the winter for heating and during the summer for cooling. If this system is to provide an adequate amount of fresh air to a house at all times, and to distribute that air throughout the house, the fan should run regularly—perhaps continuously. (For non-continuous operation, a timer with a 15-minute on/off cycle works well.) Because a longer-running fan has a higher operating cost, these systems often benefit from a high-efficiency motor, such as an ECM motor. (See Chapter 13 for a discussion of costs.)

In some cases, a supply fan is used to blow air through the fresh-air duct into the return duct. It should be wired to operate in conjunction with the furnace/air-conditioner fan and the motorized damper whenever the control calls for ventilation. By forcing air into the house with a fan, the building will be even more pressurized.

Equipment for fresh-air duct installations

Following are some companies that are currently offering equipment specifically designed for fresh-air duct applications:

American Aldes Corp. produces a fresh-air supply kit containing a Constant Airflow Regulator that can be easily added to a fresh-air duct in order to limit the amount of air entering a house. It is available in several sizes, ranging from a small 15-cfm unit up to a large model rated at 235 cfm. These regulators contain an internal bladder that adjusts itself to a wide range of pressures and restricts the amount of air passing through when the heating/air-conditioning fan is running. Constant airflow regulators are also used in conjunction with Aldes' central-exhaust ventilation equipment. (These devices require at least 0.2" w.g. of negative pressure in the return-air duct to function properly. Some return ducts don't operate at strong enough negative pressures.)

Duro Dyne Corp.'s Air-Quality Control System, **Tjernlund Products, Inc.'s** FAI-6, **Trol-A-Temp's** TMAC (also available from **Honeywell, Inc.**), and **Ztech's** Smart Vent can all be used to control the amount of air pulled through a fresh-air duct into a furnace/air-conditioner's return-air duct. They contain motorized dampers and timer controls that can activate the furnace/air-conditioner fan and the motorized damper. These devices were discussed in *Coupling a central-exhaust ventilation system with a furnace/air conditioner* in Chapter 19. The system described in that chapter has the addition of a central-exhaust fan. To pressurize a house, you simply substitute through-the-wall vents (or random holes) for the exhaust fan. The rest of the components are the same.

Skuttle Mfg. Co. has two different Make-Up Air Controls that open automatically whenever a forced-air furnace/air-conditioning fan comes on. One style opens a damper whenever there is a negative pressure in the return-air duct to which it is connected. The other style is connected to both the main supply duct and the main return duct and opens when the supply air is warmed up to a certain temperature. Because it must be warm to open, it is not suitable for use with a central air conditioner. A similar product, the Fresh Air 970, is made by **Air-King Ltd.**

Summary

Central-supply ventilation systems are less popular than central-exhaust and balanced ventilation systems, but they have some definite advantages. As long as the disadvantages (higher operating cost and the potential for hidden moisture problems in the winter) can be minimized, central-supply ventilation can be viable. This is especially true in moderate and hot climates where high operating costs and hidden moisture problems are less likely. Operating costs for modestly-sized ventilation systems are usually only high when fuel costs are high (*e.g.* with electric resistant heat).

All in all, well-engineered central-supply ventilation systems are relatively rare. Thus, experience in using them is limited in predicting when they will result in hidden moisture problems. However, because there must be both a positive pressure *and* a high indoor relative humidity for a hidden moisture problem to develop in a cold climate, you can minimize the chances of a problem by either maintaining a lower pressure and/or a lower indoor relative humidity.

In reality, if a house contains a continuously operating central-ventilation system and judiciously used local-exhaust fans it will generally have a low-enough relative humidity so that hidden moisture problems will not result.

Chapter 22

Balanced ventilation systems

Central-supply ventilation systems pressurize houses. Central-exhaust and local-exhaust ventilation systems result in depressurized houses. Pressurization and depressurization are neither good nor bad, but as we have seen, in some houses and in some climates, positive and negative house pressures can cause problems. For example, depressurization can be a potentially deadly problem in a house if it causes backdrafting or spillage. When there is the possibility of negative-pressure related problems, a balanced ventilation system (with or without heat recovery) should be considered. However, these systems also have advantages and disadvantages, which are covered below. One of the most important advantages is the fact that a balanced ventilation system gives you the most control over where air both enters and leaves the house.

A balanced ventilation system will not have an effect on the pressure in a house, but while neutral pressure is the goal, it is almost never totally achievable. Because so many natural and accidental pressures can affect a house besides the pressures induced by a ventilation system, it would be impossible to create a house that wasn't pressurized or depressurized to some extent at different times of the day, month, or year. Even a mild wind can cause pressure changes of 5 Pa. or more, and all houses contain local-exhaust fans and other mechanical equipment that affect house pressures. Even dirty filters in a ventilation system can cause the indoor air pressure to fluctuate by restricting the flow of air.

The use of a balanced ventilation system simply minimizes the intensity and duration of an imbalance in air pressures—at least those air pressures that are induced by the ventilation system. The goal with a balanced system it to have small pressure fluctuations in a house rather than large ones. In that way radon and other pollutants will be less likely to enter the living space, chimneys will tend to function correctly, and moisture problems in wall cavities will be minimized.

Understanding Ventilation

It is important to keep in mind that while a neutral pressure is often desirable, low-capacity central-supply and central-exhaust ventilation systems often don't pressurize or depressurize a house very much. Pressure-related problems are most likely the result of such things as high-capacity local-exhaust fans or leaky forced-air heating/cooling ducts.

Balanced ventilation strategies

A simple balanced ventilation system would consist of two fans. One fan could blow air into one room of a house and the other could exhaust air outdoors from another room (Figure 22–1). Fresh air would definitely

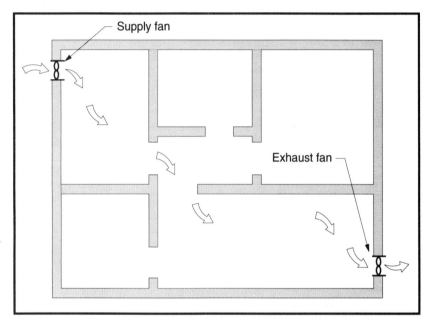

Figure 22–1. A pair of wall-mounted fans can be used to create a balanced ventilation system, but distribution will not be very effective.

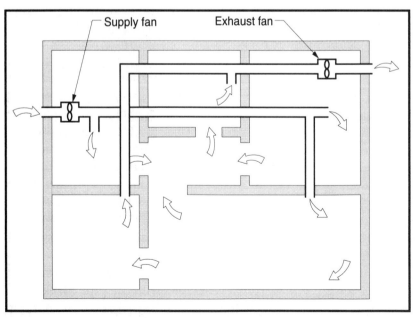

Figure 22–2. The most effective distribution involves exhausting stale air from moisture- and pollution-laden rooms and supplying fresh air to bedrooms and living areas.

be entering, and stale air would be leaving, but there wouldn't be very good distribution and there might be drafts. A better system would add individual and branch ducts with several supply and exhaust grilles throughout the house to distribute the air. (Figure 22–2). Variations might include a single exhaust grille and several supply grilles (Figure 22–3), or one supply grille and several exhaust grilles (Figure 22–4). A more complex system is more effective at removing stale air and distributing fresh air, but it is also more expensive to install. However, it will require less capacity, so there will be a lower equipment cost.

Actually, the most commonly used balanced ventilation system is the approach described in Chapter 19

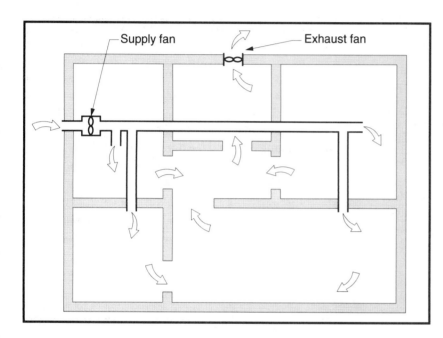

Figure 22–3. Effective distribution can often be achieved with a wall-mounted exhaust fan and fresh air ducted to several rooms.

Figure 22–4. Even though this layout has complete distribution, there may be complaints of drafts near the single supply grille—unless it is carefully located so the incoming air mixes with room air quickly above the heads of occupants and there is enough heating/cooling capacity in that room to temper the incoming air.

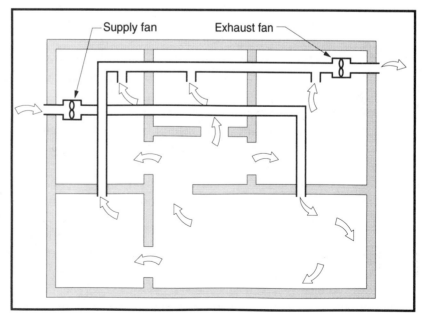

that consists of a central-exhaust fan and a fresh-air duct connected to a forced-air heating/cooling system. Although it is, in fact, a balanced-ventilation strategy, it is usually classified as a central-exhaust system. When installed and adjusted correctly, this system will result in a neutral pressure indoors, so it will have the advantages and disadvantages that are described below.

Advantages of balanced ventilation

Balanced ventilation systems create a neutral pressure in a house. This minimizes all the pressure-related problems associated with exhaust and supply ventilation systems. Of course, pressure-related problems can still occur in a house with a balanced ventilation system if strong pressures are generated in other ways—say, by leaky heating/air-conditioning ducts in the attic or crawl space. A balanced ventilation system simply reduces the chance of pressure imbalances occurring because of the ventilation system.

Minimizes hidden moisture problems

Moisture condensation within building cavities will be minimized in many climates when a house experiences neutral pressure. In order for significant amounts of moisture to get into building cavities, an air-pressure difference is necessary between the indoors and the outdoors to push moisture-laden air into those cavities. As was discussed in Chapter 8, there can still be hidden moisture problems in a house with a balanced ventilation system if there are strong natural or accidental pressures applied to the house. One of the best ways to prevent hidden moisture problems is to build a tight house and use a balanced ventilation system. Or you could use a depressurization system in a cold climate or a pressurization system in a hot, humid climate.

Keeps out unwanted pollutants

When a house is experiencing neutral pressure, a variety of pollutants, including particles and gases released from the insulation, and soil gases such as radon, will tend to remain outside the occupied living space because there is no air-pressure difference to cause them to move indoors. Of course, there still can be accidental and natural pressures that can cause pollutants to enter the living space (see Chapter 9), so the best way of preventing their entry is to combine a tight house with a balanced ventilation system.

Minimizes backdrafting and spillage

Backdrafting and spillage problems will be reduced in a house experiencing neutral pressure because there will be no ventilation-related pressures pulling air down the chimney into the house. However, because backdrafting and spillage are serious problems that can also be caused by natural and accidental pressures, the best way to keep combustion by-products out of a house is to use a heating system that is not capable of backdrafting or spillage—electric, solar, sealed combustion, etc. For a discussion, see Chapter 10.

Better control

With a balanced ventilation system, you purposefully bring air in through a single inlet and purposefully expel it through a single outlet. Such a strategy gives you very good control over where fresh air enters and stale air leaves the house. When you have better control of the air entering and leaving, the ventilation system needs less capacity to maintain comfort and freshness indoors.

When you have less control of where and when air enters a house, you need more ventilation capacity because some of the ventilation air is wasted. For example, when a ventilation system relies on an air-pressure imbalance between the house and the outdoors (either positive or negative) for there to be an exchange of air, there will be some controlled air movement through the deliberate openings, as well as some uncontrolled air movement through the random holes in the structure. (While there can still be random air movement into and out of a house due to natural and accidental pressure imbalances, at least a balanced system doesn't contribute to it.)

Filtering capability

When a fan is used to blow air into a house with a balanced ventilation system, there is often enough pressure generated by the fan to blow the incoming air through a filter. When air enters a house through a passive inlet, such as a through-the-wall vent, a filter on the inlet will simply restrict the flow of air at the inlet and cause it to enter elsewhere—often through an unfiltered, random opening. So, passive inlets can't be equipped with a filter, but the active inlet of a balanced-ventilation system can. Of course, it is important to select a fan with the correct capacity so it can overcome the resistance of the filter.

Disadvantages of balanced ventilation

The disadvantages of balanced ventilation have nothing to do with health and safety. These systems are sometimes more costly to install than the central-supply or central-exhaust approaches. And in harsh climates they can be costly to operate. In addition, they can require more complex duct systems than other strategies.

High installation cost

If a balanced ventilation system requires the purchase of two fans and has more ducts than either a central-supply or a central-exhaust ventilation system, the installed cost can be higher. However, a small offsetting savings can be realized because there is no need for through-the-wall vents. The balanced system described in Chapter 19 (central-exhaust fan coupled to a forced-air furnace/air conditioner) only requires the purchase of one fan, but it actually requires two: One is solely a ventilating fan (the central-exhaust fan) and one is used indirectly (the furnace/air-conditioner fan).

High operating cost

In a harsh climate (either very hot or very cold) when ventilation rates are excessive, a balanced ventilation system can have a high operating cost. This is partially due to the cost of running two ventilation fans, but it is also the tempering cost—the energy cost of tempering the incoming air, air that is cold in the winter and hot in the summer. However, in many parts of the U.S., where winter and summer temperatures are mild, the tempering cost will be minimal—so this may not be much of a disadvantage.

Figure 22–5. Broan's balanced ventilator has two chambers so the stale airstream and the fresh airstream don't mix and contaminate each other.

Understanding Ventilation

Because pressurization and depressurization systems interact with random leakage differently than balanced systems (see Chapter 7), an accurate comparison of operating costs should take into consideration both the controlled air exchange rate and the uncontrolled air exchange rate. Installation and operating costs are covered in Chapter 13.

Greater complexity

Because balanced ventilation systems consist of two fans and often have more ducts than central-exhaust or central-supply systems, they are perceived as being more complex, but really aren't. However, they can require a little more thought in the design stage in

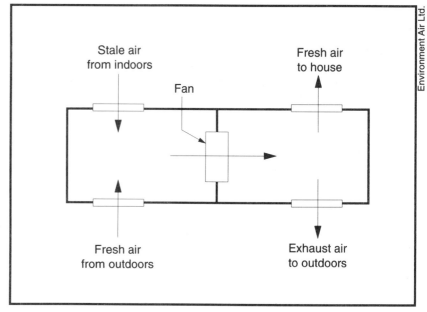

Figure 22–6. Environment Air's balanced ventilator mixes fresh air with stale air inside the unit. This tempers the incoming air so it isn't too cold in the winter or too hot in the summer. To bring in the same amount of fresh air into the living space as a balanced ventilator with separate and isolated airstreams, this approach requires a fan of greater capacity.

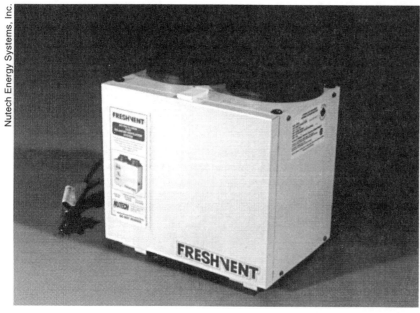

Figure 22–7. Nutech's Fresh Vent balanced ventilator uses the mixing strategy to temper incoming air.

order to plan the layout of the ducts. These systems also require the extra step of balancing (usually with airflow grids) the incoming and outgoing airflows after the equipment has been installed.

Balanced ventilation equipment

Most of the balanced ventilators on the market today are capable of heat recovery. In fact, only three manufacturers currently produce packaged balanced ventilator systems without heat recovery. Heat-recovery balanced ventilators will be covered in the next chapter. (See also Chapter 19 for a discussion of a balanced ventilation strategy that consists of a central-exhaust fan coupled to a forced-air furnace/air conditioner.) Non-heat-recovery balanced ventilators are viable options in mild climates. To provide the most effective distribution of air, these units should pull stale air from moisture- and pollution-laden rooms and blow fresh air into bedrooms and living rooms.

Broan Mfg. Co., Inc. has two very straightforward balanced non-heat-recovery ventilators (Figure 22–5). Rated at 110 cfm (85 watts) and 200 cfm (120 watts) respectively, they have an insulated cabinet, adjustable balancing dampers, and a washable filter. The Broan units are the only self-contained balanced ventilators that have two separate compartments through which the stale air and fresh air can pass, so that the airstreams don't mix and contaminate each other. If you are concerned about air entering the house that is occasionally too cold in the winter, Broan offers an electric preheater to temper the incoming air. If you are concerned about cold air entering on a regular basis, you should consider a balanced ventilator with heat-recovery (see Chapter 23).

Environment Air Ltd. has two balanced non-heat-recovery ventilators rated at 100 cfm (80 watts) and 150 cfm (113 watts) that temper the incoming air. The tempering is done by using a single fan to pull air simultaneously from both the outdoors and the indoors into one cabinet where the airstreams intermix. Then, equal amounts of the fresh/stale-air mixture are blown indoors and outdoors at the same time (Figure 22–6). With this system, some of the fresh air is blown directly outdoors before it can benefit the occupants and some of the stale indoor air is blown back indoors. This approach will temper the incoming air (so it won't be too cold in the winter or too hot in the summer) and will provide an exchange of air in the house, but the dilution factor means that it will need more overall capacity than a system with separated airstreams in order to get the same benefit.

Nutech Energy, Inc. offers three sizes (100 cfm, 190 cfm, and 255 cfm at 60, 90, and 125 watts) of its FreshVent balanced non-heat-recovery ventilator. These units also temper the incoming air by mixing it with a certain amount of stale air from the house, so the air entering the living space isn't too cold in the winter or too hot in the summer (Figure 22–7). Of the total capacity, about a third is fresh air and two-thirds is recirculated air. For example, the 100-cfm unit brings in about 35 cfm of fresh air from the outdoors and mixes it with 65 cfm of house air. At the same time, 35 cfm of air is exhausted to the outdoors. The exhaust air is also a partial mixture of fresh/stale air.

Besides these three packaged ventilators, it is very easy to create a balanced ventilation system using two fans having the same capacity. In-line fans (**Fantech, Inc., Kanaflakt, Inc., Rosenberg, Inc.,** and **Vent-Axia, Inc.**) can work very well, or you can use a pair of any of the individual fans discussed in previous chapters. With a little ingenuity, it is usually possible to create a balanced ventilation system with two quiet, low-energy fans. All you need to do is select two fans of the same size and install them so one blows air indoors and the other exhausts air to the outdoors.

A balanced ventilation system can either be interconnected with a forced-air heating/cooling system, or installed as an independent ventilating system. An independent system will need its own ducts for proper distribution, as discussed above. If coupled with a forced-air furnace or air conditioner, that system's ducts can be used to distribute the air.

Coupling with a forced-air furnace/air conditioner

A balanced ventilation system can use the ducts of a forced-air heating/cooling system to distribute fresh

Understanding Ventilation

air in a house, but as with other ventilation strategies, the furnace/air-conditioner fan must operate whenever the balanced-ventilator fan is running in order for the fresh air to be distributed throughout the house effectively. Chapter 14 describes various control strategies.

A balanced ventilation system—whether a packaged system or a pair of fans—will have connections for 4 types of ducts. Two of the ducts (a fresh-air intake and a stale-air exhaust) run to the outdoors. Chapter 16 describes the precautions you need to consider when locating them. The other two ducts run to the indoors. One picks up stale air from the house and the other supplies the house with fresh air. When interconnecting a balanced ventilator and a forced-air heating/cooling system, either one, or both, of the indoor ducts can be connected to the furnace/air conditioner's duct system.

Once a balanced ventilation system has been installed, the incoming and outgoing airflows will need to be adjusted so the house experiences a neutral pressure. If the ventilation system is connected to a forced-air heating/cooling system, the airflows should be balanced while the furnace/air-conditioner fan is running. If the furnace/air-conditioner fan operates at more than one speed, it is possible that the ventilator's airflows will be balanced at one speed but unbalanced at another. This is because the furnace/air-conditioner fan will generate different air pressures at different speeds, and those pressures can affect the airflow characteristics in the ventilator. It is also possible that a dirty filter can affect the airflows. While these are factors to consider, they are usually not significant ones. In reality, the change in flow through the balanced ventilator at different furnace/air conditioner fan speeds, or when a filter is dirty, will typically be within an allowable ±10%.

Though it is not commonly done, there is an advantage to installing motorized dampers on the balanced ventilator's incoming fresh-air and/or outgoing stale-air ducts. This will prevent air from flowing through the ventilator when the furnace/air-conditioner fan is running, but the ventilation fan isn't. In some cases, this can be done with a low-cost spring-loaded backdraft damper.

There are several possible ways of connecting a balanced ventilator to a forced-air furnace/air conditioner. When you look at ventilator manufacturers' installation literature, you will see they often prefer one method over another. Each approach has its proponents, and its own advantages and disadvantages. However, some techniques are more widely used than others. They are applicable to all balanced ventilators—those without heat recovery as well as those with heat recovery that are discussed in the next chapter.

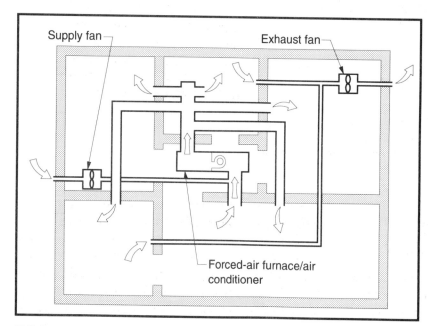

Figure 22–8. *This is the most popular method of connecting a balanced ventilator to a forced-air heating/cooling system. Fresh air is blown into the return duct of the heating/cooling system and stale air is exhausted from moisture- and pollution-prone rooms.*

Chapter 22 Balanced ventilation systems

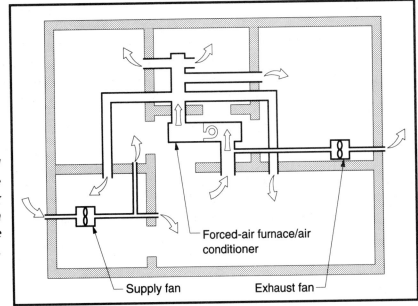

Figure 22–9. This balanced ventilation layout is not as effective at distributing fresh air to the living space or removing stale air from moisture- and pollution-prone rooms, so it is not very popular.

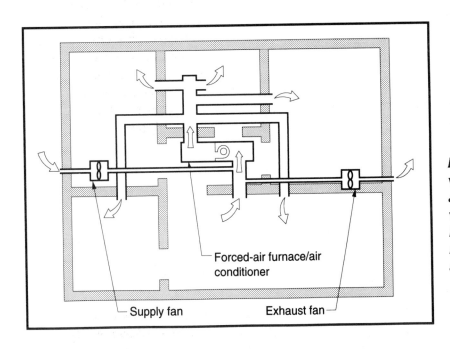

Figure 22–10. This balanced ventilation layout minimizes the amount of ducts needed for the ventilation portion of the system, but it is not very effective at removing stale air from pollution- and moisture-prone rooms.

Attaching only the fresh-air supply duct to the return duct

The most common approach is to connect the fresh-air supply from the balanced ventilator to the furnace/air conditioner's return-air duct. The stale-air duct is then routed from ceiling or high-wall grilles in the pollution-prone rooms of the house—usually the kitchen and bathroom—through the exhaust fan, then to the outdoors. During operation, the ventilation system pulls stale air from stale rooms, and simultaneously deposits fresh air in the return-air duct where it mixes with house air and is then distributed throughout the house by the heating/cooling system's air handler (Figure 22–8). A variation of this approach is used with the balanced ventilation strategy described in Chapter 19 (central-exhaust fan coupled with a forced-air furnace/air conditioner), but the forced-air furnace/air-conditioner fan pulls fresh air indoors rather than a separate supply fan.

If a multiple-speed furnace/air-conditioning fan is used, imbalanced flows can be minimized by installing a Constant Airflow Regulator (**American Aldes Corp.**) in the ventilator's fresh-air duct. These devices are sized for a certain airflow and have a flexible bladder that automatically opens and closes to maintain that airflow. They will maintain a relatively constant airflow even when the furnace/air-conditioner fan operates at different speeds. These devices can be used with a single-speed ventilator, but they render multiple speeds on a ventilator ineffective.

A variation is to attach the duct from a ventilator's supply fan to the furnace/air conditioner's supply duct. This will provide the same type of air distribution pattern (as long as the furnace/air-conditioner fan is running). The drawback to doing this is the fact that the ventilator is blowing air into a pressurized duct, so the ventilator fan must overcome more resistance than if it were blowing air into the return duct, which is under a negative pressure. An advantage of this type of setup is that a simple spring-loaded backdraft damper in the fresh-air supply duct will prevent the furnace/air-conditioner fan from blowing air backwards through the ventilator when the ventilator fan is off.

Attaching only the stale-air exhaust duct to the return duct

It is also possible to attach just the balanced ventilator's stale-air duct to the furnace/air conditioner's return-air duct and run the fresh-air supply duct to the living space, usually to heavily occupied rooms (Figure 22–9). With this type of layout, you pull stale air from one location in the house—the return-air grille—which is often mounted low on a wall. But stale air is generated in certain rooms (*e.g.* kitchen and bathrooms) where exhaust grilles high on a wall or ceiling are most effective. This approach generally doesn't mix or distribute fresh air in the house, or remove stale air, nearly as well as the first method. In fact, connecting a balanced ventilator to a furnace/air conditioner in this way is rarely done.

For a variation, you might attach the balanced ventilator's stale air duct to the furnace/air conditioner's supply duct. Distribution would still not be very good, and you would be exhausting air through the ventilator that was just heated or cooled—not a very energy-efficient thing to do. Again, balanced ventilators are rarely installed in this way.

Attaching the stale- and fresh-air ducts to the return duct

You can also connect both the fresh-air and the stale-air ducts from the balanced ventilation system to the return-air duct of the heating/cooling system. This is only viable if both ventilator ducts are attached to the return-air duct a few feet apart to prevent air from short-circuiting from one to the other. The fresh-air supply duct should be attached closer to the heating/cooling system's air handler than the stale air duct (Figure 22–10). This layout is very effective at distributing fresh air but not nearly so at removing stale air. Because it is less effective, it will require more ventilation capacity than the first layout. However, this type of installation often requires only a minimal amount of ducts.

When this layout is in operation, fresh air is introduced into the return-air duct where it mixes with house air. The heating/cooling system's air handler then blows this mixture throughout the house, and the house air finds its way back to the return-air duct where some of it is captured and blown outdoors.

A variation of this approach would be to attach the balanced ventilator's stale air duct to the furnace/air conditioner's return duct and attach the balanced ventilator's supply duct to the furnace/air conditioner's supply duct. When you do this, you are blowing air into a pressurized duct and pulling air out of a depressurized one, so both the ventilator fans must overcome a little extra resistance to airflow. However, this approach has the advantage that spring-loaded backdraft dampers on both balanced ventilator ducts will prevent unwanted air movement through the balanced ventilation fans when they are not running.

What if the furnace/air-conditioner fan is off?

If the furnace/air-conditioner fan is not running, but both of the ventilator fans are, fresh air will still enter the house and stale air will still be exhausted. But there will not be very good distribution to all the rooms in the house. This is because the furnace/air-conditioner fan must be running for air to be pushed and pulled through the entire duct system. When it isn't running, the fresh air that is in the ducts will spill into the living space through the nearest of the furnace/air-conditioner's registers, so most of the house won't be effectively ventilated. Instead, it will get less effective air exchange due to mixing and convective air currents.

The only way for some of the rooms to have an exchange of air with the furnace/air-conditioner fan turned off will be by diffusion (a very slow process) or by air currents (*e.g.* convective air currents) generated within the living space. This will occur, but it is much slower and less effective than using the furnace/air-conditioner fan to distribute the ventilation air throughout the living space. When doors between rooms are closed, air mixing will be very poor. The bottom line: For the most-effective distribution, the furnace/air-conditioner fan should run whenever the ventilation fan runs (continuously is best, but there are advantages to intermittent operation). This gives you the best fresh-air distribution and mixing and the best stale-air removal at the lowest ventilation rate.

Summary

Balanced ventilation minimizes many of the disadvantages associated with ventilation systems that pressurize or depressurize houses, but balanced ventilation systems can have a higher installation cost. They can also have a high operating cost in harsh climates (compared to a heat-recovery ventilator, especially when an expensive fuel is used for heating). While many of the balanced ventilation systems currently on the market do incorporate heat recovery (see the next chapter), the non-heat-recovery balanced ventilators described in this chapter (and the balanced system described in Chapter 19—a central-exhaust fan coupled to a forced-air heating/cooling system) can be economically viable options in many parts of the U.S. Of course, to determine what is cost-effective in your situation, you should perform some calculations (see Chapter 13).

Chapter 23

Balanced ventilation with heat recovery

A heat-recovery ventilator (HRV) is a ventilation device capable of transferring heat (and sometimes moisture) between two airstreams, or from one airstream into water. People often refer to HRVs as energy-saving devices. This isn't exactly true because HRVs are equipped with fans that require electricity to operate; therefore, they actually consume energy. In fact, HRVs, central-supply, and central-exhaust ventilators all use electricity to run their fans. Remember, all forms of ventilation consume energy. HRVs consume less energy by reducing, but not eliminating, the cost of tempering the incoming air (or by heating water, in the case of a central-exhaust heat-pump water heater). An HRV is first and foremost a ventilator and, second, offers a way to recover energy that would otherwise be wasted during the process of ventilating. (Actually, an HRV should save more energy by tempering the incoming air (or heating water) than its fans consume, but whether it does so, depends of its efficiency and the climate.)

Most HRVs are balanced ventilators in which heat is transferred from one airstream to another. These balanced HRVs are often called air-to-air heat exchangers (AAHXs). Central-exhaust heat-pump water heaters (**Therma-Stor Products**) are HRVs but they are not AAHXs because they transfer heat from one airstream into water, not between two airstreams. Another term often used is energy-recovery ventilator (ERV). All HRVs transfer *sensible heat* (see below) from one airstream to another or to water. If *latent heat* (see below) is also transferred, then an HRV can be called an ERV. While there are specific differences between these different terms, many people use HRV, AAHX, and ERV interchangeably. This chapter will only cover HRVs that are balanced ventilators with heat recovery, in other words, AAHXs. See Chapter 19 for **Therma-Stor Products'** HRV that transfers heat from air to water.

Although there is both a stale airstream and a fresh airstream passing through the same cabinet of an

Understanding Ventilation

AAHX, the airstreams (theoretically) do not intermix and contaminate each other. However, in reality there is usually a small amount (1–10%) of cross leakage between the two airstreams.

A number of manufacturers currently produce AAHXs. These devices all consist of a cabinet (galvanized metal, painted metal, or plastic) with at least one access door, one or two heat-exchange cores, and two fans. Some models come with accessories such as hanging straps, adjustable dampers for balancing, airflow grids for measuring airflows, air filters, condensate drain, defrost sensor, and controls. Most AAHXs are ducted, whole-house units but window units are available for ventilating a single room. For sensitive people,

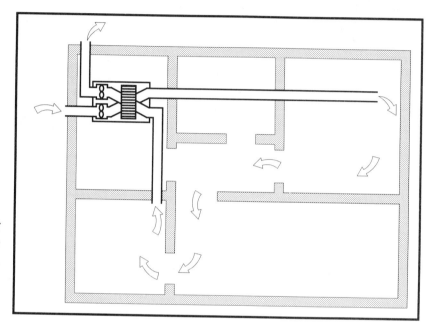

Figure 23–1. A simple AAHX duct layout will not provide an effective air exchange in all rooms.

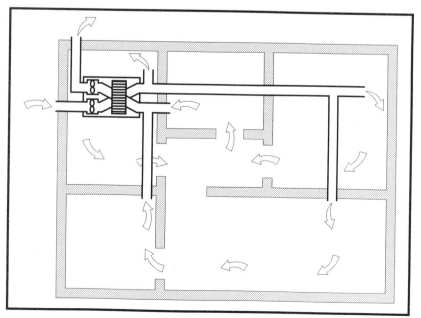

Figure 23–2. Multiple supply and exhaust grilles provide the most effective air distribution with an AAHX system.

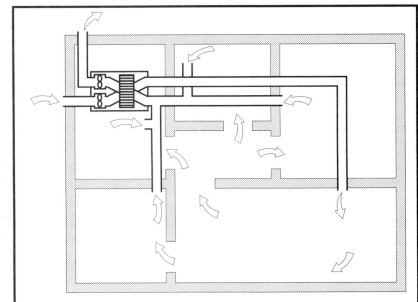

Figure 23–3. If an AAHX duct layout has only one supply grille, it must be located carefully so incoming air will mix with house air above the headspace or else the occupants will complain of drafts.

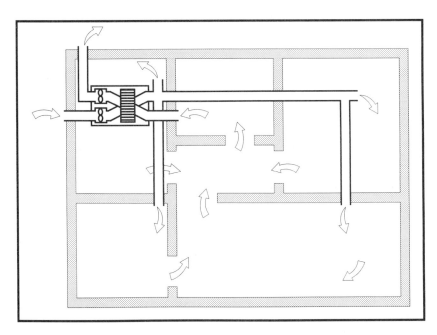

Figure 23–4. An AAHX duct layout with multiple supply grilles and a single exhaust grille can be an option if there is only one room that is generally pollution or moisture prone.

anything exposed to the fresh airstream (insulation, core, motor, gaskets, etc.) will be of concern.

Though they may seem complicated at first, an AAHX is actually a fairly simple device. Basically, it is just a box with two pathways through it—a path containing fresh air from the outdoors and a path containing stale air from the indoors—and two fans to move air through those pathways. Also inside the box is a heat-recovery core. Even though there are several different types of cores, they all do the same thing: They create multiple fresh-air paths and multiple stale-air paths that allow heat to be transferred through the core from the warmer airstream to the cooler airstream. Beyond this, some AAHXs have accessories such as con-

Understanding Ventilation

trols (timers, speed controls, and dehumidistats are common), defrost mechanisms, etc. Even though all AAHXs do the same thing—ventilate in an energy-efficient manner—the various manufacturers have different ideas about the best way to accomplish the task. That is why there are different shapes, sizes, core styles, fan types, defrost strategies, etc.

Discussions of ventilation systems often start with HRVs, but in many situations a much simpler device is all that is needed. For example, central-exhaust systems, central-supply systems, and balanced systems without heat recovery are suitable for many houses. HRVs may be among the most advanced ventilation systems available, but they can also be among the most expensive.

Balanced heat-recovery ventilation strategies

The HRVs discussed in this chapter are balanced ventilators (so they are all AAHXs), therefore the strategies for using them are exactly the same as those discussed in the last chapter. For example, a simple AAHX layout might blow air into a room at one end of a house and exhaust air from a room at the opposite end (Figure 23–1. Fresh air definitely would be entering and stale air would be leaving, but distribution would be poor and there might be drafts. A better approach would be to use a system of individual and branch ducts to distribute the air (Figure 23–2). This complicates matters a bit, but it offers much more effective distribution, thus, the capacity can be reduced. It is also possible to introduce fresh air through one supply grille and use several exhaust grilles (Figure 23–3), or use one exhaust grille and several supply grilles (Figure 23–4). The layout you choose depends on how you want the air to circulate and how complicated a system you can afford. An AAHX can also be interconnected with a forced-air heating/cooling system.

Advantages of an AAHX

AAHXs have more advantages than any other ventilation system. Because they don't affect the air pressure indoors, they minimize all the pressure-related problems that can occur. On the other hand, they can be costly and some layouts can be complex. Because AAHXs are basically balanced ventilators, many of the advantages and disadvantages below are similar to the pros and cons covered in the last chapter.

Low operating cost

Because AAHXs recover heat in the winter that would otherwise be wasted (they can also recover "cool" in the summer), they place less of a tempering burden on furnaces and air conditioners. Therefore, the heating and air-conditioning bills will be lower than using a ventilation system without heat recovery. (While the fans in an AAHX may be no more economical to operate than other similar-size fans, some other ventilation strategies are able to utilize lower-wattage fans, so it is important to consider the electricity usage of a fan, as well as the energy-efficiency of the heat-recovery process, when evaluating costs.)

AAHXs can sometimes recover enough energy to pay for themselves in a few years in a very cold climate. In a hot, humid air-conditioning climate, they can also conserve energy, especially if they have a core that is designed to transfer moisture from the incoming airstream to the outgoing airstream, but there is generally less energy to save in hot climates in the U.S. In many moderate climates where heating and cooling bills aren't very high to begin with, there may not be much of an savings in operating costs, so an AAHX may not be able to save enough energy to make its higher upfront cost worth considering. There are also climates where neither the winter savings nor the summer savings, when taken individually, are significant, but when added together, the annual savings makes an AAHX cost-effective. The costs associated with ventilation systems are covered in detail in Chapter 13.

Tempering of incoming air

Because the incoming air is warmed in heating climates and cooled in air-conditioning climates as it passes through an AAHX, it will be more comfortable than a similar system not having heat recovery. This tempering of the incoming air also minimizes the chance of a sudden temperature shock inside the furnace which could result in cracking of the furnace's heat exchanger. It also minimizes the chance of condensation or corrosion problems inside a furnace. (Actually, this type of damage to a furnace is only a problem in very cold climates when a large amount of ventilation air is brought into a furnace that has a relatively low airflow rate.)

Minimizes hidden moisture problems

Whenever a house experiences a neutral pressure, there will be no air-pressure difference between the indoors and the outdoors to push moisture-laden air through the random holes in the structure. With an AAHX, there will be no such pressures due to the ventilation system; there could still be pressures because of the wind, a clothes dryer, leaky forced-air heating/cooling ducts, or because doors between rooms are closed. An AAHX (or a balanced ventilation system without heat recovery) simply minimizes the imbalance of pressures. In most cases, low-volume central-supply or central-exhaust ventilation systems only generate fairly small pressures in a house, so this is often not a significant advantage. To eliminate hidden moisture problems completely, it is necessary to build an airtight house, eliminating the pathways for moisture to enter building cavities. See Chapter 8 for a discussion of moisture problems.

Keeps out unwanted pollutants

Pollutants that originate within the structure of a house (such as insulation) or outside the structure (such as radon or lawn chemicals) can be pulled into the living space when a house gets depressurized. When a house experiences a neutral pressure as a result of using an AAHX (or if the house is pressurized), the pollutants won't enter. Of course, there could be other things besides the ventilation system that cause depressurization, but at least an AAHX won't be a contributor. To eliminate the possibility of unwanted pollutants entering through the structure, you must build an airtight house (see Chapter 9), or pressurize the living space.

Minimizes backdrafting and spillage

Backdrafting and spillage can occur when a house gets depressurized. When a house experiences a neutral pressure as a result of using an AAHX, the chimney is more likely to function correctly—until something

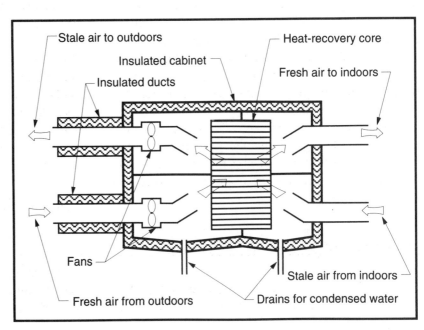

Figure 23–5. Typical features of an air-to-air heat exchanger.

such as a powerful kitchen range hood is turned on. Backdrafting and spillage are minimized in a house with an AAHX, but they may not be eliminated. To do that you need a heating system that is inherently immune from the problem, such as those discussed in Chapter 10 (*e.g.* a gas furnace with a sealed combustion chamber, or a heat pump).

Better control

With any balanced ventilation system, you bring air in through a single inlet and expel it through a single outlet. This gives you better control over where air enters and leaves. When a ventilation system relies on a pressure imbalance between the house and the outdoors (positive or negative) in order for there to be air movement, some air will enter (or leave) through deliberate openings, and some will pass through random holes in the structure of the house. Of course, to have the best control over where air enters and leaves, a house must be tightly constructed.

Filtering capability

If an AAHX uses a fan to blow air into a house (one manufacture has an AAHX without fans), there is often enough pressure generated by the fan to push or pull the incoming air through a filter. (You must use a fan to actively move air through a filter because air can't flow through most filters passively.) However, it is important to select an AAHX with a fan having enough capacity to overcome the resistance of the filter. Filters are covered in Chapter 15.

Disadvantages of an AAHX

AAHXs cost more to install than other ventilation systems because they are more complex. However, you often get what you pay for and if you want a quality ventilation system with minimal disadvantages, you must pay the price. (Of course, a competent installer is also very important; if he doesn't know what he is doing, it is possible to have an expensive system that doesn't work at all.) When used in a harsh climate an an AAHX's operating costs are often lower than other strategies—but in a mild climate they can be higher.

High installation cost

Most AAHX installations contain two fans, a core, an insulated cabinet, a defrost system, and a sys-

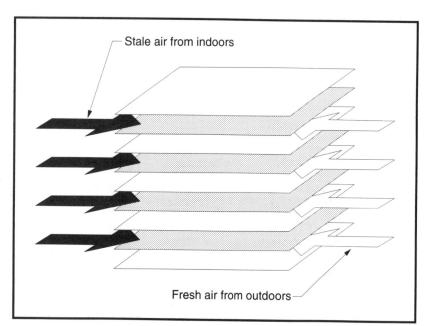

Figure 23–6. In a flat-plate counterflow AAHX, the airstreams flow in opposite directions.

tem of ducts that may extend throughout the house. There are simply more pieces than there are with many other ventilation systems. More pieces mean more money. However, the cost increase isn't as significant as it may seem at first, especially if the AAHX pays for itself in energy savings within a few years. After that, it will mean ventilation with lower heating bills for the life of the equipment.

Because much of the expense associated with an AAHX involves its system of ducts, it is often possible to simplify the duct layout and save on installation expenses. For example, if an AAHX is tied to a furnace/air conditioner's duct system, then only a few short ventilation ducts may be necessary—but you shouldn't skimp on ducts if it means less effective distribution.

High operating cost

To be cost-effective, an AAHX should save more in tempering costs than you must pay to run its fans, and that savings should pay for the extra installation cost within a few years. In a mild climate, where tempering is not a significant expense, the cost of running an AAHX's fans can exceed the cost savings of tempering the incoming air. In such a situation, an AAHX can have a higher total operating cost than a non-heat-recovery ventilation strategy that uses a lower-wattage fan (such as a central-exhaust system). If the cost of electricity is high, and the cost of heating fuel is low, an AAHX may be even less cost-effective.

Greater complexity

Many people consider AAHXs to be complicated simply because they don't know very much about them. Actually, they are far less complex than a television, or an answering machine, or an automobile. The diagram in Figure 23–5 shows the basic construction features. There are two airstreams that pass by each other inside a core (without mixing or contaminating each other). The heat from the warmer airstream passes through the core and into the cooler airstream. Most AAHXs have four duct connections—two that run to the outdoors and two that run to the indoors. Some have a fifth indoor connection as a part of a defrost system.

Not all AAHXs are created equal

AAHXs have several similar features, but there are also some differences. They all have a cabinet. Some cabinets are insulated, some are not. All have a core,

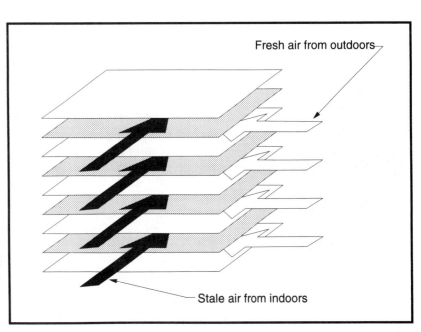

Figure 23–7. In a flat-plate crossflow AAHX, the airstreams flow at right angles to each other.

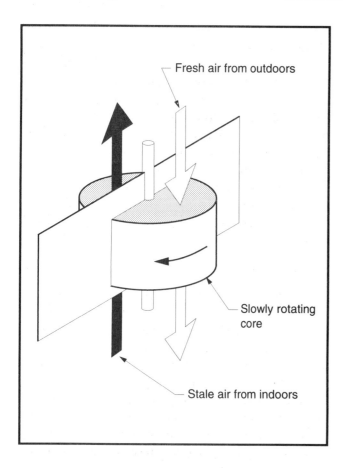

Figure 23–8. *Rotary cores have more cross leakage than other cores, but they offer the advantage of being able to transfer moisture between the two airstreams.*

but there are different types of cores, and some models have dual cores. All have two fans. Some have a single motor that turns both fans, some have two motors. Many have built-in controls, but a few come with no controls.

Most AAHXs were originally developed for use in cold-climates. Because it is possible for moisture to condense and freeze inside them in the winter, many—but not all—have defrost mechanisms. Defrosting is usually only a concern in the coldest parts of North America. In the middle and southern U.S., defrosting usually isn't necessary. Following is a rundown of the basic features of the AAHXs currently on the market.

AAHX core types

All AAHXs contain a core in which heat is transferred from one airstream to another. There are several different types of cores currently being used: flat-plate counterflow, flat-plate crossflow, rotary, and heat pipe. Some flat-plate cores are made with thin, wide passageways through which the air passes; others have passageways that are divided into multiple channels. The cores can be made from different materials: aluminum, plastic, or treated paper.

Even clean air always contains a few contaminants and air moves through a core for at least several hours each day, so all cores eventually get dirty. Many manufacturers have designed their cores so they can be removed for cleaned. These cores can be pulled out of the cabinet, taken outdoors and vacuumed or hosed off to remove accumulated debris. Some AAHXs have cores permanently mounted inside the cabinet, but in such a way that they can be cleaned with a vacuum cleaner by opening an access door. With a few AAHXs, cleaning is virtually impossible unless you dismantle the unit or unhook several connecting ducts. It is doubtful if a homeowner will dismantle a system to clean it, so partial clogging is likely.

No AAHX cores are capable of handling grease-laden air. (Actually, there are some in Sweden, but they

Chapter 23 Balanced ventilation with heat recovery

aren't being imported to North America.) This is important to remember if you are tempted to run the exhaust from a kitchen range hood through an AAHX. After all, a range hood is often expelling warm air from the house—warmth that is wasted in the winter. Unfortunately, AAHXs are not designed to handle the microscopic grease and oil droplets that a range hood blows out of a house. These pollutants can build up inside the core of an AAHX and create a fire hazard. Thus, a range hood should never be connected directly to an AAHX.

For the vast majority of people, outgassing from AAHX cores is not an issue. In fact, a study was done to determine outgassing rates from AAHX cores to see it they contributed to indoor pollution.(Piersol) The test results showed no significant outgassing of organic compounds. However, for very sensitive people, a treated-paper core or a plastic core might outgas very slightly and be bothersome. Gaskets can also be a source of outgassing that might bother hypersensitive people. Sometimes aluminum-foil tape can be used to cover offending gaskets, but care must be taken to preserve the gaskets' sealing ability.

Flat-plate cores

Most of the AAHXs currently available have flat-plate cores. These consist of a series of thin sheets of aluminum, treated paper, or plastic, laid out to separate the incoming and outgoing airstreams. Theoretically, a flat-plate core will be capable of recovering more energy if the airstreams run in opposite directions (counterflow) (Figure 23–6), than if they run perpendicular to each other (crossflow) (Figure 23–7). However, core design has gotten fairly sophisticated in recent years and crossflow cores are quite efficient; in fact, some of the most energy-efficient AAHXs contain dual crossflow cores. For a flat-plate core (counterflow or crossflow) to be efficient at transferring moisture from one airstream to the other, it must have a water-permeable treated-paper core.

Rotary cores

Currently three suppliers (**Honeywell, Inc.**, **Nutone, Inc.**, and **Stirling Technology, Inc.**) offer residential-sized AAHXs with rotary cores. In addition, **Airxchange** and **Crispaire Corp.** offer larger, commercial-sized units. Rotary-core AAHXs contain a slowly spinning, specially constructed, plastic wheel that picks up heat from one airstream and transfers it to another airstream (Figure 23–8). Since the wheel continually passes from one airstream to the other, 8–12% of the stale air is deposited into the fresh airstream and vice versa, so there is always some cross-contamination.

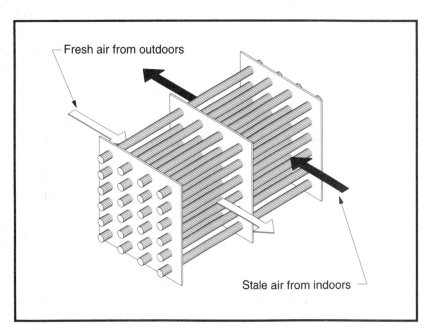

Figure 23–9. *In a heat-pipe core, heat is transferred between the two airstreams through sealed refrigerant-filled tubes.*

343

With a rotary core, some moisture is also transferred between the airstreams. In the winter, this can help prevent a house from becoming too dry. Because of their ability to transfer moisture, rotary-core AAHXs can be desirable in hot, humid climates to minimize the amount of moisture brought indoors with the fresh air. Some rotary cores are coated with a material called a desiccant which allows additional moisture to be transferred between the airstreams.

Heat-pipe cores

Though they were more popular in the past, only one company (**Environment Air Ltd.**) currently produces an AAHX with a heat-pipe core. A heat pipe is a permanently sealed tube containing a refrigerant (such as Freon). When warm air passes across one end of the tube, it boils the refrigerant, which then flows to the other end of the tube where the cold airstream captures its heat. As it gives up its heat, the refrigerant condenses back into a liquid, then flows by gravity back to the other end (the heat pipes are tilted slightly) where it is warmed up again and the cycle repeats (Figure 23–9).

Moisture control

Three AAHX manufacturers (**Altech Energy**, **Broan Mfg. Co., Inc.**, and **Research Products Corp.**) offer AAHXs with flat-plate cores that are designed to transfer both heat and moisture from the warm airstream into the cold airstream. (These AAHXs are often called ERVs.) While most AAHXs have plastic or metal cores—materials that transfer heat quite well—a flat-plate core must be made of a specially treated paper to allow both heat and moisture to pass through.

Rotary plastic cores will also transfer a certain amount of moisture from the warm airstream into the cold airstream, so they can also be called ERVs. Three manufacturers of rotary-core AAHXs (**Airxchange**,

Figure 23–10. To prevent air from moving through an AAHX's drain line, a loop can be formed in a length of plastic tubing to form a water trap. While this is a good idea, in practice the water often evaporates.

Chapter 23 Balanced ventilation with heat recovery

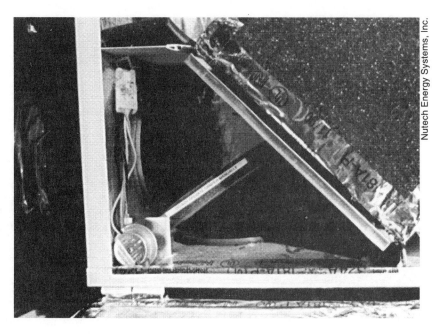

Figure 23–11. This AAHX utilizes a motorized damper to temporarily direct warm house air through the core (instead of cold outdoor air) to defrost the core.

Crispaire Corp., and **Honeywell, Inc.**) offer a desiccant-coated core that is specially designed to transfer even more moisture between the airstreams than an uncoated core. Desiccants are substances that absorb moisture. They are sometimes found in small packets inside bags of potato chips to preserve freshness, or in electronic equipment to minimize moisture damage. Desiccants are also often used inside the spacer bars of double-pane windows to keep them from fogging. When used in an AAHX, a desiccant will absorb moisture from a humid airstream and release it into a dry airstream.

Cores that transfer moisture are sometimes called *enthalpic* cores. (The word *enthalpy* refers to the total heat content of the air.) The amount of moisture transferred depends on the temperature and humidity of the incoming and exhaust airstreams, but a 40–70% transfer of moisture is typical with a treated-paper crossflow core, and 70–85% is typical with a desiccant-coated rotary core. An uncoated rotary core usually won't transfer any moisture in the summer, but as much as 50% is possible in the winter when there is a significant difference between the indoor and outdoor temperatures.

Because formaldehyde is water soluble, it has been suggested that formaldehyde can be transferred from the stale airstream to the fresh airstream along with any moisture. A study at *Lawrence Berkeley Laboratory*[(Fisk 1985)] found that only 7–15% of the formaldehyde present in the stale air was transferred to the fresh air, and it was concluded this was not enough to be considered a problem. However, individuals sensitive to formaldehyde may want to take this into consideration and opt for a different core material. In any case, it is important to keep in mind that even if a little formaldehyde does pass through an AAHX's core, a great deal more will be exhausted outdoors, so the net result will be a reduction of formaldehyde in the indoor air.

Moisture control in air-conditioning climates

In hot, humid climates, it can be very desirable to keep the indoor air as dry as possible so that the air conditioner isn't overworked. When a ventilation system brings hot, humid air indoors, the air conditioner will have to start running in order to both cool the incoming air and to remove the extra humidity.

If a ventilation system can be chosen that will transfer moisture between the incoming and outgoing airstreams, the air conditioner won't have to work as hard. This can make an AAHX more cost-effective in a hot climate. (The energy saved is actually air-conditioning energy.) Rotary cores (especially desiccant-coated rotary cores) and treated-paper cores pull moisture out of the incoming fresh air and transfer it into the stale

345

Understanding Ventilation

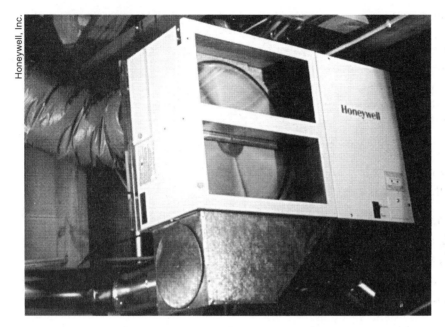

Figure 23-12. A typical rotary-core AAHX installed in a basement.

outgoing air, so these cores have a definite advantage in hot climates. However, sensitive occupants should determine if they can tolerate the core material prior to a permanent installation.

Moisture control in cold climates

When any ventilation system moves stale air out of a house, it blows a certain amount of moisture out as well. This is good, because too much moisture indoors can result in mold growth or a proliferation of dust mites. However, sometimes a house will get too dry indoors in the winter because moisture is being exhausted by the ventilation system faster than it is being generated by the occupants. This occurs primarily with oversized ventilation systems that run for extended periods of time. For example, if a ventilation system is capable of changing the air in a house several times every hour, it will tend to dry the house too much in the winter. On the other hand, a modestly sized system that provides 15 cfm of continuous ventilation per occupant will be less likely to get too dry.

Winter dryness due to overventilation can be minimized by using an AAHX that is capable of transferring moisture between the airstreams (either rotary cores or treated-paper cores). During operation, some of the moisture in the stale air leaving the house passes through the core and is released into the fresh airstream (along with some of the heat).

If a house must rely on a humidifier in the winter to maintain a comfortable indoor relative humidity, the humidifier will require energy to operate. An AAHX capable of transferring moisture between airstreams will be more cost-effective in such a situation because it minimizes the need for a humidifier. (The energy saved is humidification energy, not ventilating energy.)

Getting rid of the condensate

Many AAHXs have a small-diameter ($1/2$–$3/4$") drain connection. This is because, at different times of the year, the temperatures inside the core will result in water condensing inside the AAHX, and it must have a way to escape or else it will run onto the floor. (As was discussed in Chapter 8, the relative humidity (RH) of air will go up if the air becomes cooler. If the RH reaches 100%, condensation occurs.) If an AAHX can't be mounted in a location where its drain can be connected to a plumbing drain by gravity, a small pump must be used. Because fans generate air pressures inside an AAHX cabinet, the drain should have a trap so air isn't blown through the drain line (Figure 23–10).

When moisture condenses inside a core, mold growth is possible. Though this hasn't proven to be a serious problem in most installations, it underscores the importance of being able to clean a core periodically. Because of condensation inside AAHXs, a study was done in Canada to determine if mold growth was a possibility.(Energy) This study found that mold was generally not a problem on metal or plastic parts, but paper or cardboard air filters were susceptible. However, it was also determined that with proper and regular maintenance, mold growth is unlikely. In winter, condensation will take place in the outgoing airstream, so if mold growth does occur, it won't contaminate the fresh air.

Sweating on the outside of the cabinet

Most AAHX cabinets are insulated to prevent sweating on the outside of the cabinet when internal temperatures get too cold. (Ducts containing cold air should also be insulated to prevent sweating.) Insulating an AAHX's cabinet also makes the process of transferring heat between the two airstreams more efficient.

In the past, a few AAHXs had raw fiberglass insulation exposed to the airstreams. This is not a good idea because small glass fibers can be released into the fresh air, and the insulation can become a home for microbial contamination if it is exposed to a high relative humidity. Exposed fiberglass is also very difficult to clean without damaging it. Today, foil-faced fiberglass is widely used, with only the aluminum foil exposed to the airstreams. Other insulating materials, such as foam boards, can also be used, but ideally should also be covered with foil.

If an AAHX cabinet isn't insulated and moisture condenses outside it in the winter, a drip pan should be placed under the unit to catch the condensate and direct it to an appropriate drain.

Defrosting

In very cold climates, the temperatures inside an AAHX can get low enough so the condensed water starts

Figure 23–13. An AAHX's core must be cleaned periodically. A removable core is usually easier to clean than a non-removable core.

to freeze. The resulting ice can block the exhaust portion of the core and prevent air from freely flowing through it. Therefore, many AAHXs have defrost systems to periodically melt any ice that forms. Some mfacturers use a simple electric heater that warms up the incoming air enough to melt any ice that has formed in the exhaust section of the unit. **Honeywell, Inc.** controls its electrical-resistance heater automatically so incoming air is warmed to 5°F before it comes in contact with the core. One of the AAHXs from **Des Champs Laboratories, Inc.** uses a simple 100-watt light bulb to defrost the core periodically. During its defrost cycle, the light bulb warms up the core enough to melt any ice that has formed. While this is an interesting idea, the bulb could burn out, unnoticed by the occupants. A small electrical-resistance heater would seem to be somewhat more reliable.

Another defrost option is to have an internal motorized damper that opens and closes to allow extra warm air from the living space to pass through the core (Figure 23–11). Such a damper temporarily blocks (either completely or partially) only the cold incoming fresh air and causes extra warm house air to circulate through the cabinet, thus warming it up. When some of these AAHXs are in their defrost mode, air is still being exhausted from the house, so the house can become temporarily depressurized. This could lead to backdrafting or another negative-pressure related problem, but the defrost cycle typically only lasts 10–15 minutes, so it isn't usually considered a serious drawback. Once the defrost cycle is complete, the damper moves to its original position and the indoor/outdoor air pressures will again be in balance.

Several studies have been done on freezing in AAHXs. The precise temperature at which freezing occurs is dependent on several factors, in particular the relative humidity of the stale air leaving the house. In general, freezing has been found to only occur when fresh-air (outdoor-air) temperatures are below –23°C to –9°C (–9.4°F to +14.2°F).[Fisk 1983] If the climate is such that the outdoor temperature never drops below +14.2°F, freezing is unlikely even if the indoor RH is as high as 60%. If the indoor air is drier than 30%, freezing won't occur until the outdoor temperature dips to below –9.4°F.

Freezing occurs at lower temperatures in AAHXs that transfer moisture between airstreams than it does in other AAHXs. The reason: When moisture is transferred from the stale air to the fresh air in the winter, there is less moisture available in the exhaust air to freeze. If an AAHX is used in a climate where freezing in unlikely, defrosting is unimportant.

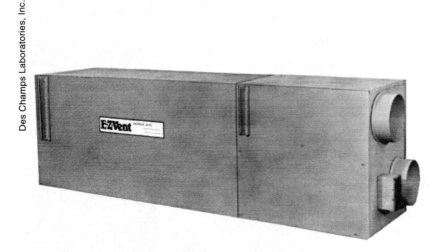

Figure 23–14. Counterflow AAHXs are often long and bulky.

Chapter 23 Balanced ventilation with heat recovery

Figure 23–15. This compact counterflow AAHX has a unique defrost mechanism in the lower right compartment—a light bulb.

Figure 23–16. A typical crossflow AAHX. This unit has all of the duct connections on the top.

Understanding Ventilation

Figure 23-17. Interior view of a typical crossflow AAHX. This unit has two duct connections on the top and one on each side.

Energy efficiency

All ventilation systems bring in outdoor air—air that is rarely at optimum room temperature—and that air must be heated or cooled. Therefore, all ventilating strategies require energy to heat or cool the incoming air. In many climates and locales in the U.S., this may not take a great deal of energy, but in harsh climates a ventilation system can place a big load on a furnace or air conditioner.

Ventilation systems that incorporate heat recovery can often save energy by reducing the cost of tempering incoming air. (Of course, a poorly designed high-capacity AAHX system can use more energy than a well-designed low-capacity non-AAHX system, so good planning and design are critical for any ventilation strategy.) AAHXs minimize the amount of energy needed to temper the incoming fresh air by removing heat from the warmer airstream and adding that heat to the cooler airstream.

While aluminum is much more efficient at transferring heat than either plastic or paper, the efficiency of an AAHX depends more on the design of the core than on the material of which the core is constructed. For example, the theoretical maximum efficiency of transferring sensible heat (see below for a description of SRE) that can be achieved with a flat-plate counterflow core is 100%, while the theoretical maximum for a flat-plate crossflow core is about 75%. (In reality, these maximums are never achieved; efficiencies in the 60–80% range are typical when a single core is used. Higher efficiencies are possible with multiple cores.)

There are actually two different kinds of heat that are transferred between the airstreams in an AAHX: sensible heat and latent heat. You measure sensible heat with a thermometer. For example, when the air temperature changes from 70°F to 50°F, that represents 20 degrees of sensible heat. However, there can be more heat in the air than the temperature indicates. This heat is "hidden" in a way, and it is called latent heat.

Latent heat is present in air only when the air contains some water vapor. If you could find some air that had 0% RH, it would have no latent heat. Because all air, even in dry climates, contains some moisture, for all practical purposes air always contains a certain amount of latent heat. The latent heat in air is the heat that was originally required to evaporate the moisture from a liquid into a gas. (It takes energy to evaporate water.) In other words, latent heat is the heat required to change something (in this case, water) from one state to another (*e.g.* liquid to gas, solid to liquid). Sometimes this is refereed to as a change of phase, meaning from liquid phase to gas phase.

Latent heat is often an important consideration. For example, when 6,000 cubic feet of air (equivalent to a 100-cfm fan running for an hour) changes temperature from 70°F to 50°F, the 20° of sensible heat contains about 2,160 Btus. If the 6,000 cubic feet of 70°F air is at 40% RH, then it contains a total of about 2,900 Btus of latent heat. The more moisture air contains (*i.e.* the higher its absolute humidity), the more latent heat it contains. Some (but not all) of this latent heat will be released when water condenses from a vapor to a liquid, inside an AAHX.

To read some companies product literature, you might think that all AAHXs are extremely energy-effi-

Company	Core type
Altech Energy (NewAire)	crossflow plastic or treated paper
American Aldes Corp.	plastic counterflow
Amerix Corp.	crossflow plastic
Aston Industries, Inc.	crossflow aluminum
Boss Aire, Inc.	crossflow aluminum
Broan Mfg. Co., Inc. (Guardian)	crossflow plastic or treated paper
Carrier Corp.*	crossflow plastic
Conservation Energy Systems, Inc. (Van EE)	crossflow plastic
Des Champs Laboratories, Inc. (E-Z Vent)	counterflow aluminum
Duro Dyne Corp. (Durovex)	crossflow aluminum
Engineering Development, Inc. (Vent-Aire)	counterflow aluminum
Environment Air Ltd. (ElitAir and Enviro)	heat pipe, crossflow plastic or aluminum
Fantech, Inc.	crossflow aluminum
Honeywell, Inc. (Perfect Window)	desiccant-coated rotary
Nutech Energy Systems, Inc. (Lifebreath)**	crossflow aluminum
Nutone, Inc. (Fresh Air Machine)	rotary
Preston-Brock Mfg. Co., Inc. (Air Changer)	counterflow plastic
Raydot, Inc.	counterflow aluminum
Research Products Corp. (PerfectAire)	crossflow treated paper
Snappy ADP (May-Aire)	crossflow aluminum
Stirling Technology, Inc. (Recouperator)	rotary
Thermax	crossflow aluminum
Venmar Ventilation, Inc. (Flair)	crossflow plastic
XeteX, Inc. (Heat-X-Changer)	aluminum crossflow

*Carrier's products are also sold under Bryant, Day & Night, Heil, and Payne brand names
Nutech's products are also marketed by **Lennox Industries and **Tradewinds**

Figure 23–18. Suppliers of residential AAHXs.

Understanding Ventilation

cient. However, once installed, they don't always perform up to their manufacturers' claims. This is because there are a variety of ways to measure energy efficiency. Sometimes an efficiency measured in a laboratory will be somewhat higher than the efficiency of the same AAHX once it is installed in a house. For this reason, if you are going to compare the efficiencies of different products, it is important that the efficiencies be measured in the same way.

In order to accurately compare products of different manufacturers, the *Canadian Standards Association* (CSA) developed a standardized test method (CSA Standard 439–88) that can be used to rate the efficiencies of AAHXs. After an approved testing labora-

Figure 23–19. This AAHX is designed to mount in the ceiling of a small house or apartment.

Figure 23–20. A crossflow-core AAHX made to mount in a window.

Chapter 23 Balanced ventilation with heat recovery

Figure 23–21. A rotary-core, window-mounted AAHX.

tory has evaluated an AAHX using CSA 439–88, the results can be certified by the ***Home Ventilating Institute*** (HVI). A list of AAHXs produced in North America that have been certified, and the test results, is available from HVI.(Home 1994)

With certified test results, it is possible to compare the performance of different AAHXs with reasonable accuracy. However, it should be kept in mind that once a ventilation system is in a house, it represents a unique installation. For example, once installed, an AAHX may have a slightly different capacity than it did during testing, or there might not be equal flows in the two airstreams. While the installed efficiency will probably not be exactly the same as the efficiency derived in a laboratory, it should be reasonably close to HVI-certified results.

By looking at an HVI-certified testing form, you can determine the air flow rates at different static pressures, the degree of cross-contamination between airstreams, how much moisture is transferred between airstreams, the power consumption of the fans, etc. The following terminology is used by CSA and HVI to compare how much energy is recovered by an AAHX.

Apparent sensible effectiveness (ASEF)

If you measure the actual temperatures of the two airstreams passing through an AAHX and compare them with each other, you can calculate the *apparent sensible effectiveness (ASEF)*. To determine the ASEF, use the following formula:

ASEF = [(Fresh air temperature after passing through the core) − (Fresh air temperature before passing through the core)] ÷ [(Temperature of stale air from house) − (Fresh air temperature before passing through the core)]

As an example, suppose 10°F fresh, outdoor air and 70°F stale indoor air pass through an AAHX's core, and after going through the core the 10°F air has been warmed up to 55°F. The ASEF will be (55 − 10) ÷ (70 −

Understanding Ventilation

Figure 23-22. A through-the-wall AAHX can simplify an installation by eliminating the two duct connections that run between the unit and the outdoors.

10), or 0.75. This can also be expressed as 75%. The ASEF is used to predict the actual temperatures of the airsteams. In this example, the incoming 55°F air will still need to be warmed up to a comfortable 70°F (remember, ventilation does cost money), but it takes considerable less energy than if the 10°F air were brought directly into the living space. An AAHX doesn't eliminate the cost of tempering, but it does minimize it. ASEF only measures sensible heat; it ignores latent heat.

Sensible recovery efficiency (SRE)

The airstream temperatures in an AAHX are affected by more than just how much energy is recovered through the core. For example, because electric motors warm up when they are running, a fan motor in an airstream will warm up the airstream. Defrost systems also add warmth to an airstream. The *sensible recovery efficiency (SRE)* takes the ASEF information and factors out the energy released by warm motors, defrost systems, electric preheaters, and cross-leakage of air between the two airstreams. As a result, the SRE is a lower number than the ASEF. A typical AAHX might have a ASEF of 0.79 (or 79%) and an SRE of 0.69 (or 69%). The SRE is useful for comparing the energy-recovery performance of different AAHXs, but it only takes sensible heat into consideration. It also ignores latent heat.

Total recovery efficiency (TRE)

The *total recovery efficiency (TRE)* is a another useful way of comparing energy-recovery performance of AAHXs. The TRE is similar to the SRE, because it factors out the influence of fan motors, cross-leakage, etc., but it also takes into consideration any latent heat that is recovered.

If the temperature inside the core of an AAHX is such that water condenses out of the air, the act of condensation releases latent heat. This is because latent heat is released whenever any substance changes from a gas to a liquid or from a liquid to a solid. For example, if water is cooled until it freezes, latent heat will be given off as it turns to a solid. More condensation means the release of more latent heat.

The TRE is said to take *enthalpy* into consideration. Enthalpy is a term often used in conjunction with latent heat. It refers to the total heat content of air, that is, the sum of the sensible heat and the latent heat. While the TRE would be very useful to know for all climates, HVI-certified evaluations only report it for a cooling climate where moisture transfer between airstreams is highly desirable. In these climates the TRE is often 20–30% for aluminum and plastic flat-plate AAHXs and a much higher 65–90% for AAHXs that are capable of moisture transfer.

How important is it to consider latent heat?

In a way, the TRE is misleading. It represents the total amount of heat transferred— both latent and sensible. But just because latent heat *can* be conserved doesn't mean it is important to do so. Sometimes it is, sometimes it isn't.

In many cases, latent heat is an important consideration. For example, it takes a certain amount of energy for an air conditioner to dehumidify the indoor air in the summer, or for a humidifier to add moisture to the air in the winter. These processes involve the transfer of latent heat. If an AAHX can minimize these expenses by transferring moisture between the two airstreams, latent heat will be conserved. But it won't necessarily be conserved inside the AAHX—it will be conserved because the air-conditioning or humidification expenses are lower.

On the other hand, just because air contains latent heat doesn't mean it is always worth considering. If dehumidification with air conditioning in the summer, or humidification in the winter, aren't necessary, then latent heat isn't a concern. So, moisture transfer (and latent-heat recovery) in an AAHX isn't always important. In fact, in many instances, it is desirable for an AAHX to remove as much moisture from a house as possible. See Chapter 8 for a discussion of how ventilation relates to moisture control.

If the ventilation process saves some energy associated with dehumidification (with the air conditioner) or humidification, you have conserved energy—but you have done so at the dehumidifier (or air conditioner) or at the humidifier. Sometimes, when you transfer moisture inside an AAHX, you conserve heat inside the AAHX itself. This occurs when the temperatures and humidities inside the AAHX result in condensation. If condensation occurs inside an AAHX, heat is liberated and that heat can be absorbed by the opposing airstream. (Latent heat is *released* when moisture changes from a

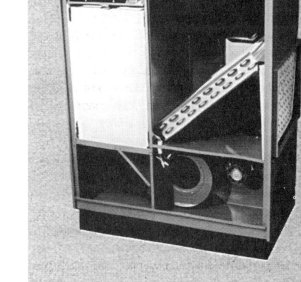

Figure 23–23. This unit combines the functions of a forced-air furnace/air conditioner and an AAHX into a single piece of equipment.

Understanding Ventilation

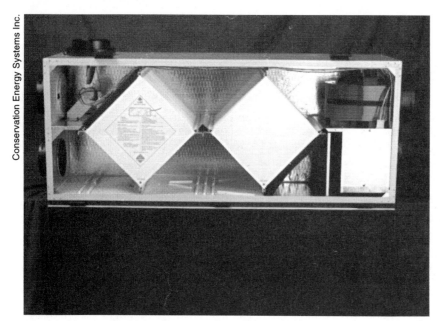

Figure 23–24. Some manufacturers produce AAHXs with dual cores for extra energy conservation.

gas to a liquid.) Condensation only takes place occasionally inside some AAHXs (*e.g.* in a plastic or aluminum core in the winter when the stale airstream has an RH above 50% and the fresh airstream is below 40°F.). Though possible, condensation in an AAHX is less likely in an air-conditioned house during the summer.

Fan wattage

SRE and TRE are very useful for comparing AAHXs, but they aren't the only factors to consider. For example, suppose two AAHXs have similar SREs but one requires 200 watts of power to run the fans and the other only consumes 100 watts. They may transfer heat from one airstream to another comparably, but one will cost more to operate. An HVI report will tell you how much power is consumed (in watts) at several different airflows. But the airflows listed for one manufacturer's AAHX are often different from those of a different AAHX. A useful way to compare power consumption at different airflow rates is to divide the number of watts consumed by the motor by the cfm of the fan. This will give you the number of watts/cfm, a figure that can be compared between different units. Most AAHXs consume about 1 watt/cfm total for both fans. The lower the watt/cfm figure, the more efficiently air is moved through the unit.

Maximizing energy efficiency

It doesn't matter how energy-efficient an AAHX is, if it isn't installed conscientiously, it can lose its economic advantage. (Of course, non-heat-recovery systems can be installed poorly too.) For example, if an AAHX is mounted in a cold crawl space and the ducts are leaky and uninsulated, enough heat can be lost that the overall efficiency of the system will be considerably lower than the efficiency of the AAHX core itself.

To minimize heat losses from the cabinet, an AAHX should be mounted in a conditioned part of the house (not in an attic or crawl space). To minimize heat losses from the ducts between the AAHX and the outdoors, they should be as short as possible, well-sealed, and well-insulated. Ideally, the ducts between the AAHX and the living space should be within the conditioned space and they should also be well-sealed.

Balanced heat-recovery ventilation equipment

There are numerous AAHX manufacturers in the U.S. and Canada. Many are able to ship directly to contractors or homeowners, but a few only sell through distributors. Figure 23–18 contains a listing of the various North American producers of AAHXs for residential installations. Most offer several models in different sizes and most offer a variety of accessories.

Most AAHXs on the market are made to be part of a whole-house, ducted, ventilation system; however, there are a few models that can be used to ventilate a single room. **Altech Energy** has a small ducted unit that mounts in a ceiling; it is advertised for use in apartments (Figure 23–19). **Thermax** and **Stirling Technology Inc.** both produce AAHXs that can be mounted in a window (much like a window air conditioner) in order to ventilate one room (Figures 23–20, 23–21).

For an office installation, **Crispaire Corp.** has a 450-cfm Marveair AAHX with a desiccant-coated rotary core that is designed to fit into a 2' x 4' suspended ceiling. **Crispaire Corp.** also produces an E-Tech heat-pump water heater that removes heat from the indoor air and transfers it to a hot water tank. Though its main function is to produce hot water, it can be hooked to a duct system to extract heat from the outdoor air and act as a balanced heat-recovery ventilator as well as a water heater.(Next) (Although they are actually central-exhaust ventilators, the Therma-Vent and Envirovent heat-recovery water heaters by **Therma-Stor Products** (see Chapter 19) are specifically designed to supply hot water and ventilate in an energy-efficient manner.)

Conservation Energy Systems, Inc. offers a Thruwall AAHX that hangs on an exterior wall with the inlet and outlet passing directly through the wall (Figure 23–22).

Besides a conventional stand-alone AAHX, **Engineering Development Inc.** has several unique pieces of Vent-Aire equipment. One model is an integral forced-air heating/cooling/AAHX unit that is suitable for a well-insulated house with low heating and air-conditioning requirements (Figure 23–23). They also have

Figure 23–25. Large-capacity AAHXs are specially designed for ventilating indoor swimming pools.

models with modules that can add heating and/or cooling capabilities to an AAHX, and a ClasVent that is a one-piece through-the-wall heating/cooling/AAHX unit that looks similar to the heaters used in schools and motel rooms.

Most AAHXs on the market consist of a single self-contained piece of equipment. The AAHX from **American Aldes Corp.** has two modules: a fan unit and a heat-recovery core unit. With this approach, the fan can be mounted some distance from the living space—say, in an attic—to minimize any fan noise inside the house. Then the heat-recovery core unit can be mounted within the conditioned part of the house with branch ducts leading between it and the various rooms. This allows for easy cleaning.

Environment Air Ltd. offers conventional AAHXs with crossflow cores as well as an Enviro passive AAHX that has a heat-pipe core *and no fans*. This is the only AAHX on the market that does not rely on its own fans to move air. Instead, it uses the pressures generated in both the supply and return ducts of a forced-air furnace/air-conditioning system to move fresh air and stale air through the core. The main supply duct (under positive pressure) pushes air through the core to the outdoors, and the main return duct (under negative pressure) pulls fresh in from the outdoors. Though this setup saves on the cost of buying the fans, a certain amount of fresh air that is brought into the return duct is exhausted from the supply duct before it has a chance to benefit the occupants. This unit comes with a motorized damper that can be wired to open whenever the furnace/air-conditioning fan is running.

Boss Aire, Inc., **Conservation Energy Systems, Inc.**, and **Nutech Energy Systems, Inc.** offer both single-core AAHXs and higher-efficiency, dual-core residential units (Figure 23–24).

Venmar Ventilation, Inc. offers an AAHX with a built-in negative-ion generator, but it is doubtful if it will have a significant effect on air quality. This is because air moving through a ducted system usually produces positive ions which would counteract the negative ones being generated.

Manufacturers of higher-capacity AAHXs for light-commercial applications, large houses, swimming pools, and spas include **Airxchange** (500–3,000 cfm, desiccant-coated rotary core (not appropriate for pools and spas)), **Altech Energy** (900 cfm, crossflow plastic or treated paper core), **Conservation Energy Systems, Inc.** (700 cfm, crossflow plastic core), **Engineering Development Inc.** (600–3,000 cfm, crossflow aluminum core), **Nutech Energy Systems, Inc.** (700–1,200 cfm, crossflow aluminum cores), and **Raydot, Inc.** (custom-made sizes up to 40,000 cfm, counterflow aluminum core). **Boss Aire, Inc.** has aluminum crossflow core AAHXs up to 3,500 cfm and can custom-make them to handle up to 60,000 cfm, **Des Champs Laboratories, Inc.** has aluminum counterflow units from 600–4,000 cfm, **Environment Air Ltd.** offers a 600-cfm model, **Thermax** produces AAHXs with plastic or aluminum counterflow cores up to 10,000 cfm, and **Venmar Ventilation, Inc.** has units up to 1,300 cfm with either aluminum or plastic crossflow cores. **XeteX, Inc.** has large crossflow aluminum-core AAHXs specifically designed for indoor swimming pools called Pool-X-Changers (Figure 23–25), and they can custom-make sizes up to 38,000 cfm.

Coupling an AAHX with a forced-air furnace/ air conditioner

To combine an AAHX with a forced-air heating/cooling system, the heating/cooling fan and the AAHX fans should run simultaneously for the best distribution of air throughout the house. The fans can be interconnected in a variety of ways (see *Controlling a ventilation fan and a furnace/air-conditioning fan simultaneously* in Chapter 14). One advantage of interconnecting the ventilation system and the heating/air-conditioning system is that fewer ducts are needed than if the ventilation system were installed independently. In other words, the furnace/air-conditioning ducts handle both ventilation air and heated or cooled air. Another advantage is the fact that one filter can be used for both systems. (Filters are covered in Chapter 15.)

A disadvantage to interconnecting these two systems is that you must run the larger heating/cooling fan as well as the AAHX fan to obtain the most effective distribution of air to all rooms of the house. One of the primary advantages of an AAHX is the fact that it is an

Chapter 23 Balanced ventilation with heat recovery

energy efficient way to ventilate houses. By operating an extra fan, this advantage is lessened—sometimes completely. However, the cost of running the larger fan can be minimized by using a furnace/air-conditioner equipped with an energy-efficient, variable speed (ECM) fan motor.

If the systems are interconnected and the AAHX is running but the furnace/air-conditioning fan is off, some rooms won't be ventilated as effectively as others. This means more ventilation capacity is necessary to have the same effect. See *What if the furnace/air-conditioner fan is off* in Chapter 22.

There are several methods of attaching an AAHX's ducts to a forced-air heating/cooling system. The options are identical to the methods that are used to connect a balanced ventilator without heat recovery. See Chapter 22 for a discussion.

Summary

AAHXs have a number of definite advantages, but they can also be costly to install. In some climates, this cost can be recouped in a few years because of a savings in energy, but this is not true in all parts of the U.S. Sometimes, another ventilation strategy (*e.g.* central supply, central exhaust, or balanced without heat recovery) will work just as well at a lower installed cost. But just because an AAHX isn't always cost-effective doesn't mean that it should be ruled out. After all, it will temper incoming air, making it more comfortable—and we are all seeking comfort. AAHXs also offer the advantage of being available in a wide variety of styles and sizes, so there are more pieces of equipment to choose from than if you decide on a different strategy.

Chapter 24

Passive ventilation systems

With most controlled ventilation systems, two things are controlled: 1) the airflow (by using one or two fans) and 2) the size and location of deliberate openings in the building (inlets and outlets). With passive ventilation you place deliberate inlets and outlets in a house—but use no fans—so the air movement is actually only half-controlled. In other words, you purposely place openings in a house, then you let the natural and accidental pressures move air through those holes.

The drawback to this approach to ventilation is the fact that the air pressures you are relying on to move the air (natural and accidental pressures) are inconsistent. Air will only pass through the openings when pressures are applied to the building. Natural pressures will exert no pressure when the wind isn't blowing, or when the indoor and outdoor temperatures are the same. The mechanical devices that create accidental pressures will only exert a pressure on the building when they are operating. However, there are many times when these pressures can supply a house with sufficient fresh air. For example, in the winter months there may be enough continual pressure being applied to the building due to stack effect for there to be sufficient air moving through the inlets and outlets. On the other hand, there will always be other times when there are no natural or accidental pressures available to move air through the house. And there will probably be occasions when too much air is blown through the house. Still, even though a passive ventilation strategy is less controlled and predictable than other strategies, it is better than having no ventilation system at all.

The idea of passive ventilation is intriguing, but it is often easier said than done. As a result, passive ventilation is often better suited to buildings that do not require continuous ventilation, such as barns, storage buildings,(Westlake) and garages. Attics and crawl spaces are often ventilated passively. Usually this works reasonably well, but if an attic or crawl space contains an

Understanding Ventilation

unusually large amount of moisture, a fan is generally necessary because the passive pressures are not sufficient to provide enough air movement. (Actually, if moisture is a problem in a crawl space, a better control strategy is to prevent it from entering in the first place.) As with other ventilation strategies, you will be able to minimize air movement through random holes, and maximize air movement through deliberate openings, if a house is tightly constructed.

Passive ventilation strategies

Open windows have been used for years as a means of passive ventilation. But to effectively provide fresh air to all parts a house and remove stale air from all the rooms, there will need to be an open window in each room. Then, the natural and accidental pressures will push air inward through some of the windows, and outward through others. When the outdoor temperature is moderate and the outdoor air is clean, you may have all of the windows wide open and receive plenty of air movement through the house. But during a hot summer or a cold winter, it can be very difficult to determine precisely how much to open the windows. If they are open too wide, the furnace or air conditioner will need to run continuously to keep up. If the windows aren't open enough (or if they are closed), you won't have enough air exchange. Open windows are a legitimate way of ventilating a house, especially in moderate climates, but they have some definite drawbacks.

Another approach to passive ventilation simply involves placing deliberate openings in a house, such as through-the-wall vents, and then letting the natural and accidental pressures push air through the openings. But how many openings do you need, and how big should they be? We learned in Chapter 2 that the amount of air flowing into and out of a house depends on two variables: the size of the openings and the amount of pressure. When you use a fan to ventilate, the fan manufacturer's literature will tell you precisely how much pressure you will have at a certain airflow. With this information, you can accurately select an opening size. When you rely on passive ventilation, the pressures can vary widely from day to day. So, whatever size openings you select, there will be times when a great deal of pressure leads to a great deal of air exchange in the house, and periods of no pressure and no air exchange. Again, this is better than nothing, but it is far from perfect.

Another way to passively ventilate a house is to combine a *ventilating chimney* with through-the-wall

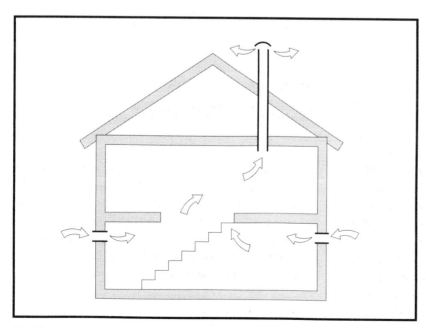

Figure 24–1. *In the winter, stack effect causes air to leave a house through a ventilating chimney and enter through through-the-wall vents.*

Chapter 24 Passive ventilation systems

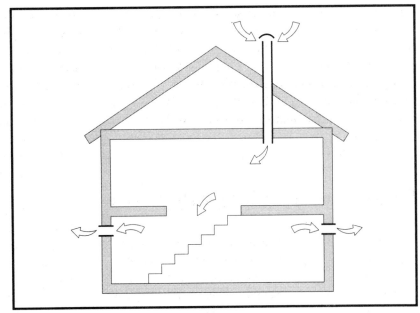

Figure 24–1. In the summer, stack effect causes air to enter a house through a ventilating chimney and leave through through-the-wall vents.

vents. A ventilating chimney is only used for ventilation (not for combustion by-products) and in cold weather it will have air moving up it and out of the house due to stack effect (Figure 24–1). Ventilating chimneys must have a cap on the top to prevent rain entry, and while they can be made to function, they are very dependent on temperature differences to function well. When it is very cold outdoors and warm indoors, the flow rates can be high, but during mild weather, the flow rates will decrease substantially. During the air-conditioning season when it is cool indoors and hot outdoors, stack effect will cause fresh air to *enter* a house through a ventilating chimney (Figure 24–2).

Though passive ventilation is not widely relied on in the U.S. and Canada, work is being done in other countries to develop more viable strategies for utilizing passive ventilation systems.(Knoll) For example, a school in Europe was designed to receive fresh air through a passive ventilation system—but the results were less than successful. The school had a south facing conservatory that was warmed by the sun. It was hoped that the solar warming would enhance the stack effect pressures enough to ventilate the building. The system worked, but not well enough to provide the recommended ventilation rate. Only when there were periods of high temperature and solar radiation, and fire doors were left open, was the recommended ventilation rate able to be met.(Palmer) Actually, passive ventilation is not as suitable in schools as it is in houses because of the greater population density. For example, in a school there may be ten students per 500 sq. ft. of floor space, but in a house there may only be one person per 500 sq. ft. When there are more people per square foot, a ventilation system needs to be more powerful and more effective, and passive ventilation generally isn't reliable enough to deliver the correct amount of air at all times.

Advantages of passive ventilation

The big advantage of passive ventilation is simplicity. The basic components of a passive ventilation system are a few inlets and outlets (such as through-the-wall vents) placed in the walls or ceilings. This makes it a low-cost system to install. There are no motors, no ducts, and no controls. However, even with a low installed cost, there is still a cost associated with heating or cooling the incoming air, so passive ventilation isn't free. In fact, if the airflows are excessive, the cost of tempering may exceed the cost of running a small mechanical venting fan. The simplicity of passive ventilation means there is very little for the occupant to understand, maintain, or repair. Because there is no fan, passive ventilation systems are somewhat quieter than mechanical systems.

Backdrafting and spillage might be minimized in a house with a passive ventilation system if the use of inlets and outlets prevents large air-pressure differences from occurring. But if the system is not carefully designed, inappropriate airflows may still result in a problem. So the best way to prevent backdrafting and spillage is to use a furnace, boiler, or water heater that doesn't require a natural draft chimney, such as an electric heat pump or a gas appliance with a sealed combustion chamber.

Disadvantages of passive ventilation

There are three basic drawbacks to passive ventilation. First, it can have a high operating cost, especially in harsh climates. Second, the amount of air passing through the building is less controlled. And third, the air movement is less predictable.

High operating cost

The air entering a building that is ventilated by passive means will be the same temperature as the outdoor air. It will be cold in the winter and hot in the summer. While there is no actual cost to run the ventilating fan, there will be a tempering cost associated with bringing hot or cold air indoors. There are only two ways to minimize tempering costs: 1) reduce the ventilating capacity (controlling a passive system can be difficult) or 2) use a ventilation system that is capable of heat recovery (there aren't any truly passive HRVs). In mild climates, the cost of heating or cooling incoming air is often not excessive, so this disadvantage only applies to harsh climates.

Minimal control

Because passive ventilation uses no fans, air will only enter and leave the house when a pressure acts on the building. Because the pressures involved are the result of uncontrolled pressures or activities, you can't turn them on and off, so there is little that can be done to control such a system. It is possible to use motorized dampers or manually-operable inlets and outlets. Or, differential pressure sensing devices (see Chapter 14)

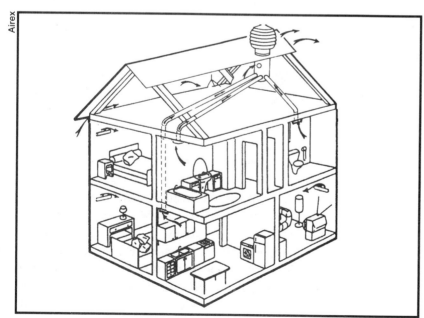

Figure 24–3. The Aereco passive ventilation system distributed by Airex combines a static fan with through-the-wall vents.

Chapter 24 Passive ventilation systems

Figure 24–4. Trickle Ventilators mount in a window sash to provide a pathway for air to enter and leave a house passively.

could automatically open and close dampers to vary the incoming and outgoing airflow, but such a control system must be engineered carefully. It is also possible to design a combined system that relies on passive pressures for part of the year, and a fan at other times. These approaches tend to make a fairly simple ventilation system more complex—and a passive ventilation system's main appeal is its simplicity.

Minimal predictability

A variety of different ventilation strategies, including passive ventilation, have been used in France since a 1969 regulation required ventilation in all buildings. When the passively ventilated buildings are analyzed, airflows are difficult to predict because wind and stack-effect pressures vary considerably from day to day. Therefore, in practice, it is difficult to determine the size of the openings for optimum airflow rates. When used in multistory buildings, some installations have had air entering through outlets during unfavorable climatic conditions.(Beinfait) When a passive ventilation system works well, it is often a matter of luck rather than good design.

All in all, it is difficult to predict precisely what effect passive ventilation will have on pollutants entering a house outside the living space, moisture migration, or backdrafting and spillage. This makes the design of a reliable passive ventilation system very difficult. If there is a great deal of pressure, there can be a great deal of ventilation. If there is little or no pressure, there will be little or no ventilation. But then, this is better than nothing. After all, occasionally the ventilation rate will be "just right."

Passive ventilation equipment

While passive ventilation isn't as popular as mechanical ventilation in North America, a few companies are actively marketing passive ventilation systems for houses. These systems consist of inlets and outlets that are carefully located in walls or ceilings.

Airex has an Aereco passive ventilation system that relies on natural pressures. This system combines humidity-sensitive through-the-wall air inlets with humidity-controlled exhaust grilles to vary the airflow through a house. The exhaust grilles are connected to either a wind turbine or a Static Fan mounted on the roof. During operation the inlets automatically vary their opening size depending on indoor humidity levels, natural pressures, and accidental pressures so that the avail-

Understanding Ventilation

Figure 24–5. A roof turbine utilizes both wind and stack effect to pull air from a building passively.

able air pressures will push the right amount of air through the openings (Figure 24–3).

Environment Air Ltd. has an Enviro passive HRV (see Chapter 23) that uses no fans of its own. Instead, it relies on the fan pressures generated in the ducts by a forced-air furnace/air-conditioning system. Because air will only move through the ventilator when the air handler is running, this really shouldn't be considered a passive ventilation system. When calculating the operating cost, a portion of the forced-air furnace/air-conditioning fan's operating cost should be taken into consideration.

Thinking Vents, Inc. has a passive ventilation system that consists of several different ceiling vents and two different wall vents. One style of wall vent can be used as an inlet, the other as an outlet. The company's literature shows ceiling vents connected to an attic so that stale house air will be expelled directly into the attic, then presumably outdoors through roof vents. This is very poor practice because stale house air is often very humid, and humid air in the winter can condense on cold surfaces in the attic, resulting in mold growth and/or decay.

Titon Inc. markets a very complete line of Trickle Ventilators as a passive ventilation system. Attractive, inconspicuous, well-made, and available in a variety of styles and sizes, they are made to be mounted in window frames and sashes (Figure 24–4). Trickle ventilators can be used as either inlets or outlets. Some models are made to remain open all the time while other styles can be opened and closed. They have been used in Europe for over 20 years.

A standard roof turbine ventilator (Figure 24–5) can be used as a passive-exhaust ventilation device if it is connected to the living space with a duct. But these devices are probably most suitable for ventilating non-residential buildings or uninhabited attic spaces. Roof turbines are driven by either the wind or by stack effect. If moist house air passes through them, they can be subject to freezing in the winter.

Summary

As with other ventilation strategies, passive ventilation has both advantages and disadvantages. If the advantages can be maximized, and the disadvantages minimized, this can be a viable approach to ventilating a house. For example, passive ventilation can work in a small, tightly constructed, two-story house in a climate where it is cold most of the year because stack effect will provide a relatively continuous pressure. It might also be an option for a remote vacation cabin where electricity isn't available. However, in most parts of the country, a small central-supply or central-exhaust fan can substantially increase the system's effectiveness without adding much to the cost.

Chapter 25

A nine-step design and installation process

In this chapter, we will go through the basic steps involved in planning and installing a central ventilation system. Because every house is unique and occupant requirements vary, the best ventilation system for one house may be completely different from the system next door. Or, if two houses have identical floor plans but are in different climates, one may have a central-supply system and the other a central-exhaust system.

All the various ventilation strategies have both advantages and disadvantages. With new construction, you should consider all the pluses and minuses of a particular system while the house is still in the design phase. That way, the negatives can be minimized and the positives maximized. For example, it is easier to build an airtight house if you plan to do so from the beginning. If you would like to conceal ventilating ducts inside building cavities, you may need to lay out the floor joists in a different direction or move a partition wall a few inches to make room to run a duct. Or it may become apparent there isn't enough money in the budget for a hot tub, a sealed-combustion gas furnace, *and* a fully ducted heat-recovery ventilation system.

Too often, a house is completely designed, then the ventilation system is fit in—with difficulty. As the plumbers and electricians are roughing in the plumbing and wiring, someone is just getting around to suggesting some ideas for ventilation. At this stage, there will no doubt be many more compromises than if the system was designed much earlier. The resulting installation will very likely end up being less effective and more expensive. So, you should plan out the entire ventilation system very early in the design process. This doesn't mean you should work out all the small ducting details before you determine where the kitchen will be located. But if you have a general concept of what you want your ventilation system to actually accomplish, you may decide to lay out the kitchen somewhat differently to allow for better airflow.

It generally involves some extra thought, planning, and expense to install a ventilation system in an existing house—compared to a new house—because of the difficulty in installing ducts in walls that have already been closed up. Therefore, there may need to be more compromises when working with an older home. But it can still be done, and with a little forethought, it can usually be done well. Planning ahead is also important for the builder once the house in under construction, especially if a particular piece of ventilation equipment must be special ordered. An out-of-stock ventilator can mean a delay of at least a few days, occasionally a few weeks. A builder also needs to make sure his heating/air-conditioning subcontractor is familiar with installing ventilation equipment.

The following steps are not overly detailed, but they will give you an idea of the basic process of designing and installing a central ventilation system.

Step 1
Fix any problems with the house

First, you should analyze the house to determine if there are any existing problems that should be taken care of. If a chimney is susceptible to backdrafting or spillage, do whatever is necessary to remedy the situation because this can lead to serious health problems—even death. If radon, or other dangerous soil gases, are entering the house in excessive amounts, you should take remedial steps to prevent their entry. If hidden moisture problems exist, you should analyze what is causing them, repair any damage, and prevent the situation from recurring. If a house is excessively leaky, it should be weatherized to improve comfort and energy efficiency. If leaky heating ducts are causing excessive infiltration or exfiltration, they should be sealed.

Step 2
Determine the capacity

The capacity of a ventilation system is measured in cfm and will generally be based either on a certain number of cfm per person (based on average occupancy) or on a certain number of air changes per hour (ACH). To convert between cfm and ACH, you will need to calculate the volume of the house. Large houses generally need fewer air changes per hour than smaller houses. If a central ventilation system is going to operate intermittently when the house is occupied, the equipment should be oversized so the *average* ventilation rate is suitable. For more information on sizing a ventilation system, see Chapter 11.

Step 3
Select a strategy

Once the capacity has been determined, the basic strategy should be selected: central exhaust, central supply, balanced, or passive. If you have decided on a balanced system, you should determine whether or not you require heat recovery capability. Before selecting a heat-recovery strategy, you should always compare all of the costs with alternative strategies. See Chapter 7 for a discussion of basic ventilation strategies.

There are two basic approaches to installing controlled ventilation systems: They can be connected to a forced-air heating/air-conditioning system or they can be installed as a stand-alone system. In many houses there will be both a central ventilation system as well as one or more local-exhaust fans.

There are several key questions to ask when evaluating a house to determine the best central ventilation strategy. For example, could negative pressures cause backdrafting or spillage to occur? Will pollutants be drawn indoors from outside the living space? Are hidden moisture problems likely? Is the climate harsh enough to result in a high operating cost? What is the budget for equipment installation?

Step 4
Select the equipment

There are many companies producing ventilation equipment in North America. Some utilize similar technologies, while others have very different design ideas. Few systems are inherently better than others; they are just different. Before selecting a particular ventilator, you should define your specific needs. For example, a sensitive person might look for a unit that has well-sealed cabinet insulation, that doesn't have many plastic parts with the potential to outgas, and that doesn't have the motor in the fresh airstream. Someone in a hot, humid climate may lean toward an HRV capable of transferring moisture between the airstreams. Someone in a very cold climate may opt for an HRV with maximum energy efficiency.

If a filter is to be coupled with a ventilator, the two should be selected together. Because high-efficiency filters designed specifically for use with ventilation systems aren't very common, it is important to select equipment that is compatible so that adequate airflows can be maintained.

During the equipment-selection process, you should determine how the incoming air will need to be tempered. For example, will through-the-wall vents provide enough mixing to be successfully temper the air, will an HRV be necessary, or is the climate mild enough that tempering won't be a significant consideration?

Whatever equipment is selected, it should be reliable and designed for continuous operation. It should also be quiet and easy to maintain. Ventilation equipment is covered in Chapters 19–24; filters are discussed in Chapter 15.

Step 5
Select a control

Controls should be accessible, reliable, and easy to understand. They can be either manual (operated by the occupants) or automatic, and it is possible to use a combination of several different controls on the same system. The types of controls available include: on/off switch, speed control, timer, motion switch, carbon monoxide monitor, etc.

If heating/air-conditioning ducts are also used to distribute ventilation air, the ventilation control will usually need to activate both the ventilation fan and the forced-air furnace/air-conditioner fan. One of the simplest controls is no control—in other words, let the ventilation system run continuously. For more information on controls, see Chapter 14.

Step 6
Plan the general flow of air

Locate the fresh air inlet(s) where the outdoor air is the cleanest, and the outlet(s) where the stale air won't create problems outdoors. Also select effective locations for supply and exhaust grilles. The airflow from the outdoors, through the house, and back outdoors can take a variety of different paths. For example, air can enter and leave the house through a duct or through several through-the wall vents, it can pass from room to room by way of ducts or by passing under doors, it can move from closets to rooms, or from rooms to closets, etc.

There can be a system of ducts that only contains ventilation air, or ventilation air can be combined with air passing through a forced-air furnace/air-conditioning system. See Chapter 16 for inlet and outlet locations and Chapter 17 for distributing air around the house. When planning the pathways for air to follow, be sure to consider the consequences of air moving through the random holes because of pressurization or depressurization.

Step 7
Design the ducts and grilles

Select a duct material (metal or plastic, rigid or flexible). Plan the locations for the ducts (inside or outside the conditioned space) and determine where insulation will be required. Avoid excessive resistance to airflow by minimizing the use of elbows and keeping runs as short and straight as possible. Don't move ventilation air through building cavities if it can be helped. Determine the sizes for the duct runs and the grilles. In laying out the duct system, consideration should be given to future cleaning and maintenance. These issues are covered in Chapters 16 and 17.

Step 8
Install the equipment

At this stage, the entire system should be carefully planned out, and it is time to assemble the pieces in their proper locations. A variety of issues are covered throughout this book that should be taken into consideration. For example, equipment should be mounted where it can be accessed easily for servicing; vibration noise should be minimized; controls should be easy to find and labeled with basic instructions; seams in the ducts should be sealed. Be sure to always follow the manufacturer's installation instructions carefully.

Step 9
Balance and measure the airflows

Central-supply and central-exhaust ventilation systems generally create a pressure imbalance between the indoors and outdoors. Balanced ventilation systems (with or without heat recovery) are designed to operate at a neutral pressure, so the airflows should be measured and adjusted to ensure that the indoors is neither pressurized or depressurized.

With all ventilation systems, the actual airflow rate to or from each individual rooms should be measured and adjusted where necessary. If everything has been designed and installed carefully, the actual airflows should be within 10% ($\pm$) of the design calculations. At this point, the system should be explained carefully to the occupants and they should be left with the operation and installation manuals.

If you have done anything to change the tightness of the house, or the degree of pressurization or depressurization it experiences, it is very important to determine if your changes will result in any pressure-related problems. For example, it is always advisable to check combustion appliances for backdrafting or spillage under a worst-case depressurization. You may also want to perform a radon test or consider the possibility of hidden moisture problems.

Chapter 26

Typical central ventilation systems

There are literally hundreds of possible layouts for well-designed central ventilation systems. However, in practice, most designers and installers either opt for an approach that depressurizes a house or operates under a neutral pressure. Because of their potential to cause hidden moisture problems in cold climates, pressurization systems are not as common.

This chapter outlines six central ventilation systems—two depressurization systems: three balanced systems that have become popular in recent years, and one pressurization system that would be suitable in a mild climate. The balanced systems shown in Figures 26-5 and 26-2 are most often installed with HRVs.

All ventilation systems have a number of advantages and disadvantages that should be considered prior to selecting a particular strategy. For example, depressurization systems (Figure 26-1 and 26-2) have the potential to cause backdrafting and spillage, so they are generally not recommended in houses containing combustion-fired furnaces, boilers, water heaters, or fireplaces that are connected to a natural-draft chimney.

Though not shown in the examples, most houses should also have local-exhaust fans in the moisture- and pollution-prone rooms (*e.g.* bath fans, kitchen range hood, and possibly a laundry-room fan) to clear them of contaminants quickly. In Figures 26-1, the exhaust fan in Bath 2 functions as both a local-exhaust fan and a central-exhaust fan. With proper controls and motorized dampers or grilles, the systems in some of the other examples can also be made to function as both central and local systems, but separate local-exhaust fans are somewhat more common.

When a house has a forced-air heating/cooling system, well-sealed ducts are important for energy efficiency, comfort, health, and safety. Well-sealed ventilation ducts are also important for similar reasons.

With any ventilation system, you will be able to minimize overventilation and have a better idea of where

Understanding Ventilation

air is entering and leaving if the house is tightly constructed. In other words, air will be more likely to enter and leave through the deliberate inlets and outlets if there aren't many random holes in the structure. Tight houses are highly desirable because they are more comfortable, less expensive to heat and cool, and quieter.

In tightly constructed houses having pressurization and depressurization ventilation systems, through-the-wall vents can be useful to provide overall pressure relief and to direct airflows into or out of specific rooms. However, in most cases, through-the-wall vents are only effective in small houses (less than 1,500 sq. ft.) and apartments, because they are inherently tighter than larger houses. The depressurization systems shown in Figures 26-1 and 26-2, and the pressurization system shown in Figure 26-4 would benefit from through-the-wall vents in a smaller or very tightly constructed house.

In these six examples, the capacity of the fan isn't listed because that will depend on the house size, the number of occupants, and how many polluting materials are in the house. Similarly, the duct sizes aren't shown because they will depend on the fan capacity. However, typical central ventilation systems move less than 200 cfm through 3-7" diameter ducts.

The floor plan shown in the following examples contains about 1,600 sq. ft. A ventilation rate of about 75 cfm would be based on ASHRAE's recommendation of .35 air changes per hour. Because the plan has three bedrooms, it would be reasonable to assume an occupancy of two adults and two children. Based on ASHRAE's recommendation of 15 cfm per person, this would translate to a total ventilation rate of 60 cfm. These rates should satisfy the needs of average people (not hypersensitive people) if the fresh air is distributed effectively, the stale air is removed effectively, the house is not constructed of polluting materials, and excessive amounts of pollution are not sucked indoors by air-pressure differences.

Of the six examples, Figure 26-1 has the least effective distribution, so it would need more capacity than the other examples. The systems in Figures 26-3, 26-5, and 26-6 have the best distribution, so they would need the least capacity. If a rate of 60-75 cfm has been deemed appropriate, but the ventilator will operate only half the time (*e.g.* on a 15-min. on/off cycle), it should have a 120-150 cfm capacity. If a ventilation system is designed to be operated 50% of the time under normal conditions, you have the option of doubling the rate by running it full-time when occupancy or indoor-pollution levels increase (*e.g.* when guests are visiting). A long on/off cycle (*e.g.* 12 hr. on, 12 hr. off) provides less effective distribution, so the overall ventilation rate will need to be higher.

Specific controls aren't included in the examples because they are often a matter of personal preference. Some people prefer a ventilation system that runs 24 hours per day; others opt for controls that allow the system to be shut off periodically. In general, simple controls that are low in cost and easy to understand are the most popular.

The six examples that follow should be viewed as schematics. The actual layout and the specific details will depend on the house plan. For example, the equipment and the ducts might be located in an attic, basement, or crawl space, or some ducts might be run vertically within wall cavities. Any of these approaches can be adapted to virtually any house style: multi-level, split-level, basement, slab-on-grade, etc.

Chapter 26 Typical central ventilation systems

Upgraded bathroom exhaust fan

This is one of the lowest-cost central ventilation systems available, but it also provides the least effective distribution. Consisting of a high-quality, quiet, bathroom exhaust fan, this approach works best in small houses and apartments. The fan should be sized to provide a ventilation rate that is appropriate for the whole house. During operation, the fan will do two things: First, it will exhaust any stale or moisture-laden air from the bathroom it is located in, and second, it will depressurize the whole house so make-up air will enter from the outdoors through the random holes in the structure. If a house is tightly constructed (or smaller—small houses are inherently tighter than large houses), there will not be enough random holes for make-up air to enter easily, so through-the-wall vents should be placed in the bedrooms and living areas. The through-the-wall vents will relieve some of the depressurization and improve the distribution of fresh air. The fan functions as both a local-exhaust fan in the bathroom and a central-exhaust fan for the whole house. In order for make-up air to be able to get into the bathroom from the rest of the house when the door is closed, there should be a pass-through installed in the wall between the bathroom and the rest of the house (for a high-capacity fan), or the bathroom door should be undercut at least $1/2$" (for a low-capacity fan). A variation of this approach would be to mount the exhaust fan in the ceiling of a central hallway, in which case the fan should only be considered a central-exhaust fan, and another fan should be used in the bathroom for local exhaust. Another variation would be to connect the fan to some ducts so it pulls air from two bathrooms simultaneously. Often a two-speed fan is used—operated on low speed for general ventilation and on high speed to clear the bathroom(s) quickly.

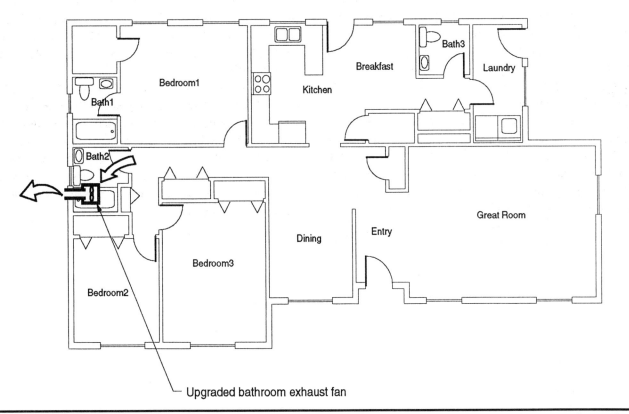

Figure 26-1. Upgraded bathroom exhaust fan.

Understanding Ventilation

Multi-point central-exhaust fan

This design uses a central-exhaust fan with multiple suction ducts running from several moisture- or pollution-prone rooms. A variation would be to use an individual fan and connect it to a single duct with branches to various rooms. This system will be quieter and will have better exhaust effectiveness than the system in Figure 26-1. When the fan is operating, the house will be depressurized and make-air will enter from the outdoors through random holes in the structure. It will then mix with house air, move through the living space, and leave though the various exhaust grilles. If a house is tightly constructed (or smaller), through-the-wall vents should be used in the bedrooms and other principal rooms that are the most likely to be occupied. The combination of tight construction and through-the-wall vents will improve the effectiveness of fresh-air distribution. Two-speed or variable-speed fans and/or motorized grilles or dampers can be used to provide for different ventilation rates at different times. This approach was originally developed for small houses and apartments without forced-air heating/cooling systems, and there are some well-designed packaged systems available. It can also be used in houses with forced-air heating/cooling, but since the pressures induced on a house by leaky forced-air heating/cooling system ducts can overpower the pressures induced by this type of ventilation system, well-sealed ducts are very important for maximum ventilation effectiveness.

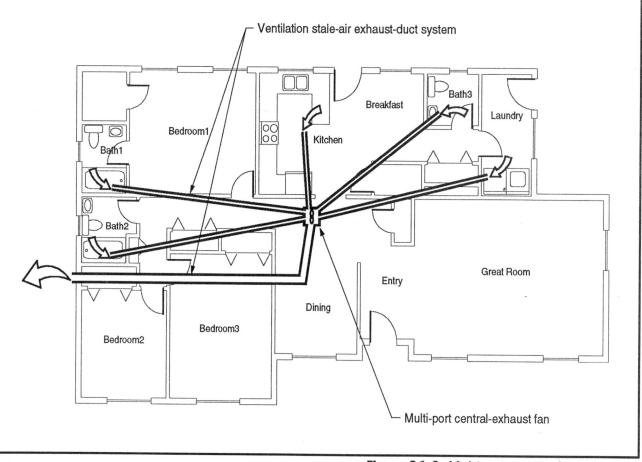

Figure 26-2. Multi-point central-exhaust fan.

Chapter 26 Typical central ventilation systems

Multi-point central-exhaust fan coupled with a forced-air furnace/air conditioner

With this approach, stale- and moisture-laden air is exhausted from pollution-prone rooms by a multi-port central-exhaust fan—the same as in Figure 26-2. Fresh air is pulled into the house through a fresh-air duct that is connected to the return-air duct of a forced-air heating/cooling system and is distributed throughout the living space by the forced-air heating/cooling system's fan. This design is very effective at both distributing fresh air and exhausting stale air, but the forced-air heating/cooling system's fan must operate whenever the central-exhaust fan is running. Extended operation of the forced-air heating/cooling system's fan will mean an increased operating cost, so it can be important to use an energy-efficient fan motor (such as an ECM motor). The volume of air leaving through the multi-port central-exhaust fan should be measured, then the damper in the incoming fresh-air duct should be adjusted so the incoming and outgoing airflows are equal. Once balanced, the ventilation system will neither pressurize nor depressurize the house—it will operate under a neutral pressure. This means that the ventilation system will not contribute to pressure-related problems such as backdrafting and spillage. If the exhaust fan is running, but the furnace fan is off, the house will be depressurized just as it is in the layout shown in Figure 26-2. This type of installation is discussed more fully in Chapter 19. It works well in all houses having a forced-air heating/cooling system, but when airflows are excessive, in very harsh climates, or when the cost of heating fuel is high, there can be a substantial cost associated with tempering the incoming air.

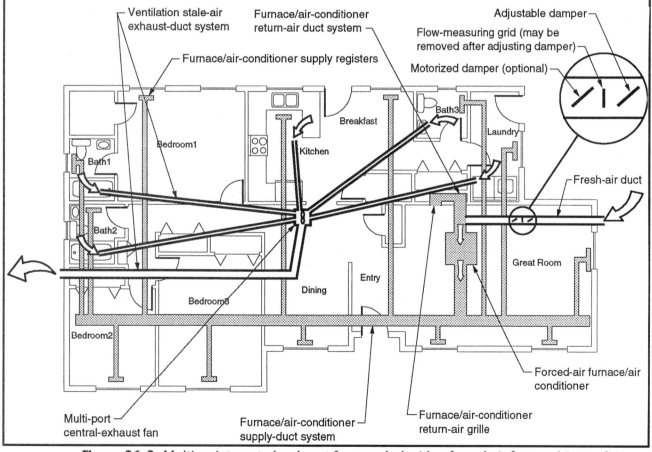

Figure 26-3. Multi-point central-exhaust fan coupled with a forced-air furnace/air conditioner.

Understanding Ventilation

Fresh-air duct coupled with a forced-air furnace/air conditioner

This system is quite simple: A fresh-air supply duct is connected between the outdoors and the return-air duct of a forced-air furnace/air conditioner. The adjustable damper is used to regulate the amount of air entering the fresh-air duct. It is usually adjusted one time only, after which the flow-measuring grid can be removed. When the furnace/air-conditioner fan is operating, fresh outdoor air will be pulled into the duct system where it will mix with house air and be distributed throughout the living space. Fresh air distribution will be very effective, but stale air will not be exhausted effectively. Because the house will be slightly pressurized, stale air will leave the house through the random holes in the structure. If a house is tightly constructed (or smaller), there will not be enough random holes through which air can escape easily, so through-the-wall vents should be used in moisture- and pollution-prone rooms as stale-air outlets. This will improve the exhaust effectiveness. Because this system only provides ventilation when the furnace/air-conditioning fan is running, the fan must run for extended periods (often continuously). This can lead to a high operating cost—unless a high-efficiency fan motor is used. This system is suitable for any house having a forced-air heating/cooling system, but in cold climates there is the possibility of hidden moisture condensation, especially when the indoor relative humidity is high (e.g. in a small house having many occupants), and a high tempering cost.

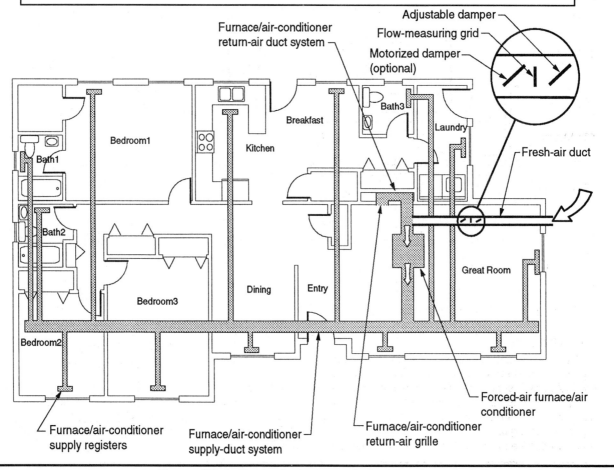

Figure 26-4. Fresh-air duct coupled with a forced-air furnace/air conditioner.

Chapter 26 Typical central ventilation systems

Dual-fan, independently ducted system

In this approach, a pair of identical ventilating fans are installed with their own series of ducts. This type of system can be used in a house with or without forced-air heating or cooling. Because this ventilation strategy operates under a neutral pressure, it will not contribute to any pressure-related problems (*e.g.* hidden moisture problems, backdrafting, etc.). The ventilation system itself can consist of two fans of equal capacity, as shown, or a ventilator containing two fans in a single cabinet. The ducts are laid out to pull stale air from near the ceiling of moisture- or pollution-prone rooms and exhaust it outdoors, and to pull fresh air from the outdoors and blow it into bedrooms and living areas. A much simpler layout might consist of a single fresh-air grille at one end of the house and a single stale-air exhaust grille at the opposite end, but this would mean less effective distribution, so it would require more capacity. Though not shown in the illustration below, to balance the incoming and outgoing airflows, there should be an adjustable damper and a flow-measuring grid in the fresh-air and the stale-air ducts. The flow-measuring grid can be removed once the airflows have been adjusted.

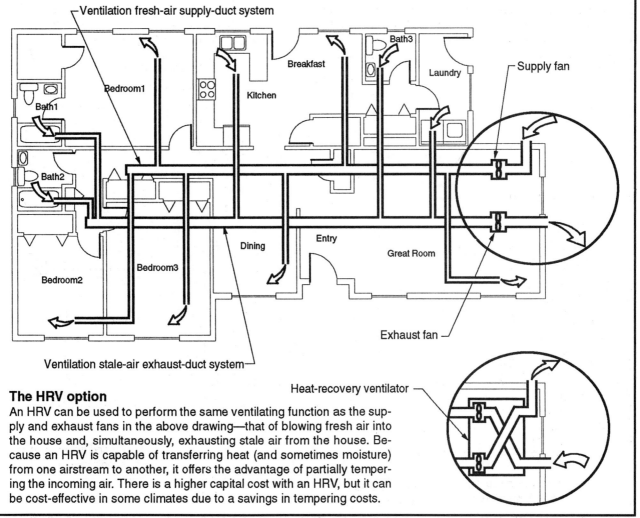

The HRV option
An HRV can be used to perform the same ventilating function as the supply and exhaust fans in the above drawing—that of blowing fresh air into the house and, simultaneously, exhausting stale air from the house. Because an HRV is capable of transferring heat (and sometimes moisture) from one airstream to another, it offers the advantage of partially tempering the incoming air. There is a higher capital cost with an HRV, but it can be cost-effective in some climates due to a savings in tempering costs.

Figure 26-5. *Dual-fan, independently ducted system.*

Understanding Ventilation

Dual-fan system coupled with a forced-air furnace/air conditioner

With this approach, a supply fan blows fresh air indoors into the return-air duct of the forced-air furnace/air conditioner. At the same time, an identically sized exhaust fan pulls stale air from moisture- and pollution-prone rooms and blows it outdoors. For effective distribution of fresh air, the furnace/air conditioner fan must be running whenever the ventilator fans are running. (With extended operating times, an energy-efficient fan motor will be more economical to operate.) If the furnace/air conditioner fan is not running, the supply fan will not deliver fresh air effectively because it will enter the living space primarily through the furnace/air conditioner's return air grille(s). A lower-cost installation would use a single stale-air exhaust grille in the most used bathroom, but this would mean less effective exhaust in the rest of the house. With less effective ventilation, more capacity is needed. This system is suitable for any house having a forced-air heating/cooling system. Because it results in a neutral pressure in the house, it will not contribute to pressure-related problems. As with the system in Figure 26-5, there should be an adjustable damper and flow-measuring grid in the supply and exhaust ducts to balance the airflows.

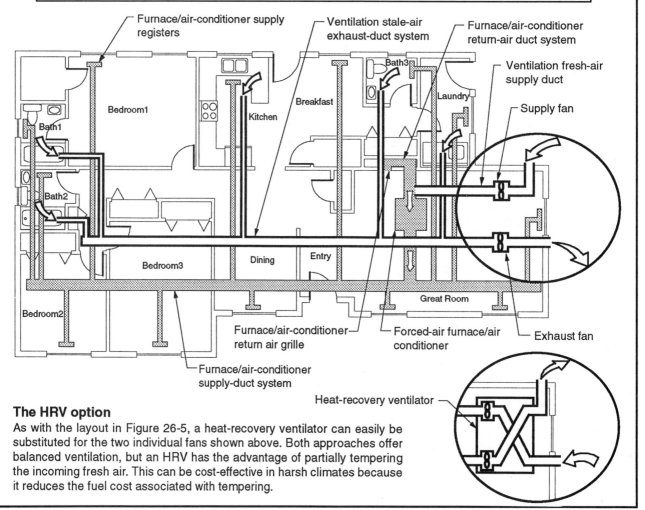

The HRV option
As with the layout in Figure 26-5, a heat-recovery ventilator can easily be substituted for the two individual fans shown above. Both approaches offer balanced ventilation, but an HRV has the advantage of partially tempering the incoming fresh air. This can be cost-effective in harsh climates because it reduces the fuel cost associated with tempering.

Figure 26-6. Dual-fan system coupled with a forced-air furnace/air conditioner.

Part 6

Resources

Appendix A Abbreviations
Appendix B Glossary
Appendix C Bibliography
Appendix D Organizations
Appendix E Equipment sources

Appendix A
Abbreviations

AAHX	air-to-air heat exchanger
ACH	air changes per hour
ACH$_{50}$	air changes per hour at 50 Pascals of depressurization
ADA	airtight drywall approach
ASEF	apparent sensible effectiveness
ASHRAE	American Society of Heating, Refrigerating, and Air-Conditioning Engineers
Btu	British thermal unit
ccf	hundred cubic feet
c.f.	cubic foot
cfm	cubic feet per minute
CFM$_{50}$	cubic feet per minute at 50 Pascals of depressurization
CO	carbon monoxide
CO$_2$	carbon dioxide
CSA	Canadian Standards Association
dB	decibel
DD	degree day
DOP	di-octyl phthalate
DCV	demand-controlled ventilation
ECM	Electrically commutated motor

EI	environmental illness
ELA	effective leakage area (in the U.S.); equivalent leakage area (in Canada)
EMR	electromagnetic radiation
ERV	energy-recovery ventilator
fpm	feet per minute
HEPA	High Efficiency Particulate Air
HRV	heat-recovery ventilator
HVI	Home Ventilating Institute
in. hg. (or " hg.)	inches of mercury on a pressure gauge
in. w.g. (or " w.g.)	inches of water on a pressure gauge
KISS	keep it simple stupid
kWh	kilowatt-hour
l/s	liters per second
m.c.	moisture content
MCS	multiple chemical sensitivity
NCAMP	National Coalition Against the Misuse of Pesticides
nfa	net free area
Pa.	Pascal
pC/l	picoCurie per liter
ppm	parts per million
psi	pounds per square inch
RH	relative humidity
s.p.	static pressure
sq. ft.	square foot
sq. in.	square inch
SRE	sensible recovery efficiency
TD or ΔT	temperature difference
TRE	total recovery efficiency
TSP	trisodium phosphate
UFFI	urea-formaldehyde foam insulation
USEPA	United States Environmental Protection Agency
VOC	volatile organic compound
W	watt
w.g.	water gauge
Wh	watt-hour

Appendix B

Glossary

Absolute humidity: The weight of water in a given volume of air, often expressed in pounds of water per pound of dry air. See also **relative humidity**.

Accidental pressure: An air-pressure difference between the indoors and the outdoors that is induced by mechanical devices whose primary purpose is not ventilation.

Accidental ventilation: Random air movement into and out of a house caused by accidental pressures.

Activated alumina: A form of aluminum oxide used as an adsorption material in air filters, more correctly called activated alumina impregnated with potassium permanganate. Brand names include Purafil and Carusorb.

Activated carbon: A form of carbon used as an adsorption material in air filters, often derived from coconut, wood, or coal.

Active: Caused by a fan, as in active ventilation or active make-up air supply. See also **passive**. (When used to describe a chimney, *i.e.* an active chimney, active means the chimney is in use, as opposed to *inactive* or *not in use*.)

Adsorption: The adhering of a gas onto the surface of a substance without changing chemically.

Adsorption filter: A filter that removes gases from the air by the process of adsorption.

Air changes per hour (ACH): The number of times the volume of air in a house is replaced with outdoor air during an hour, sometimes expressed at a 50 Pa. pressure difference (ACH_{50}) between the indoors and outdoors.

Airflow grid: A device that mounts inside a duct (often temporarily) and can be connected to an air-pressure gauge to determine the quantity of air flowing in the duct.

Understanding Ventilation

Air handler: A cabinet containing a fan, usually used to describe a fan cabinet that moves air through a system of heating/cooling ducts.

Air-pressure difference: The dissimilarity in air pressures between two locations that can cause air to move between them. See also **positive pressure** and **negative pressure**.

Air silencer: See **sound attenuator**.

Airtight construction: A building technique that results in a house with a small ELA.

Airtight drywall approach (ADA): See **airtight construction**.

Air-to-air heat exchanger (AAHX): A balanced ventilation device capable of transferring heat (and sometimes moisture) between two airstreams. See also **heat-recovery ventilator**.

American Society of Heating, Refrigerating, and Air-Conditioning Engineers (ASHRAE): A professional organization that sponsors research and sets voluntary standards for the ventilation industry.

Apparent sensible effectiveness (ASEF): A measurement useful in determining the temperatures of airstreams in a heat-recovery ventilator.

Atmospheric draft: See **natural draft**.

Atmospheric-spot-dust test: A method of testing medium-efficiency air filters.

Axial fan: A type of ventilating fan having propeller-like blades, often used for window fans.

Backdraft damper: A device that allows air to only flow in one direction through a duct.

Backdrafting: Complete reversal of flow in a chimney, usually due to negative pressures indoors.

Balanced ventilation: A general ventilation strategy that results in the house experiencing a neutral pressure, consisting of either a supply fan and an exhaust fan, or a central-exhaust fan coupled with a fresh-air duct on a forced-air furnace/air conditioner.

Balancing: The process of adjusting a balanced ventilation system so that the house experiences a neutral pressure.

Biological pollutants: Air pollutants, such as mold and pollen, that either are or once were alive, or **by-products of metabolism**.

Bio-nasties: See **biological pollutants**.

Blower door: A device used to pressurize or depressurize a house to evaluate its tightness.

Breathing zone: See **headspace**.

British thermal unit (Btu): The amount of heat required to raise the temperature of 1 pound of water 1°F, used in calculating heating and air-conditioning loads.

Building envelope: The exterior portion of a structure, usually insulated, that surrounds the conditioned space.

Bypass: A pathway through a building, usually hidden within the structure, through which air can move (and heat loss can occur) without directly causing an air exchange in the living space.

By-products of metabolism: Solids, liquids, and gases given off by living creatures as a normal part of the life process.

Canadian Standards Association (CSA): An organization that maintains standards used to rate ventilation equipment.

Capital cost: The installed cost of equipment (*i.e.* the equipment itself, incidentals expenses, and labor). See also **operating cost**.

Carbon dioxide (CO_2): A colorless, odorless gas released in exhaled breath, or by combustion processes.

Carbon-dioxide sensor: A device that senses the carbon dioxide concentration in the air and can be used to control a ventilation system.

Carbon monoxide (CO): A colorless, odorless, poisonous gas released during the incomplete combustion of carbon-containing fuels.

Carbon-monoxide detector: A device that senses carbon monoxide in the air and sounds an alarm when the concentration reaches a certain point.

Carcinogen: A substance that tends to produce cancer.

Category A room: A room, such as a bedroom or living room, that is often occupied for long periods and requires general ventilation.

Category B room: A room, such as a kitchen or bathroom, in which moisture or pollutants are generated, that is often in need of special ventilation requirements.

Centrifugal fan: A ventilating fan with blades resembling a squirrel cage, often used in furnaces and ventilating devices.

Central-exhaust ventilation: A general-ventilation strategy that uses an exhaust fan to remove air from a house. An equal volume of make-up air enters either by way of through-the-wall vents, through a fresh-air duct connected to a forced-air heating/cooling system, or through random holes in the structure of the house.

Central-supply ventilation: A general ventilation strategy that uses either a supply ventilation fan or a forced-air heating/cooling fan to draw air into a house. An equal volume of air leaves the house either by way of through-the-wall vents or through random holes in the structure of the house.

Central ventilation: Any general-ventilation strategy, *i.e.* balanced, central-supply or central-exhaust; with or without heat recovery.

Chimney: A vertical structure or conduit that usually carries combustion by-products out of a house. See also **ventilating chimney**.

Circulating air: Air that moves from room to room in a house.

Combustion air: Air that enters a house specifically to be used in the combustion of a fuel. See also **ventilation air**.

Combustion-air duct: A duct through which combustion air travels from the outdoors to a combustion device.

Combustion by-products: Gases and particulates released during the burning of a fuel.

Condensation: The changing of a gas or vapor into a liquid, accompanied by the release of heat. See also **evaporation**.

Conditioned space: The part of a house, within the **building envelope**, that is kept at a comfortable temperature and humidity.

Continuous duty: An electrical safety rating for motors, not the same as **continuous use**. Most motors (except elevator motors) are rated for continuous duty.

Continuous use: A designation that a fan is capable of running for indefinite periods of time, not the same as **continuous duty**. Not all motors are rated for continuous use.

Controlled pressure: An air-pressure difference between the indoors and the outdoors, purposefully induced by ventilation equipment.

Controlled ventilation: Purposeful air movement into and out of a building in a regular way, usually caused by a fan moving air through deliberate openings.

Convective air currents: The movement of air resulting from heat transference through the air by convection.

Cubic feet per minute (cfm): A unit of measurement used to rate a fan's capacity.

Damper: A device, often motorized or manually adjustable, used to vary or control the airflow in a duct. See also **backdraft damper** and **modulating damper**.

Decibel (dB): A logarithmic unit of measurement used to express sound intensity. Two decibels are 10 times as loud as 1 decibel.

Defrost mechanism: A device used in HRVs to melt any ice that builds up in the core in cold climates.

Understanding Ventilation

Degree day: A unit of measurement used to estimate fuel consumption and heating or cooling costs, based on temperature and time.

Dehumidistat: A control device that can be used to activate a ventilation system as the relative humidity rises.

Deliberate holes: Openings (**inlets** or **outlets**) that are purposely placed in a house through which air can enter or leave. See also **random holes**.

Demand-controlled ventilation (DCV): The process of automatically supplying air to, and removing air from, a house whenever needed by the occupants.

Depressurization: When the air pressure inside a house is less than the atmospheric pressure outside the house.

Desiccant: A substance that absorbs moisture, used in some heat-recovery ventilators to enhance moisture transfer between airstreams.

Dew point: The temperature at which air is saturated with moisture (100% relative humidity), below which condensation will occur.

Diffuser: A grille designed to direct airflow in a specific pattern into a room for proper mixing.

Diffusion: The migration of molecules of a gas or a vapor (or a liquid) from an area of high concentration to an area of low concentration.

Diffusion barrier: See **diffusion retarder**.

Diffusion retarder: A material that slows down the amount of diffusion through a solid material, often used in insulated walls.

Dilution: The mixing of fresh air into stale air to reduce the concentration of pollutants; also applies to liquids.

Dilution air: Air that mixes with combustion by-products prior to their being expelled from a house through a natural-draft chimney.

Di-octyl phthalate (DOP): An inert gas used to test high-efficiency (HEPA) particulate air filters.

Displacement ventilation: A method of very effectively moving air through a room, generally only used in specialized applications.

Distribution: Movement of air through a house with a ventilation system or heating/cooling system.

DOP-smoke-penetration test: A method of testing high-efficiency (HEPA) air filters

Double-flow ventilation: See **balanced ventilation**.

Downdraft: A term often applied to kitchen exhaust fans that pull air downward.

Draft: A breeze or current of air in a room, often uncomfortable. See also **Natural draft**.

Drying potential: The ability of a substance to dry out after it becomes wet.

Duct: A conduit for moving air, made from a variety of materials and in a variety of shapes.

Ductboard: A semirigid fiberglass material with an aluminum-foil facing on one side, used to construct ducts.

Dynamic wall: See **permeable wall**.

Earth tube: A duct for moving air from the outdoors, through the soil, into the living space, used for tempering incoming air.

Effective leakage area (ELA): The size of a single round-edged opening that a house would have if all the small random holes between the indoors and the outdoors were combined into one hole. Used in the U.S. and based on a pressure of 4 Pa. It is the same concept as **equivalent leakage area**, but is calculated differently.

Effectiveness: The degree to which fresh air is introduced where the occupants are located and stale air is removed from near pollution sources.

Electret: A plastic material that carries a permanent static charge, used in some air filters.

Electrically-commutated motor (ECM): A higher-cost type of variable-speed motor that utilizes electricity efficiently, especially at low speeds.

Electromagnetic radiation (EMR): The electrical and magnetic energy surrounding electrical wires and appliances.

Electrostatic air filter: An air filter composed of plastic materials that capture particulate pollutants using static electricity. See also **electret**.

Electrostatic precipitator: An air filter that generates a high voltage which causes particulates to become electrically charged and cling to metal plates having an opposite charge.

Energy-Recovery Ventilator (ERV): A **heat-recovery ventilator** that recovers both **latent heat** and **sensible heat**.

Enthalpy: The total amount of heat contained in air, the sum of the **sensible heat** and the **latent heat**.

Envelope: See **building envelope**.

Environmental illness (EI): See **multiple chemical sensitivity (MCS)**.

Equivalent length: The length of straight duct that would have the same resistance to airflow as a fitting.

Equivalent leakage area (ELA): The size of a single square-edged opening that a house would have if all the small random holes between the indoors and the outdoors were combined into one hole. Used in Canada and based on a pressure of 10 Pa. This is the same concept as **effective leakage area**, but is calculated differently.

Evaporation: The changing of a liquid to a gas or vapor, requiring heat. See also **condensation**.

Exfiltration: Air leaving a house through random holes in the structure. See also **infiltration**.

Exhaust air: The air leaving a house through a ventilation system. See also **supply air**.

Exhaust grille: A grille through which stale air leaves a room. See also **supply grille**.

Exhaust ventilation: See **central-exhaust ventilation**.

Exhaust-only ventilation: See **central-exhaust ventilation**.

Extended-surface filter: An air filter configured to have a large amount of surface area to minimize resistance to airflow. See also **medium-efficiency filter**.

Exchange rate: The rapidity at which indoor air is replaced with outdoor air.

Fan: An electrically powered device that moves air.

Fan door: See **blower door**.

Filtration: The process of removing pollutants from air.

Flame rollout: The pushing of flames outside a combustion chamber as a result of backdrafting.

Flat-plate core: A metal, plastic, or treated-paper device that is used in some HRVs to transfer heat (and sometimes moisture) between two airstreams.

Forced draft: A system utilizing a fan to blow air into the combustion chamber of a combustion appliance.

Formaldehyde: A common gas found in indoor air, often released from building materials, consisting of 1 carbon atom, 2 hydrogen atoms, and 1 oxygen atom.

Fresh air: Outdoor air that is brought indoors.

Fresh-air duct: A duct through which air travels from the outdoors to the indoors to benefit the occupants, often integrated with a forced-air heating/cooling system.

Gas: A substance that will fill a container of any size or shape, as opposed to a solid or liquid. See also **vapor**.

General ventilation: Supplying air to, and removing air from, all parts of a house simultaneously for the ordinary needs of the occupants. See also **local-exhaust ventilation**.

Grille: An often decorative covering on a wall or ceiling through which air moves to or from the conditioned space.

Headspace: The breathing zone in a room, in residences it usually extends from a few inches off the floor to about 6 feet off the floor.

Understanding Ventilation

Heat pipe: A sealed tube partially filled with a refrigerant, used in some HRVs to transfer heat between two airstreams.

Heat-recovery ventilator (HRV): A ventilation device capable of transferring heat (and sometimes moisture) between two airstreams, or between water and air. See also **air-to-air heat exchanger (AAHX)** and **energy-recovery ventilator (ERV)**.

HEPA filter: A very high-efficiency particulate air filter, often used in hospitals and laboratories.

Home Ventilating Institute (HVI): An organization that certifies that ventilation equipment has met certain criteria.

Humidistat: See **dehumidistat**.

Humidity: See **absolute humidity** and **relative humidity**.

Hygrometer: A device that senses relative humidity.

Inches of mercury (in. hg. or " hg.): A unit of pressure measurement. About 3,386 Pa. equals 1" hg.

Inches water gauge (in. w.g. or " w.g.): A unit of pressure measurement. About 250 Pa. equals 1" w.g.

Induced draft: A system utilizing a fan to exhaust air from the combustion chamber of a combustion appliance.

Infiltration: Air entering a house through random holes in the structure. See also **exfiltration**.

Inlet: An opening through which fresh air enters a house.

In-line fan: A ventilating fan made to attach a duct at each end, often called a tube fan.

Insulation: A material that inhibits the flow of heat.

Ionizer: A device that generates ions (usually negatively charged), sometimes for the purpose of cleaning the air. See also **negative ions**.

Irritant: A substance that causes a part of the body to become overly excited or sensitive.

Jump duct: A short duct connecting two rooms through which air can move when pressure imbalances occur between the rooms. Also called a jump-over duct or a transfer duct.

Kilowatt hour (kWh): A unit of measuring electrical energy equal to 1,000 watts of power consumed over an hour.

KISS: Keep It Simple Stupid.

Latent heat: The amount of heat that must be removed from air to change the water vapor from a gas to a liquid. See also **enthalpy** and **sensible heat**.

Liters per second (l/s): A metric unit of measurement used to rate a fan's capacity. One liter per second is equal to approximately 2 cubic feet per minute.

Living space: The part of a house to which the occupants normally have access, or habitable rooms; also called the occupied space.

Local-exhaust ventilation: The quick removal of air pollutants from near the source to the outdoors. See also **general ventilation**.

Magnahelic gauge: A type of pressure-measuring device.

Make-up air: Outdoor air that enters a house to replace air that is exhausted from the house.

Manometer: A gauge for measuring air-pressure differences, of which there are several types.

Mechanical ventilation: The process of supplying air to, and removing air from, a house using a fan. See also **controlled ventilation**.

Media filter: A filter that relies on a fibrous material, usually polyester or fiberglass, to physically strain particulates out of the air. See also **electrostatic air filter** and **electrostatic precipitator**.

Medium-efficiency filter: A particulate filter, usually in the 30–40% efficiency range when measured on the **atmospheric-spot-dust test**.

Micron: A unit used to measure air pollutants; a millionth of a meter.

Mixed-gas sensor: A device that senses oxidizable gases in the air and can be used to control a ventilation system.

Modulating damper: A motorized damper that can be opened and closed a variable amount.

Moisture content (m.c.): The amount of moisture in wood, expressed as a percentage of the wood's oven-dry weight.

Motion sensor: A device that senses movement in a room and can be used to control a ventilation system.

Multiple chemical sensitivity (MCS): A medical condition involving hypersensitivities to a wide variety of common air and food contaminants at very low levels.

Multi-port ventilator: An exhaust device having several suction connections for running ducts from different rooms. See also **single-port ventilator**.

Mutagen: A substance that causes changes in chromosomes or genes.

National Coalition Against the Misuse of Pesticides (NCAMP): A nonprofit organization devoted to pesticide safety and alternative, less toxic methods of controlling pests.

Natural aspiration: See **natural draft**.

Natural draft: The negative pressure in a chimney caused by rising, warm combustion by-products. Often referred to simply as draft.

Natural pressure: An air-pressure difference between the indoors and the outdoors, induced by natural phenomena such as wind, stack effect, and diffusion.

Natural ventilation: Random air movement into and out of a house caused by natural pressures.

Negative ions: Negatively charged atoms or groups of atoms that will cling to oppositely charged surfaces and can induce a feeling of well-being.

Negative pressure: See **depressurization**.

Net free area (nfa): The amount of unobstructed open area of a grille, inlet, or outlet.

Neurotoxin: A substance that is poisonous to the nervous system, spinal cord, brain, etc.

Neutral pressure: Air pressure inside a house that is equal to the atmospheric pressure outside the house.

Neutral-pressure plane: That part of a house that experiences neutral pressure as a result of stack effect, or as a result of combined pressures.

Occupied space: The part of a house to which the occupants normally have access, or habitable rooms; also called the living space.

Olf: A unit of odor measurement equal to the amount of body odor produced by one person performing normal activities in one day.

On-purpose ventilation: See **controlled ventilation**.

Operating cost: The cost of running equipment (*i.e.* maintenance expenses, electricity for fans, and tempering costs). See also **capital cost**.

Outgassing: The release of volatile gases from a solid material as a part of aging, decomposition, or curing.

Outlet: An opening through which stale air leaves a house.

Ozone: A highly reactive gas consisting of three oxygen atoms, produced in small amounts by electric motors, electrostatic precipitators, ionizers, etc.

Ozone generator: An air-cleaning device that creates ozone for the purpose of reacting with air pollutants to neutralize them.

Partial-bypass filter: A filter that removes some contaminants from the air, but allows some air to pass through without contacting the filter, usually to minimize the resistance to airflow.

Particulates: Solid (or liquid) air pollutants, as opposed to gases.

Particulate filter: An air filter designed to remove particulates from the air.

Parts per million (ppm): A small unit of measurement, often used to express the concentration of pollutants in the air.

Pascal (Pa.): A small unit of pressure measurement, useful in diagnosing houses. One pound per square inch equals about 7,000 Pascals.

Understanding Ventilation

Passive: Caused by house pressurization or depressurization, not directly caused by a fan, as in passive ventilation. Infiltration and exfiltration involve passive air movement.

Passive ventilation: A semi-controlled general-ventilation strategy, utilizing natural or accidental pressures, that involves air movement through deliberate openings into and out of a house. See also **active** and **passive**.

Pass-through: An opening between two rooms through which air can move when pressure imbalances occur between the rooms.

People pollutants: See **by-products of metabolism**.

Permanent-split-capacitor motor: A type of electric motor characterized by a main winding and an auxiliary winding wired in series with a capacitor.

Permeable wall: An insulated part of a house's structure through which outdoor air can pass from the outdoors to the indoors and is tempered in the process.

Pesticides: Chemical compounds formulated to kill living creatures.

PicoCurie per liter (pC/l): A unit of measuring the concentration of radon in the air.

Porosity: Capacity of a material to be penetrated through its pores by gases or vapors.

Positive pressure: See **pressurization**.

Pounds per square inch (psi): A unit of pressure measurement.

Pressurization: When the air pressure inside a house is greater than the atmospheric pressure outside the house.

Primary ventilation: See **general ventilation**.

Psychometric chart: A graph used by engineers to determine the moisture content of air at different temperatures.

Radon: A naturally occurring radioactive gas, often released from soil and rocks.

Radon daughters: The natural radioactive decay products of radon, some of which release harmful radiation.

Radon progeny: See **radon daughters**.

Random holes: The miscellaneous cracks, gaps, and openings in a house between the indoors and the outdoors through which air can move. See also **Deliberate holes**.

Recirculate: To move air through the living space without adding fresh air or removing stale air.

Recirculating range hood: A ductless range hood, one that is not connected to the outdoors.

Relative humidity (RH): The amount of moisture in air compared to the maximum amount of moisture that air at that temperature can contain, usually expressed as a percentage. See also **absolute humidity**.

Return air: Air from the living space that enters a furnace/air conditioner to be conditioned.

Rotary core: A slowly spinning plastic wheel that is used in some HRVs to transfer heat and moisture between two airstreams.

Sealed combustion: A system used in some furnaces, boilers, fireplaces, and water heaters that is immune from sensing pressure imbalances in a house because it draws combustion air into a combustion chamber from the outdoors and expels combustion by-products to the outdoors, all within totally enclosed conduits and chambers.

Secondary ventilation: See **spot ventilation**.

Sensible heat: The amount of heat involved in raising or lowering the temperature of air, not including any heat required to cause water vapor to change state (*e.g.* from a gas to a liquid). See also **enthalpy** and **latent heat**.

Sensible recovery efficiency (SRE): A measurement, that does not include latent heat recovery, useful in comparing the amount of energy passed between airstreams in a heat-recovery ventilator.

Appendix B Glossary

Separation: The principle of isolating pollutants from the living space.

Shaded-pole motor: A type of low-cost, usually inefficient, motor whose starting torque is provided by a permanently short-circuited auxiliary winding.

Short circuiting: When fresh air moves through a room quickly without adequate mixing.

Single-port ventilator: An exhaust device having only one suction connection for a duct from the living space. See also **multi-port ventilator**.

Sone: A linear unit of sound measurement used to express sound intensity. Two sones are twice as loud as one sone.

Sound attenuator: A device, similar to an automotive muffler, that can be used to absorb some of the sound generated in a ventilation system.

Source control: The principle of using nonpolluting materials in the living space, or of removing polluting materials from the living space.

Spillage: A situation where some of the combustion by-products spill into the living space, rather than go up the chimney, usually due to insufficient draft.

Spot ventilation: Ventilation for the quick removal of air pollutants from near a source. See also **local-exhaust ventilation**.

Squirrel-cage fan: See **centrifugal fan**.

Stack effect: The naturally occurring phenomena of warm air exerting pressure on cooler air that results in warm air rising.

Stale air: See **exhaust air**.

Static pressure: The amount of pressure exerted against the walls of a duct or airway, created by the friction and impact of air as it moves.

Static pressure drop: The change in pressure resulting from resistance to airflow.

Strain hygrometer: A relative-humidity sensing device containing a spiral-wound material that is sensitive to changes in relative humidity.

Supply air: The air entering a house through a ventilation system. See also **Exhaust air**. Also, the conditioned air passing from a furnace/air conditioner into the living space.

Supply grille: A grille through which fresh air enters a room. See also **Exhaust grille**.

Surrogate: A substitute.

Supply ventilation: See **central-supply ventilation**.

Supply-only ventilation: See **central-supply ventilation**.

Synergism: The interaction of two substances such that the combined effect is greater than the sum of the individual effects.

Temperature difference (TD or ΔT): The difference in temperature between two air masses.

Tempering: The conditioning of incoming outdoor air so it will more closely match the temperature and humidity of the indoor air.

Teratogen: A substance that interferes with the development of a fetus or causes birth defects.

Through-the-wall vent: A deliberate opening in an exterior wall through which air moves. The direction of air movement can be inward or outward, depending on the pressure difference between the indoors and the outdoors

Timer: A control that can automatically turn off (or on) a ventilation system.

Total recovery efficiency (TRE): A measurement that includes latent heat recovery, useful in comparing the amount of energy passed between airstreams in a heat-recovery ventilator.

Tracer gas: An gas, often inert, that is used to evaluate the air-exchange rate in a house.

Transfer duct: See **jump duct**.

Transfer grille: See **pass-through**.

Transparency: The concept that a ventilation system should not be detectable by any of the five senses.

Trisodium phosphate (TSP): A granular cleaning product sold in hardware and paint stores.

Tube fan. See **in-line fan**.

Urea-formaldehyde foam insulation (UFFI): A type of insulation widely used in the 1970s that sometimes released formaldehyde gas, now rarely used.

United States Environmental Protection Agency (USEPA): A federal agency charged with monitoring and regulating environmental quality, including indoor air quality, and providing research-based information to the public.

Vapor: The gaseous form of a substance that is normally a liquid at room temperature. See also **gas**.

Vapor pressure: The small pressure exerted by vapor molecules in a mixture of gases.

Vapor retarder: See **diffusion retarder**.

Ventilating chimney: A vertical structure or conduit that carries ventilation air into or out of a house passively.

Ventilation: The process of supplying air to, or removing air from, a house, most often with a fan. See also **controlled ventilation**.

Ventilation air: Air that enters a house specifically for the purpose of ventilating. See also **combustion air**.

Volatile organic compound (VOC): A class of hundreds of different molecular compounds containing carbon that easily evaporates, often released from building materials and found as contaminants in indoor air. See also **outgassing**.

Water gauge (w.g.): A device used to measure pressure differences in inches of water.

Weight-arrestance test: A method of testing very low-efficiency air filters.

Wetting potential: The ability of a substance to become wet after it has dried out. See also **drying potential**.

Whole-house fan: A fan (usually an exhaust fan) used to move air through an entire house, usually in high volumes for the purpose of summer cooling.

Wind: A naturally occurring phenomena that can cause a house to experience positive or negative air-pressure differences.

Appendix C

Bibliography

Abrams, Robert. *Lawn Care Pesticides: A Guide for Action*. Albany NY: New York State Department of Law, May 1987.

Affordable Comfort, Inc. *A Draft Affordable Comfort Policy Statement, Indoor Air Quality, Ventilation and Combustion Safety*. Philadelphia, PA: Affordable Comfort, Inc., n.d.

Air Conditioning Contractors of America. *Manual D: Duct Design for Residential Winter and Summer Air Conditioning and Equipment Selection*. Washington, DC: Air Conditioning Contractors of America, 1984.

Air Infiltration and Ventilation Centre. *AIRBASE, a Bibliographical Database*. Coventry, Great Britain: Air Infiltration and Ventilation Centre, n.d.

_____. *Literature List, 5: Domestic Air to Air Heat Exchangers*. Coventry, Great Britain: Air Infiltration and Ventilation Centre, n.d.

_____. *Literature List, 16: Sick Buildings*. Coventry, Great Britain: Air Infiltration and Ventilation Centre, n.d.

_____. *Literature List, 19: Locations of Exhausts and Inlets*. Coventry, Great Britain: Air Infiltration and Ventilation Centre, n.d.

_____. *Technical Note 17: Ventilation Strategy—A Selected Bibliography*. Coventry, Great Britain: Air Infiltration and Ventilation Centre, July 1985.

_____. *Technical Note 35: Advanced Ventilation Systems—State of the Art and Trends*. Coventry, Great Britain: Air Infiltration and Ventilation Centre, March 1992.

_____. *Technical Note 42: Current Ventilation and Air Conditioning Systems and Strategies*. Coventry, Great Britain: Air Infiltration and Ventilation Centre, February 1994.

_____. *Technical Note 43: Ventilation and Building Airtightness: an International Comparison of Standards, Codes of Practice and Regulations.* Coventry, Great Britain: Air Infiltration and Ventilation Centre, February 1994.

Alfano, Sal. "Pop-Up Downdraft Vents." *Journal of Light Construction,* June 1993, 30–33, 34.

Allen Associates. *Efficient and Effective Residential Air Handling Devices.* Ottawa, Canada: Canada Mortgage and Housing Corp., March 1993.

Allison, Tanya. "Beware the Smell." *The Southface Journal,* Summer-Fall 1991, 18–19.

American College of Allergy & Immunology. *Advice from Your Allergist.* Palatine, IL: American College of Allergy & Immunology, n.d.

American National Standards Institute. *ANSI Z223.1–1992, National Fuel Gas Code, Appendix H: Recommended Procedure for Safety Inspection of an Existing Appliance Installation.* New York: American National Standards Institute, 1992.

American Society of Heating, Refrigerating and Air-Conditioning Engineers. *ASHRAE Standard 52–76 (1976), Method of Testing Air Cleaning Devices Used in General Ventilation for Removing Particulate Matter.* Atlanta, GA: American Society of Heating, Refrigerating and Air-Conditioning Engineers, 1976.

_____. *ASHRAE Standard 62 (1989), Ventilation for Acceptable Indoor Air Quality.* Atlanta, GA: American Society of Heating, Refrigerating and Air-Conditioning Engineers, 1989.

Anachem, Inc. and Sandia National Laboratories. *Indoor Air Quality Handbook.* Albuquerque, NM: Sandia National Laboratories, September 1982.

"An HVAC Systems Perspective on IAQ." *Indoor Air Quality Update,* February 1991, 1–9.

Anne Arundle County Public Schools. *Indoor Air Quality Management Program.* Baltimore, MD: Maryland State Department of Education, n.d.

"Arsenic in Lumber." *Environmental Health Monthly,* September 30, 1988, 1–10.

Ashford, Nicholas and Claudia Miller. *Chemical Exposures, Low Levels, and High Stakes.* New York: Van Nostrand Reinhold, 1991.

Barringer, C.G. and C.A. McGugan. "Effect of Residential Air-to-Air Heat and Moisture Exchangers on Indoor Humidity." *ASHRAE Transactions* 95–2 (1989): 461–474.

"BC Advanced House Ventilation." *Solplan Review,* February-March 1993, 10–11.

Best, Don. "Ventilation in the '90s." *Practical Homeowner,* February 1898, 24–25.

Bienfait, Dominique. "Design of Ventilation Systems in Residential Buildings." in *Proceedings of the 5th International Conference on Indoor Air Quality and Climate, Volume 4 in Toronto, ON, Canada, 29 July to 3 August 1990,* by the International Conference on Indoor Air Quality and Climate, 1990, 47–52.

Blandy, Thomas. "Finding Hidden Heat Leaks." *Journal of Light Construction,* August 1992, 25–28.

Bonneville Power Administration. *Super Good Cents Construction Manual.* #DOE/BP–688. Portland, OR: Bonneville Power Administration, September 1986a.

_____. *Super Good Cents Technical Reference Manual.* #DOE/BP–689. Portland, OR: Bonneville Power Administration, September 1986b.

_____. *Issue Backgrounder: Energy-Efficient New Homes & Indoor Air Pollutants.* Portland, OR: Bonneville Power Administration, August 1987.

_____. *Builder's Field Guide.* Portland, OR: Bonneville Power Administration, 1991.

Borders, Paul. "Evaluating Filters—Our Tools for Cleaning the Air." *Indoor Air Review,* September 1994, 33–34.

BossAire, Inc. *Installation Techniques for the BossAire.* Produced by BossAire, Inc., n.d. VHS videocassette.

Bower, John. *The Healthy House: How to Buy One, How to Build One, How to Cure a Sick One.* Secaucus, NJ: Lyle Stuart Inc., 1989a.

_____. "Use Fiberglass with Caution." *Environ* # 9 (1989b), 17–20.

_____. "Air Filters." *East West,* September/October 1991a, 28–35.

_____. "Cellulose Insulation: Handle with Care—If at All." *Environ* # 11 (1991b), 11–13.

Appendix C Bibliography

_____. "Healthy House Construction." in *Proceedings of the Excellence in Housing Conference in Research Triangle Park, NC March 4–7, 1992,* by the Energy-Efficient Building Association, 1992a, D7–D12.

_____. *Your House, Your Health: A Non-Toxic Building Guide.* Produced by John Bower. 27.5 min. The Healthy House Institute, 1992b. VHS videocassette.

_____. "Healthy House Construction: What to Do, What Not to Do—A 14-Point Plan." in *Proceedings of the Building Solutions Conference in Boston, MA March 3–6, 1993,* by the Energy-Efficient Building Association and the Northeast Sustainable Energy Association, 1993a, D1–D6.

_____. *Healthy House Building: A Design and Construction Guide.* Unionville, IN: The Healthy House Institute, 1993b.

_____. "Healthy House Construction." in *Proceedings of the Innovative Housing '93 Conference, Volume 1, in Vancouver, BC, Canada June 21–25, 1993,* by Canada Mortgage and Housing Corp. and Energy, Mines and Resources Canada, 1993c, 441–445.

_____. "Healthy Construction Recommendations for Healthy People—Building a Generically Healthy House." in *Proceedings of the Excellence in Housing Conference in Dallas, TX February 23–26, 1994,* by the Energy-Efficient Building Association, 1994a, C1–C8.

_____. "Improved Indoor Air Quality in an Energy-Efficient Demonstration House." in *Proceedings of the Excellence in Housing 1994 EEBA Conference in Dallas, TX February 23–26, 1994,* by the Energy-Efficient Building Association, 1994b, C9–C16.

Bower, Lynn Marie. *The Healthy Household.* Bloomington, IN: The Healthy House Institute, 1995.

Breecher, Maury and Shirley Linde. *Healthy Homes in a Toxic World.* New York: John Wiley & Sons, Inc., 1992.

Broan Mfg. Co., Inc. *Guardian Indoor Air Quality Systems.* Produced by Broan Mfg. Co., Inc., n.d. VHS videocassette.

Buchan, Lawton, Parent Ltd. *Improved Make-Up Air Supply Techniques.* Ottawa, Canada: Canada Mortgage and Housing Corp., June 1989 (revised).

Canada Mortgage and Housing Corp. *Builders' Series, Indoor Air Quality.* Ottawa, ON, Canada: Canada Mortgage and Housing Corp., 1988.

_____. *Bibliography on Wind Pressure and Buildings.* Ottawa, ON, Canada: Canada Mortgage and Housing Corp., November 1990.

_____. *Ventilation: A Bibliography.* Ottawa, ON, Canada: Canada Mortgage and Housing Corp., September 1991.

_____. *Investigating, Treating and Diagnosing Your Damp Basement.* Ottawa, ON, Canada: Canada Mortgage and Housing Corp., 1992.

_____. *Soil Gases and Housing: A Guide for Municipalities.* Ottawa, ON, Canada: Canada Mortgage and Housing Corp., 1993a.

_____. *The Clean Air Guide: How to Identify and Correct Indoor Air Problems in Your Home.* Ottawa, ON, Canada: Canada Mortgage and Housing Corp., 1993b.

Canata Research Inc. *The Effectiveness of Low-Cost Continuous Ventilation Systems.* Ottawa, ON, Canada: Canada Mortgage and Housing Corp., March 1990.

_____. *Ventilation File: A Compendium of Canadian Ventilation Research and Demonstration Project Results.* Ottawa, ON, Canada: Canada Mortgage and Housing Corp., April 1992.

Carter, Jon. "Why Ventilate?" *The Southface Journal,* 1993–1, 22–24.

Centers for Disease Control. "Pentachlorophenol in Log Homes—Kentucky." *Morbidity and Mortality Weekly Reports* 29 (1980): 431–432, 437.

_____. "Mercury Exposure from Interior Latex Paint—Michigan." *Morbidity and Mortality Weekly Reports* 39 (1990): 125–126.

_____. *Preventing Lead Poisoning in Young Children.* Atlanta, GA: Centers for Disease Control, October 1991.

"Childhood Asthma Linked to Indoor Pollution." *The Delicate Balance,* #1–2, 1992, 3.

Clarke, Sherrill. "The Everything Box." *Popular Science,* August 1987, 76.

Coffel, Steve and Karyn Feiden. *Indoor Pollution.* New York: Fawcett Columbine, 1990.

Combustion Air, Minnesota House Warming Guide Series. St. Paul, MN: Minnesota Department of Public Service, January 1994.

"Combustion Safety for Residential Equipment." *Solplan Review*, June-July 1992, 8–10.

Cone, James and Michael Hodgson, eds. "Problem Buildings: Building-Associated Illness and the Sick Building Syndrome." *Occupational Medicine: State of the Art Reviews* 4 (October-December 1989).

"Continuous Ventilation Gives Builder Confidence." *Northwest Builder*, April 1991, 1, 8.

Cook, Gary and Virginia Peart. "The Moisture/Sick Building Connection: Issues, Relationships, and Solutions. in *Proceedings of the Excellence in Housing Conference in Indianapolis, IN March 21–23, 1991,* by the Energy-Efficient Building Association, 1991, C9–C23.

Corman, Rita. *Air Pollution Primer.* New York: American Lung Association, 1978.

Courpas, Mira. *CRS Report for Congress, Indoor Air Pollution: Cause for Concern.* Washington, DC: Congressional Research Service, December 1, 1988.

Creech, Dennis. "Ventilation for Your Home—Don't Leave it to Chance." *Southface Journal of Energy and Building Technology*, Winter 1990, 7–9.

Cutter Information Corp. *Psychometrics for Builders and Designers.* Arlington, MA: Cutter Information Corp., 1993.

Dadd, Debra Lynn. *The Nontoxic Home.* Los Angeles, CA: Jeremy Tarcher, Inc., 1986.

Davis, Bruce. "The Impact of Air Distribution System Leakage on Heating Energy Consumption in Arkansas Homes." in *Proceedings of the Excellence in Housing Conference in Research Triangle Park, NC March 4–7, 1992,* by the Energy-Efficient Building Association, 1992, F1–F17.

Dietz, Dennis. "Ventilation and Moisture" (letter). *Custom Builder*, May 1988, 48.

————. "Residential Ventilation Systems: Proposed Energy Labeling System by the Home Ventilating Institute." Paper presented as a part of the "Advanced Ventilation Workshop" at the Energy-Efficient Building Association annual conference, Dallas, TX, February 26, 1994.

Dietz, Dennis and Mark Jackson. "Multi-Family Ventilation Systems: A Review of Continuous Multi-Point Exhaust-Only Ventilation Systems in Washington and Oregon." in *Proceedings of the Building Solutions Conference in Boston, MA March 3–6, 1993,* by the Energy-Efficient Building Association and the Northeast Sustainable Energy Association, 1993, D7–D25.

"Dirty Ventilators and Moisture Problems in Swedish Houses." *Energy Design Update*, September 1992, 7.

Dolan, Michael C., Thomas L. Haltom, George H. Barrows, Craig S. Short, and Kathleen M. Ferriel. "Carboxyhemoglobin Levels in Patients with Flu-like Symptoms." *Annals of Emergency Medicine* 16 (July 1987): 782/87–786/91.

Downing, Andrew Jackson. *The Architecture of Country Houses.* New York: D. Appleton & Co., 1850.

Drerup Construction Ltd. *Housing for the Environmentally Hypersensitive, Survey and Examples of Clean Air Housing in Canada.* Ottawa, ON, Canada: Canada Mortgage and Housing Corp., July 1990.

Dunford, Randall and Kevin May. *Your Health and the Indoor Environment.* Dallas, TX: NuDawn Publishing, 1991.

du Pont, Peter and John Morrill. *Residential Indoor Air Quality & Energy Efficiency.* Washington, DC: American Council for an Energy-Efficient Economy, 1989.

"Earthtubes Temper Make-up Air." *Northwest Builder*, December 1992, 2.

Ecotope, Inc. *Modeled and Measured Infiltration: Phase II, A Detailed Case Study of Three Homes.* EPRI TR–102511. Palo Alto, CA: Electric Power Research Institute. January 1994.

Eich, Bill. "Water Hazards & Your Building's Health." *Journal of Light Construction*, August 1989, 28–30.

Elson, Clive M. *Development of Air Purifiers.* Ottawa, ON, Canada: Canada Mortgage and Housing Corp., August 1983.

Energy, Mines & Resources Canada. *Studies of Mold Growth Potential in Heat-Recovery Ventilators.* Ottawa, ON, Canada: Energy, Mines & Resources Canada, n.d.

————. "Mold Growth in Heat-Recovery Ventilators." in *The Air You Breathe.* Ottawa, ON, Canada: Energy, Mines & Resources Canada, n.d.

Etheredge, Mike. "North Carolina Cautions Against Use of Ozone-Generating Air Purifiers, Recommends Addressing Pollution Source." *Indoor Air Review*, July 1992, 11–12.

"Exhaust Ventilation Systems and Radon—Rethinking Common Wisdom." *Energy Design Update*, June 1993, 8–9.

Fahlén, Per. "Demand Controlled Ventilating Systems—Sensor tests." *Air Infiltration Review* 14 (June 1993): 4–7.

Fannin, Kerby. "Respirable Particulates Pose Challenges to Evaluating IAQ Filters." *Indoor Air Review*, January 1993, 13–14.

"Fans for Central-Exhaust Ventilation Systems." *Energy Design Update*, April 1993, 7–13.

Farr Co. *Filtration and Indoor Air Quality: A Two-Step Design Solution*. Los Angeles, CA: Farr Co., 1992.

Feustel, Helmut E., Mark P. Modera, and Arthur H. Rosenfeld. *Ventilation Strategies for Different Climates*. LBL-20364. Berkeley, CA: Lawrence Berkeley Laboratory, June 1986.

Fisk, W.J., K.M. Archer, R.E. Chant, D. Hekmat, F.J. Offerman, and B.S. Pedersen. *Freezing in Residential Air-to-Air Heat Exchangers*: An Experimental Study. Berkeley, CA: Lawrence Berkeley Laboratory, September 1983. LBL-16783.

Fisk, W.J., B.S. Pedersen, D. Hekmat, R.E. Chant, and H. Kaboli. "Formaldehyde and Tracer-Gas Transfer Between Airstreams in Enthalpy-Type Air-to-Air Heat Exchangers." *ASHRAE Transactions* 91-1B (1985): 173–186. LBL-18149.

Fitzgerald, J., G. Nelson, and L. Shen. "Sidewall Insulation and Air-Leakage Control." *Home Energy Magazine*, January/February 1990, 13–20.

Fry, Jason. "Canadian Study: Cheaper Humidity Meters are Best." *Indoor Air Review*, August 1994, 6.

Gammage, Richard and Steven Kaye. *Indoor Air and Human Health*. Chelsea, MI: Lewis Publishers, Inc., 1985.

Garbesi, Karina. "Experiments and Modeling of the Soil-Gas Transport of Volatile Organic Compounds into a Residential Basement." MS Thesis. LBL-25519 Rev. Berkeley, CA: Lawrence Berkeley Laboratory, December 1988.

"Gas Sensors to Control Indoor Pollution." *Healthy Buildings International Magazine*, March/April 1991, 8–9.

Gebefugi, I. and F. Korte. "Indoor Contamination of Household Articles Through Pentachlorophenol and Lindane." in *Proceedings of the 3rd International Conference on Indoor Air Quality and Climate, Volume 4, in Stockholm, Sweden, August 20–24, 1984*, by the International Conference on Indoor Air Quality and Climate, 1984, 317–322.

Geddes Enterprises, Scanada Consultants, Ltd., National Energy Conservation Association, and Energy Building Group. *Combustion Venting Training Course*. Ottawa, Canada: Canada Mortgage and Housing Corp., January 1991.

Geddes Enterprises, Union Gas Ltd., Northwestern Utilities, B.C. Gas Ltd., Scanada Consultants Ltd., and Howell Mayhew Engineering, Inc. *Evaluation of the Effectiveness of a Hard-Connected Duct into the Return-Air System of a Furnace Forced Air Duct System as a Means for Providing Ventilation and Make-up Air*. Ottawa, ON, Canada: Canada Mortgage and Housing Corp., December 21, 1991.

Gehring, Ken. "Year-Round Humidity Control in the Energy-Efficient Home: Creating a Healthy Allergen-Free Environment." in *Proceedings of the Excellence in Housing Conference in Indianapolis, IN March 21–23, 1991*, by the Energy-Efficient Building Association, 1991, C24–C38.

_____. "Energy Conservation, Comfort and Indoor Air Quality in a High-Humidity Climate." in *Proceedings of the Excellence in Housing Conference in Research Triangle Park, NC March 4–7, 1992*, by the Energy-Efficient Building Association, 1992, F18–F27.

_____. "Minimal Air Conditioning, Maximum Dehumidification." in *Proceedings of the Building Solutions Conference in Boston, MA March 3–6, 1993*, by the Energy-Efficient Building Association and the Northeast Sustainable Energy Association, 1993, D27–D36.

_____. "Ventilation and Humidity Control of a Home in a Humid Climate." in *Proceedings of the Excellence in Housing Conference in Dallas, TX February 23–26, 1994*, by the Energy-Efficient Building Association, 1994, B34–B48.

Giannini, James A., Donald A. Malone and Thaddeus A. Piotrowski. "The Seratonin Irritation Syndrome—A New Clinical Entity?" *Journal of Clinical Psychiatry* 47 (January 1986): 22–25.

Godish, Thad. *Indoor Air Quality Notes: Residential Formaldehyde Control*. Muncie IN: Ball State University, Department of Natural Resources, Summer 1986.

———. "Botanical Air Purification Studies Under Dynamic Chamber Conditions." in *Proceedings of the 81st Annual Meeting of the Air Pollution Control Association* in Dallas, TX, 1988, by the Air Pollution Control Association, 1988.

———. *Indoor Air Pollution Control*. Chelsea, MI: Lewis Publishers, Inc., 1989.

———. "Impact of Infiltration-Reducing Energy Conservation Measures on Indoor Air Quality." in *Proceedings of the Innovative Housing '93 Conference, Volume 1, in Vancouver, BC, Canada June 21–25, 1993*, by Canada Mortgage and Housing Corp. and Energy, Mines and Resources Canada, 1993, 391–403.

Goldbeck, David. *The Smart Kitchen*. Woodstock, NY: Ceres Press, 1989.

Good, Clint. *Healthful Houses*. Bethesda, MD: Guaranty Press, 1988.

Gots, Ron. "Carbon Dioxide—Is It a Good Litmus Test for IAQ?" *Indoor Air Review*, September 1992, 9.

Gregerson, Joan. "High-Efficiency Residential Blower Motor Replacements: Capturing a Lost Opportunity." in *Proceedings of the Excellence in Housing Conference in Dallas, TX February 23–26, 1993*, by the Energy-Efficient Building Association, 1994, B49–B60.

Habitat Design and Consulting Ltd. *A Survey of Ventilation Systems for New Housing*. Ottawa, ON, Canada: Canada Mortgage and Housing Corp., September 1988.

"Half of New Swedish Homes have Heat-Recovery Ventilators." *Energy Design Update*, November 1993, 2.

Hamlin, T., J. Forman, and M. Lubun. *Ventilation and Airtightness in New Detached Canadian Housing*. Ottawa, ON, Canada: Canada Mortgage and Housing Corp., 1990.

Hayson, J.C. and J.T. Reardon. "1989 Survey of Airtightness of New Merchant Builder Houses." in *Proceedings of the 5th International Conference on Indoor Air Quality and Climate, Volume 4, in Toronto, ON, Canada, 29 July to 3 August 1990*, by the International Conference on Indoor Air Quality and Climate, 1990, 263–268.

"Heat-Recovery Ventilation in Warm Humid Climates." *Energy Design Update*, September 1989, 8–10.

Hedge, Alan, William Ericson, and Gail Rubin. "Effects of Man-Made Mineral Fibers in Settled Dust on Sick Building Syndrome in Air-Conditioned Offices." in *Proceedings of the 6th International Conference on Indoor Air Quality and Climate, Volume 1, in Helsinki Finland, July 4–8, 1993*, by the International Conference on Indoor Air Quality and Climate, 1993, 291–296.

Heller, Jonathon (Ecotope, Inc.). *Residential Construction Demonstration Project: Cycle III, Analysis of Innovative Ventilation Systems in Multifamily Buildings, Final Report*. Spokane, WA: Washington State Energy Office, December 1992.

Hill, David. *Ventilation Solutions for Health and Comfort*. Paper presented as a part of the symposium "NAHB Energy Update: New Products, Technologies, and Research Findings" at the National Association of Home Builders Annual Convention, Las Vegas, NV, February 20, 1993.

———. *Ventilation*. Paper presented as a part of the workshop "Ventilation Advanced Session" at the Building Solutions Energy-Efficient Building Association and the Northeast Sustainable Energy Association joint Conference in Boston, MA March 3–6, 1993.

Hill, David and Yvonne Kerr, "Concerns about Venting." *Solplan Review*, April-May 1993, 9–10.

Hodgson, Michael, Geoffrey Block, and David Parkinson. "Organophosphate Poisoning in Office Workers." *Journal of Occupational Medicine* 28 (June 1986): 434–437.

Home Ventilating Institute. *Home Ventilating Guide*. Arlington Heights, IL: Home Ventilating Institute, 1986.

———. *Installation Manual for Heat-Recovery Ventilators*. Arlington Heights, IL: Home Ventilating Institute, 1990.

———. *Certified Home Ventilating Products Directory*. Arlington Heights, IL: Home Ventilating Institute, 1994.

Honeywell, Inc. *Electronic Air Cleaner Theory and Fundamentals*. Minneapolis, MN: Honeywell, Inc., n.d.

———. *Clearing the Air: The Real Story about Indoor Air Quality*. Minneapolis, MN: Honeywell, Inc., 1986.

Appendix C Bibliography

Honicky, Richard, J. Scott Osborne III, and C. Amechi Akpom. "Symptoms of Respiratory Illness in Young Children and the Use of Wood-Burning Stoves for Indoor Heating." *Pediatrics* 75 (March 1985): 587–593.

HOT2000 Version 7 (Ottawa, ON, Canada: Canadian Homebuilders' Association).

Housing Resource Center. *Housemending Notebook*. Cleveland, OH: Housing Resource Center, n.d.

_____. *Healthy House Catalog and Conference Proceedings*. Cleveland, OH: Housing Resource Center, 1988.

_____. *Healthy House Catalog*. Cleveland, OH: Housing Resource Center, 1990.

HRAI Technical Services Division, Inc. and Energy Systems Centre. *Residential Exhaust Equipment*. Toronto, ON, Canada: Canada Mortgage and Housing Corp., September 12, 1988.

"HRV System Efficiency." *Solplan Review*, June-July 1993, 6–7.

Hughes, John. "Heat-Recovery Ventilators." *Fine Homebuilding*, August-September 1986, 30–34.

IAQ Publications, Inc. *Indoor Air Quality, Product Manufacturers & Distributors*. Bethesda, MD: IAQ Publications, Inc., 1993.

Interior Concerns Publications. *Interior Concerns Resource Guide*. Mill Valley, CA: Interior Concerns Publications, 1993.

Iris Communications, Inc. *Energy Source Directory, A Guide to Products Used in Energy-Efficient Residential Buildings*. Eugene, OR: Iris Communications, Inc., 1994.

Iris Communications, Inc. *Advanced Air Sealing*. Eugene, OR: Iris Communications, Inc., n.d.

Jackson, Mark. "Ventilation for Manufactured Houses: An Evaluation of Technologies for Small Residences." in *Proceedings of the Building Solutions Conference in Boston, MA March 3–6, 1993*, by the Energy-Efficient Building Association and the Northeast Sustainable Energy Association, 1993, D37–D46.

_____. "Integrated Heating and Ventilation: Double Duty for Ducts." *Home Energy Magazine*, May-June 1993, 27–33.

Janssen, John. "Ventilation for Acceptable Indoor Air Quality." *ASHRAE Journal*, October 1989, 40–48.

Johnson, Duane. "Can Your House Run Out of Breath?" *The Family Handyman*, November-December 1988, 26–32.

Jones, Don, James LaRue, Helene Roussi, and Alan Wasco. *Weatherization & Indoor Air Quality*. Cleveland, OH: Housing Resource Center, 1992.

Kadulski, Richard. *Residential Ventilation: Achieving Indoor Air Quality*. Vancouver, BC, Canada: The Drawing Room Graphic Services Ltd., 1988.

_____. "Ventilation Case Study: The Importance of Noise Levels." *Solplan Review*, April-May 1990, 7–8.

_____. "Clean Air: Getting the Most Out of Filters." *Solplan Review*, December-January 1991, 3–6.

_____. "Residential Ventilation." *Solplan Review*, February-March 1991, 3–8.

_____. "Indoor Air: Is it Dangerous to Your Health." *Solplan Review*, April-May 1993, 8–9. (This is a reprint of an October-November 1986 *Solplan Review* article.)

Karg, Richard. "Tips and Cautions about Air Exchangers." *Solar Age*, October 1984, 15–19.

Kaufman, David. "Ventilation Secrets from Sweden." *Journal of Light Construction*, September 1992, 40–43.

King, J. Gordon. "Air for Living." *Respiratory Care* 18 (March/April 1973), 160–164.

"Kitchen Fans." *Solplan Review*, December-January 1990, 15.

"Kitchen Ventilation." *The Practical Homeowner*, September-October 1992, 14.

Knoll, Bas. *Advanced Ventilation Systems—State of the Art and Trends*. Coventry, Great Britain: Air Infiltration and Ventilation Centre, March 1992.

Kormos, Robin. *Environmental Medicine & An Environmental Home*. Produced by Robin Kormos. Healthy Homes Video Series, 1993. VHS videocassette.

Lawson, Lynn. *Staying Well in a Toxic World*. Chicago, IL: Noble Press, Inc., 1993.

Leclair, Kim and David Rousseau. *Environmental by Design, Volume 1: Interiors*. Point Roberts, WA: Hartley & Marks, Inc., 1992.

Lepage, Michael F. and Glenn D. Schuyler. "How Fresh is Fresh Air?" in *Proceedings of the 5th International Conference on Indoor Air Quality and Climate, Volume 4, in Toronto, ON, Canada, 29 July to 3 August 1990,* by the International Conference on Indoor Air Quality and Climate, 1990, 311–316.

Levin, Hal. "Ventilation Requirements Based on Subjective Responses." *Indoor Air Quality Update*, October 1990, 1–7.

Lillie, Thomas H., and Edward S. Barnes. "Airborne Termiticide Levels in Houses on United States Air Force Installations." in *Proceedings of the 4th International Conference on Indoor Air Quality and Climate, Volume 1, in Berlin, Germany, August 17–21, 1987,* by the International Conference on Indoor Air Quality and Climate, 1987. 200–205.

Linddament, Martin W. *Energy-Efficient Ventilation Strategies*. in Proceedings of the Innovative Housing '93 Conference, Volume 1, in Vancouver, BC, Canada June 20–25, 1993, by Canada Mortgage and Housing Corp. and Energy, Mines and Resources Canada, 1993, 417–26

Lischkoff, James K. and Joseph Lstiburek. *The Airtight House*. Ames, IA: Iowa State University Research Foundation, 1985.

Lloyd Publishing. *Safe Home Digest House Building Resource Guide*. New Canaan, CT: Lloyd Publishing, 1993.

Lossnay™ Heat and Moisture (Enthalpy) Exchange Cores for Commercial and Industrial Ventilation. Madison, WI: Altech Energy, May 1994.

Lotz, William. "Choosing Kitchen and Bath Exhaust Fans." *Journal of Light Construction*, February 1992, 38–39.

Lstiburek, Joseph. *Building Science*. Chestnut Hill, MA: Building Science Corp., n.d.

_____. "Moisture Woes in the South." *Custom Builder*, January/February 1994, 28–32.

_____. "Multiple Chemical Sensitivity." *Energy-Efficient Building Association Excellence* IX (Special Conference Issue 1994), 2–4.

Lstiburek, Joseph, and John Carmody. *Moisture-Control Handbook: Principles and Practices for Residential and Small Commercial Buildings*. New York, NY: Van Nostrand Reinhold, 1993.

Lubliner, Michael and Marvin Young. "Is it All a Lot of Hot Air?—Mechanical Ventilator Performance." *Home Energy Magazine*, November-December 1990, 25–29.

Mahajan, B.M. *A Method for Measuring the Effectiveness of Gaseous Contaminant Removal Filters* NBSIR 89–4119. Gaithersburg, MD: National Institute of Standards and Technology, 1989.

Maker, Tim. "Blower Door Basics." *Journal of Light Construction*, April 1994, 71–72.

Malin, Nadav. "Fresh Air Supply for Exhaust-Only Ventilation." *Environmental Building News*, March-April 1993, 13–14.

Man-Made Mineral Fibers, Position Paper. Washington DC: Safety and Health Committee, AFL-CIO, June 1991.

Massachusetts Audubon Society. *Contractor's Guide to Finding and Sealing Hidden Air Leaks*. Lincoln, MA: Massachusetts Audubon Society, 1993.

Maunder, Jim. "Residential Construction Demonstration Project: Cycle III: Single Family Ventilation Case Study Investigations." in *Proceedings of the Excellence in Housing Conference in Research Triangle Park, NC March 4–7, 1992,* by the Energy-Efficient Building Association, 1992, F28–F39.

Mayell, Mark. "Smoke Detectors: The Non-Nuclear Alternative." *East West*, April 1990, 80–81.

Mayk, Gary. "Furnace Add-Ons That Clean the Air." *Journal of Light Construction*, October 1989, 24–26.

McCarthy, Jeff. "Ventilation System Design Made Easy: The Layman's Guide to Residential Ventilation." in *Proceedings of the Building Solutions Conference, in Boston MA March 3–6, 1993,* by the Energy-Efficient Building Association and the Northeast Sustainable Energy Association, 85–90.

McLeister, Dan. "A Low-Tech Solution to a 20th Century Problem." *Professional Builder*, June 1989, 55.

Meier, Alan. "Infiltration: Just ACH_{50} Divided by 20?" *Energy Auditor & Retrofitter*, July-August 1986, 16–19.

Meier, Simon. "Mixed-Gas or CO_2 Sensors as a Reference Variable for Demand Controlled Ventilation." in *Proceedings of Indoor Air '93, The 6th International Conference on Indoor Air Quality and Climate, July 4–8, 1993 in Helsinki, Finland, Volume 5,* by the International Conference on Indoor Air Quality and Climate, 1993, 85–90.

Minnegasco. *Combustion Air.* Minneapolis, MN: Minnegasco, 1990.

Minnesota Department of Public Service. *Combustion Air, Housewarming Guide Series.* Minneapolis, MN: Energy Information Center, May 1994.

Minnesota Extension Service. *Building for Performance: Ventilation Series.* Produced by Minnesota Extension Service. 39 min. University of Minnesota, 1991. VHS videocassette.

Moffatt, Peter. "The Bermuda Triangle: Airtightness, Ventilation, and Insulation." *Solplan Review*, February-March 1993, 6–8.

Moffatt, Sebastian and Peter Moffatt. "House Depressurization Limits for Mid-Efficiency Gas-Fired Furnaces." *Solplan Review*, February-March 1990, 7–9.

"Moisture Recovery Ventilators." *Energy Design Update*, September 1989, 12–14.

Molhave, Lars. "Indoor Air Pollution Due to Organic Gases and Vapours of Solvents in Building Materials." *Environment International* 8 (1982): 117–127.

Morris, Dan. *Healthy Buildings Resource Guide.* Clinton, WA: Healthy Buildings Associates, 1993.

Muller, Christopher O. "A Comparison of Packed-Bed and Partial-Bypass Gas-Phase Air Filters." *Indoor Air Review*, January 1993, 15, 26.

"Multi-Family Ventilation Systems—Cost Comparison and Builders' Guide." *Energy Design Update*, September 1990, 10–12.

Nagda, N. and M. Koontz. "Detailed Evaluation of Range Exhaust Fans." in *Proceedings of the 4th International Conference on Indoor Air Quality and Climate, Volume 3 in Berlin, Germany, August 17–21, 1987,* by the International Conference on Indoor Air Quality and Climate, 1987, 324–328.

Nair, Indira, Granger Morgan, and Keith Florig. *Biological Effects of Power Frequency Electric and Magnetic Fields—Background Paper.* # OTA–BP–E–53. Washington, DC: Office of Technology Assessment, U.S. Congress, 1989.

National Association of Home Builders. *Controlling Moisture in Homes.* Washington, DC: Home Builder Press, 1987.

National Center for Appropriate Technology. *Heat-Recovery Ventilation for Housing.* Butte, MT: National Center for Appropriate Technology, n.d.

_____. *Moisture and Home Energy Conservation.* Butte, MT: National Center for Appropriate Technology, n.d.

National Coalition Against the Misuse of Pesticides, *Safety at Home.* Washington, DC: National Coalition Against the Misuse of Pesticides, 1991.

Nelson, Gary, Robert Nevitt, Neal Moyer, and John Tooley. "Measured Duct Leakage, Mechanical System Induced Pressures and Infiltration in Eight Randomly Selected Minnesota Homes." in *Proceedings of the Building Solutions Conference in Boston, MA March 3–6, 1993,* by the Energy-Efficient Building Association and the Northeast Sustainable Energy Association, 1993, F1–F12.

Nelson, Gary, Robert Nevitt, and Gary Anderson. "Are Your Houses Too Tight?" *Journal of Light Construction*, August 1994a, 29–33.

Nelson, Gary. *Notes on Mechanical Ventilation Workshop.* Paper presented as a part of the session "Mechanical Ventilation" at the 1994 Region 5 Weatherization Conference, Indianapolis, IN, August 19, 1994b.

Nelson, Harold S. and others. "Recommendations for the Use of Residential Air-Cleaning Devices in the Treatment of Allergic Respiratory Diseases." *Journal of Allergy and Clinical Immunology* 82 (October 1988): 661–669.

"Next Generation(?) Heat-Pump Water Heater." *Energy Design Update*, December 1993, 7–8.

Nisson, J.D. Ned. *Air-to-Air Heat Exchanger Directory and Buyer's Guide.* Arlington, MA: Cutter Information Corp., 1987

_____. "Building a Better Bathroom Fan." *Journal of Light Construction*, August 1989, 41–42.

_____. "Don't Ignore Leaky Ducts." *Journal of Light Construction*, April 1990, 40–41.

Nisson, J.D. Ned, and Gautam Dutt. *The Superinsulated Home Book*. New York: John Wiley and Sons, 1985.

Nuess, Mike. "Beating Indoor Air Pollution." *Fine Homebuilding*, December 1992/January 1993, 68–71.

Oboe Engineering, Inc. *Medium-Efficiency Filtration: Improved Filters for Residential Forced-Air Furnaces*. Ottawa, ON, Canada: Canada Mortgage and Housing Corp., September 1986.

Offerman, Francis J., III. "Ventilation Standard Was Misunderstood." *Indoor Pollution Law Report*, September 1992, 3.

Olkowski, William, Sheila Daar, and Helga Olkowski. *Common-Sense Pest Control*. Newton, CT: The Taunton Press, 1991.

Olson, Wanda. "Clearing the Kitchen Air." *Journal of Light Construction*, August 1991, 38–41.

Palmer, J., P. Shaw, and M. Trollope. "Stack-Effect Ventilation of an Infant's School." In *Energy Impact of Ventilation and Air Infiltration: Proceedings of the Air Infiltration and Ventilation Centre, 14th AIVC Conference held in Copenhagen, Denmark, September 21–23, 1993*, by the Air Infiltration and Ventilation Centre, 1993, 157–166.

Parfitt, Yvette. *Ventilation Strategy—A selected Bibliography*. Coventry, Great Britain: Air Infiltration and Ventilation Centre, July 1985.

Penny, Rob. "The Energy Doctor." *Southface Journal of Energy and Building Technology*, Winter 1990, 25–26.

"Performance Guidelines for Controlling IAQ by Ventilation." *Indoor Air Quality Update*, September 1991, 4–5.

Performance Woodburning, Inc. *That Nice "Woodsy" Smell, Combustion Spillage from Residential Wood Heating Systems*. Ottawa, ON, Canada: Canada Mortgage and Housing Corp., March 1991a.

_____. *That Nice "Woodsy" Smell, Combustion Spillage from Residential Wood Heating Systems, Appendices*. Ottawa, ON, Canada: Canada Mortgage and Housing Corp., March 1991b.

Peterson, Roger Alan. "Clearing the Air." *Custom Builder*, July 1988, 19–22.

Pfeiffer, Guy O. and Casimir M. Nikel. *The Household Environment and Chronic Illness*. Springfield, IL: Charles C. Thomas Publisher, 1980.

Pholbrick, David, Martin Thompson, and Mike O'Brien. *Installing Non-Heat-Recovery Ventilation Systems*. Corvallis, OR: Oregon State University Extension Energy Program, June 1989.

Piersol, P. and K. Matsummura. *Development of a Procedure to Assess Organic Outgassing from Heat-Recovery Ventilators*. Ottawa, ON, Canada: Energy, Mines & Resources, January 1987.

Proctor, John, Michael Blasnik, Bruce Davis, Tom Downey, Mark P. Modera, Gary Nelson, and John Tooley Jr. "Diagnosing Ducts, Finding the Energy Culprits." *Home Energy Magazine*, September-October 1993, 26–31. (This issue also has several other good articles dealing with leaky duct issues on pp. 32–67.)

Product Knowledge Center. *House Ventilation and Air Quality Reference Guide*. Toronto, ON, Canada: Ontario Hydro, 1988.

Pure Air Systems, Inc. *600 HEPA Shield Air Filtration System*. Produced by Pure Air Systems, Inc. 3:44 min. n.d. VHS videocassette.

Raab, Karl H. *Upgrading Residential Forced Air Filtration*. Ottawa, ON, Canada: Canada Mortgage and Housing Corp., May 1982.

Randolph, Theron G. *Human Ecology and Susceptibility to the Chemical Environment*. Springfield, IL: Charles C. Thomas Publisher, 1962.

Randolph, Theron G. and Ralph Moss. *An Alternative Approach to Allergies*. New York: Harper & Row, 1980.

Raydot, Inc. *Improve Your Home Air Quality with a Raydot HRV System*. Produced by Raydot, Inc. 8 min. 1992. VHS Videocassette.

"Rebalancing Energy and Indoor Air Quality." *Building Design & Construction*, May 1993, 61–62.

Reiss, Chuck. "Wet-Spray Cellulose Insulation." *Journal of Light Construction*, August 1994, 24–27.

"Residential Ventilation." Solplan Review, April-May 1994, 6–8.

Röben, Jürgen. "Review of Sick Building Syndrome." *Air Infiltration Review* 14 (June 1993): 11–15.

Rosenbaum, Marc. "Simple Ventilation for Tight Houses." *Journal of Light Construction*, May 1991, 26–29.

Rosenbaum, Marc. "Testing Hones for Air Leaks." *Fine Homebuilding*, February/March 1994, 51–53.

Rousseau, David and William Rea. *Your Home, Your Health and Well-Being*. Berkeley, CA: Ten Speed Press, 1988.

Ruggles, Rick and Mary Witt. *Guidelines for Non-Toxic Construction*. Atlanta, GA: The Southface Institute, 1990.

Samet, Jonathan, Marian Marbury, and John Spengler. "Health Effects and Sources of Indoor Air Pollution. Part I." *American Reviews of Respiratory Disease* 136 (1987): 1486–1508.

_____. "Health Effects and Sources of Indoor Air Pollution. Part II." *American Reviews of Respiratory Disease* 137 (1988): 221–242.

Scanada-Shletair Consortium Inc. *Chimney Safety Tests Users' Manual* (Second Edition). Ottawa, ON, Canada: Canada Mortgage and Housing Corp., January 12, 1988.

Schaplowsky, A.F., F.B. Oglesbay, J.H. Morrison, R.E. Gallagher, and W. Berman Jr. "Carbon Monoxide Contamination of the Living Environment, A National Survey of Home Air and Children's Blood." *Journal of Environmental Health* 36 (1974): 569–573.

Schell, Michael. "The Case for HRV's." *Indoor Air Review*, April 1992, 14, 20.

Schell, Mike. "Understanding the Difference: CO_2 Sensors and Air Quality Sensors." *Indoor Air Review*, December 1994, 31–32.

Sheldon, L., H. Zelon, J. Sickles, C. Eaton, T. Hartwell, and L. Wallace. *Project Summary, Indoor Air Quality in Public Buildings: Volume II*. #EPA/600/S6–88/009b. Washington, DC: U.S. Environmental Protection Agency, September 1988.

Sheltair Scientific Ltd. *Barriers to the Use of Energy-Efficient Residential Ventilation Equipment*. Ottawa, ON, Canada: Canada Mortgage and Housing Corp., June 1992.

Sheltair Scientific Ltd. and SAR Engineering Ltd. *Demand Controlled Ventilation: Final Report*. Ottawa, ON, Canada: Canada Mortgage and Housing Corp., 1991.

Shumate, Monroe W. and John E. Wilhelm. *Evaluation of Fiber Shedding Characteristics Under Laboratory Conditions and in Commercial Installations*. Denver, CO: Schuller International, Inc., n.d.

Shurcliff, William. *Air-to-Air Heat Exchangers for Houses*. Andover, MA: Brick House Publishing Co., 1982.

_____. "Air-to-Air Heat Exchangers for Houses." *Solar Age*, March 1982, 19–22.

Small, Bruce. *Indoor Air Pollution and Housing Technology*. Ottawa, ON, Ontario: Canada Mortgage and Housing Corp., 1985.

Smith, Bill Rock. "Heat-Recovery Ventilators." *Journal of Light Construction*, March 1994, 31–36.

Special Legislative Commission on Indoor Air Pollution. *Indoor Air Pollution in Massachusetts, Final Report*. Boston, MA: Commonwealth of Massachusetts, Special Legislative Commission on Indoor Air Pollution, April 1989.

Spengler, John, Harriet Burge, and Jenny Su. "Biological Agents and the Home Environment." in *Proceedings of Bugs, Mold & Rot, a Workshop on Residential Moisture Problems, Health Effects, Building Damage, and Moisture Control, in Washington, DC, May 20–21, 1991,* ed. E. Bales and W.B. Rose (Washington, DC: National Institute of Building Sciences, 1992), 11–18.

Steege, Douglas. "Controlling Moisture with Heat-Recovery Ventilators." *Custom Builder*, March 1989, 27–30.

Stum, Karl. "Designing and Installing Residential HVAC Duct Systems that Minimize Duct Leakage and Its Effects." in *Proceedings of the Building Solutions Conference in Boston, MA March 3–6, 1993,* by the Energy-Efficient Building Association and the Northeast Sustainable Energy Association, 1993, F31–F47.

Sullivan, Bruce. "Study Raps Bath Fans." *Journal of Light Construction*, October 1989, 9.

_____. "Making Mastic Stick." *Journal of Light Construction*, December 1992, 38.

_____. "Tightening Up With Air Sealing." *Journal of Light Construction*, January 1993, 33–35.

_____. "Using Heating Ducts for Ventilation." *Journal of Light Construction*, September 1993, 51.

_____. "Choosing a Furnace Filter." *Journal of Light Construction*, May 1994, 57–58.

Summary of the Minnesota Attorney General's Lawsuit against Alpine Air Products. St. Paul, MN: Minnesota Attorney General's Office, n.d.

Swedish Council for Building Research. Several English-Language publications dealing with ventilation. Swedish Council for Building Research, Box 128 66, 112 98 Stockholm, Sweden.

"The Cost of Clean Air." *Energy Design Update*, October 1992, 9–10.

"The Cost of Ventilation." *Energy Design Update*, October 1992, 8–9.

"The Effectiveness of Outdoor Air for Indoor Humidity Control." *Energy Design Update*, December 1990, 8–9.

The Energy Conservatory. *Minneapolis Blower-Door Operation Manual*. Minneapolis, MN: The Energy Conservatory. n.d.

"The Ventilation Rate Myth." *Energy Design Update*, September 1993, 9–10.

Tiller, Jeff. "Put a Stop to MAD-Air and Leaky Ducts." *The Southface Journal*, Winter 1990, 12–13.

_____. "Southface Finds MAD-AIR Problems are Driving People Crazy." *The Southface Journal*, Summer-Fall 1991, 4–7.

Tolpin, Jim. "Builder's Guide to Moisture Meters." *Tools of the Trade*, Summer 1994, 41–45.

Traudt, Jon. "Control of Indoor Air Pressure for Protection of Health, Preservation of Buildings, and Conservation of Energy." in *Proceedings of the 16th World Engineering Congress and 3rd Environmental Technology Exposition in Atlanta, GA, October 26–28*, by the World Engineering Congress, 1993.

_____. "Protecting People and Houses while Saving Energy." *Energy Exchange*, November 1993, 29–32.

Trethewey, Richard and Don Best. *This Old House Heating, Ventilation, and Air Conditioning*. New York: Little, Brown & Co., 1994.

Tsongas, George. "Dehumidification as a Needed Complement to Ventilation for Indoor Moisture Control in New Residences in Northern Climates." in *Proceedings of the Building Solutions Conference in Boston, MA March 3–6, 1993*, by the Energy-Efficient Building Association and the Northeast Sustainable Energy Association, 1993, D57–D68.

Turiel, Isaac. *Indoor Air Quality and Human Health*. Stanford, CA: Stanford University Press, 1985.

Ulness, Amy. "Air Apparent." *Building Products*, Winter 1994, 77–80.

Uniacke, Michael and John Proctor. "Getting the Most from Mechanical Cooling." *Journal of Light Construction*, August 1993, 18–22.

U.S. Consumer Product Safety Commission. *Consumer Product Safety Alert: Dirty Humidifiers May Cause Health Problems*. Washington, DC: U.S. Consumer Product Safety Commission, December 1988.

U.S. Department of Defense. *MIL Standard 282 Filter Units, Protective Clothing, Gas Masks*. Irvine, CA: U.S. Department of Defense, Global Engineering Documents, 1956 with notices 1974 and 1989.

U.S. Department of Energy. *Indoor Air Quality Environmental Information Handbook: Radon*. # DOE/PE/72013–2. Washington, DC: U.S. Department of Energy. January 1986.

_____. *Indoor-Air Quality Environmental Information Handbook: Building System Characteristics*. # DOE/EV/10450–H1. Washington, DC: U.S. Department of Energy, January 1987.

_____. *Indoor-Air Quality Environmental Information Handbook: Combustion Sources, 1989 Update*. # DOE/EH/79079–H1. Washington, DC: U.S. Department of Energy, June 1990.

_____. *Improving the Efficiency of Your Duct System*, # DOE/EE–0015. Washington, DC: U.S. Department of Energy, April 1994.

U.S. Environmental Protection Agency. *Guidance for Controlling Asbestos-Containing Materials in Buildings*. #EPA 560/5–85–024. Washington, DC: U.S. Environmental Protection Agency, June 1985.

_____. *Radon Reduction Techniques for Detached Houses, Technical Guidance.* #EPA/625/5–86/019. Washington, DC: U.S. Environmental Protection Agency, June 1986.

_____. *Indoor Air Facts No. 3: Ventilation and Air Quality in Offices.* Washington, DC: U.S. Environmental Protection Agency, February 1988.

_____. *Indoor Air Facts No. 7: Residential Air Cleaners.* #20A–4001. Washington, DC: U.S. Environmental Protection Agency, February 1990.

_____. *Residential Air-Cleaning Devices.* #EPA 400/1–90–002. Washington, DC: U.S. Environmental Protection Agency, February 1990.

_____. *Radon-Resistant Construction Techniques for New Residential Construction, Technical Guidance.* #EPA/625/2–91/032. Washington, DC: U.S. Environmental Protection Agency, February 1991.

_____. *Sub-Slab Depressurization for Low-Permeability Fill Material, Design & Installation of a Home Radon Reduction System.* #EPA/625/6–91/029. Washington, DC: U.S. Environmental Protection Agency, July 1991.

_____. *Building Air Quality: A Guide for Building Owners and Facility Managers.* #EPA/400/1–91/033. Washington, DC: U.S. Environmental Protection Agency, December 1991.

_____. *Home Buyer's and Seller's Guide to Radon.* #402–R–93–003. Washington, DC: U.S. Environmental Protection Agency, March 1993.

_____. *The Inside Story: A Guide to Indoor Air Quality.* #EPA 402–K–93–007. Washington, DC: U.S. Environmental Protection Agency, September 1993.

"Using Central Heating System Ducts for Fresh-Air Ventilation." *Northwest Builder*, December 1992, 4.

Vance, Mary. *Ventilation: A Bibliography.* Monticello, IL: Vance Bibliographies, May 1986.

Ventilation: Health and Safety Issues. Ottawa, ON, Canada: Canada Mortgage and Housing Corp., 1987.

"Ventilation Systems vs. House Occupants." *Energy Design Update*, November 1993, 6.

"Ventilation vs. Source Control: Witnesses Spar on Best IAQ Strategy." *Indoor Pollution News*, July 11, 1991, 1, 4.

Washington Energy Extension Service. *Indoor-Air Pollutant: Particles, Factsheet.* Seattle, WA: Washington Energy Extension Service, December 1987.

_____. *Indoor-Air Quality Factsheet.* Seattle, WA: Washington Energy Extension Service, December 1987.

Westlake, Joan Kay. "Condensation Control—Stopping the Indoor Rain." *Mini-Storage Messenger*, August 1991. 63–64.

White, Jim. "Ventilation Compliance Tests for Houses," in *Proceedings of the 77th Annual Meeting of the Air Pollution Control Association in San Francisco, CA, June 24–29, 1984*, by the Air Pollution Control Association, 1984.

_____. "Moisture, Mould, & Ventilation." *Solplan Review*, April-May 1991a, 3–6.

_____. *The Energy Efficiency of Residential Ventilation Fans and Fan/Motor Sets.* Ottawa, ON, Canada: Canada Mortgage and Housing Corp., 1991b.

_____. "Ventilation Compromised by Poor Fan Quality." *Indoor Air Review*, January 1993, 5–6.

Whitmore, Roy W., Janice E. Kelly, and Pamela L. Reading. *National Home and Garden Pesticide Survey, Final Report.* RTI/5100/17–03F. Research Triangle Park, NC: Research Triangle Institute, March 1992.

Williams, Gurney. "Allergy Proofing Your Home." *Practical Homeowner*, February 1991, 22–24

Wolbrink, David W. *Common Sense Ventilation.* Hartford, WI: Broan Mfg. Co., Inc., May 27, 1993.

_____. *Energy Labeling for Residential Ventilation Equipment* (Draft). Hartford, WI: Broan Mfg. Co., Inc., September 13, 1993.

Wolverton, B.C. *The Role of Plants and Microorganisms in Assuring a Future Supply of Clean Air and Water: A Summary of NSTL Research.* NSTL, MS: National Aeronautics and Space Administration, April 1985.

Wolverton, B.C., Rebecca C. McDonald, and E.A. Watkins, Jr., "Foliage Plants for Removing Indoor Air Pollutants from Energy-Efficient Homes." *Economic Botany* 38 (1984): 224–228

Yates, Alayne, Frank B. Gray, John I. Misiaszek, and Walter Wolman. "Air Ions: Past Problems and Future Directions." Environment International 12 (1986): 99–108.

Yuill, G.K. and Associates Ltd. *Investigation of the Indoor Air Quality, Airtightness, and Infiltration Rates of a Statistically Random Sample of 78 Houses in Winnipeg*. Ottawa, ON, Canada: Canada Mortgage and Housing Corp., December 14, 1987.

Zamm, Alfred V. *Why Your House May Endanger Your Health*. New York: Simon and Schuster, 1980.

Appendix D

Organizations

Following is a list of the most prominent organizations that regularly deal with residential ventilation issues both from a research and a practical standpoint. Most produce a variety of publications and they also hold periodic conferences.

Air Infiltration and Ventilation Centre (AIVC)
University of Warwick Science Park
Sovereign Court, Sir William Lyons Rd.
Coventry CV4 7EZ Great Britain
Phone +44 (0) 203 692050

American Society of Heating, Refrigerating, and Air-Conditioning Engineers (ASHRAE)
1791 Tullie Circle NE
Atlanta, GA 30329–2305
404–636–8400

Bonneville Power Administration (BPA)
P.O. Box 3621
Portland, OR 97208
503–230–3000

Canada Mortgage and Housing Corp. (CMHC)
700 Montreal Rd.
Ottawa, ON, Canada K1A 0P7
613–748–2000

Canadian Standards Association (CSA)
178 Rexdale Blvd.
Rexdale, ON, Canada M9W 1R3
416–747–4000

Energy-Efficient Building Association (EEBA)
1829 Portland Ave.
Minneapolis, MN 55404–1898
612–871–0413

Understanding Ventilation

Home Ventilating Institute (HVI)
Div. of Air Movement and Control Assoc., Inc.
30 West University Dr.
Arlington Heights, IL 60004–1893
708–394–0150

Lawrence Berkeley Laboratory (LBL)
University of California
#1 Cyclotron Rd.
Berkeley, CA 94720
510–486–4000

U.S. Environmental Protection Agency (USEPA)
Public Information Center
401 "M" Street
Washington, DC 20460
202–382–2080
or
U.S. Environmental Protection Agency (USEPA)
Indoor Air Quality Information Clearinghouse (IAQ INFO)
P.O. Box 37133
Washington, DC 20013–7133
800–438–4318

Appendix E

Equipment sources

Many of the ventilation equipment manufactures listed below also offer a wide variety of accessories such as pressure balancing kits, flow-measuring devices, duct silencers, controls, grilles, intake or exhaust hoods, ducts and fittings, duct insulation, dampers, filters, duct heaters, etc. Some of these companies sell directly to end users, but some only sell through distributors.

3-M Do-It-Yourself Division
3M Center
St. Paul, MN 55144–1000
800–388–3458
612–733–1110
Electrostatic air filters

ACS Filter Division
186 El Camino Real
Milbrae, CA 94030
800–633–4007
415–697–2761
Activated-carbon filters

Aero Hygenics, Inc.
5245 San Fernando Rd. West
Los Angeles, CA 90039
800–346–4642
818–246–4006
Air filters

Airex
5 Sandhill Ct. Unit C
Brampton, ON, Canada L6T 5J5
905–790–8667
Central-exhaust ventilation equipment, humidity-sensing controls, through-the-wall vents.

Air Filtration Products
639 Broadway
N. Amityville, NY 11701
516–789–2350
Full line of air filters

Understanding Ventilation

Airguard Industries, Inc.
P.O. Box 32578
Louisville, KY 40232–2578
502–969–2304
HEPA air filters

Air-King Ltd.
110 Glidden Rd.
Brampton, ON, Canada L6T 2J3
905–456–2033
Thermostatic fresh-air control

Air Kontrol
221 Pearson
Batesville, MS 38606
800–647–6192
601–563–4736
Electrostatic air filters

AirPro, Inc.
8541 Meredith Ave.
Omaha, NE 68134
402–572–0404
Pressure controllers and gauges

Airxchange
401 VFW Dr.
Rockland, MA 02370
617–871–4816
Commercial-sized HRVs

Allermed Corp.
31 Steel Rd.
Wylie, TX 75098
214–442–4898
Central air-filtration systems

Altech Energy
7009 Raywood Rd.
Madison, WI 53713
800–627–4499
608–221–4499
HRVs, controls

American Aldes Corp.
4539 Northgate Ct.
Sarasota, FL 34234–2124
800–255–7749
813–351–3441
Central-exhaust equipment, through-the-wall vents, grilles, HRVs, etc.

American Sensors, Inc.
30 Alden Rd. Unit 4
Markham, ON, Canada L2R 2S1
416–477–3320
Carbon-monoxide monitors

Amerix Corp.
2796 5th Ave. South
Fargo, ND 58103
800–232–4116
HRVs

Artis Metals Co.
3323 Chindin Blvd.
Boise, ID 83714
208–336–1560
Inlets and outlets

Aston Industries, Inc.
P.O. Box 220
St. Leonard d'Aston, PQ, Canada J0C 1M0
819–399–2175
HRVs, central-exhaust systems

Barneby & Sutcliffe
P.O. Box 2526
Columbus, OH 43216–2526
614–258–9501
Full line of adsorption filters

Boss Aire
2109 SE 4th St.
Minneapolis, MN 55414
800–847–7390
612–378–9652
HRVs, grilles, dampers

Appendix E Equipment sources

Bowers/Thermador
8685 Bowers Ave.
South Gate, CA 90280
800–669–2626
213–566–2111
Remote-mount kitchen fans

Broan Mfg. Co., Inc.
P.O. Box 140
Hartford, WI 53027
800–548–0790
414–673–4340
HRVs, central and local ventilation equipment, through-the-wall vents, controls, flow-measuring equipment

Carrier Corp.
P.O. Box 70
Indianapolis, IN 46206
800–227–7437
HRVs, filters

Carus Chemical Co.
1001 Boyce Memorial Dr.
Ottawa, IL 61350
815–223–1500
Carusorb adsorption material

Chim-A-Lator Co.
8824 Wentworth Ave. S.
Bloomington, MN 55420
612–884–7274
Chimney dampers

Columbus Industries
2938 State Road 752
Ashville, OH 43103
614–983–2552
Polysorb activated-carbon filters

Conserval Systems, Inc.
4242 Ridge Lea Rd. Suite 1
Buffalo, NY 14226–1051
716–835–4903
or

Conserval Systems, Inc.
200 Wildcat Rd.
Downsview, ON, Canada M3J 2N5
416–661–7057
Solarwall ventilation-air preheater

Conservation Energy Systems, Inc.
2525 Wentz Ave.
Saskatoon, SK, Canada S7K 2K9
306–242–3663
or
Conservation Energy Systems, Inc.
Box 582416
Minneapolis, MN 55458–2416
800–667–3717
HRVs, controls, dampers, grilles, airflow-measuring equipment

Crispaire Corp.
3570 American Dr.
Atlanta, GA 30341–2493
404–458–6643
HRVs, heat-pump water heaters

Des Champs Laboratories, Inc.
Box 220
Natural Bridge Station, VA 24579
703–291–1111
HRVs

Dornback Furnace and Foundry
9545 Granger Rd.
Garfield Heights, OH 44125
216–662–1600
Sealed-combustion oil furnace

Duro-Dyne Corp.
130 Route 110
Farmingdale, NY 11735
800–899–3876
516–249–9000
HRVs, fresh-air intake control package

413

Understanding Ventilation

Dust Free, Inc.
P.O. Box 519
Royse City, TX 75189
800-441-1107
214-635-9566
Electrostatic air filters

Dwyer Instruments, Inc.
P.O. Box 373
Michigan City, IN 46360
219-879-8000
Differential pressure transducers and air-pressure gauges

Econsys, Inc.
2590 E. Devon Ave.
Des Plaines, IL 60018
708-299-4460
Pressure controller

Eder Energy
7535 Halstead Dr.
Mound, MN 55364
612-446-1559
Blower doors

Emerson Electric Co. (Electro-Air)
White-Roders Division
9797 Reavis Rd.
St. Louis, MO 63123
314-577-1300
Electrostatic precipitators

Eneready Products Ltd.
6860 Antrim Ave.
Burnaby, BC, Canada V5J 4M4
604-433-5697
Ventilation equipment distributor, manufacturer of zone-control timers and grilles

Engineering Development Inc.
Vent-Aire Systems
4850 Northpark Dr.
Colorado Springs, CO 80918-3872
719-599-9080
HRVs, some with heating and cooling capability

Enviro-Energy Marketing Services Ltd.
P.O. Box 45057
Victoria, BC, Canada V8Z 7G9
604-595-7283
Inflatable fireplace draftstopper

Environment Air Ltd.
P.O. Box 10
Cocagne, NB, Canada E0A 1K0
506-576-6672
HRVs, balanced ventilators, exhaust fans

Exhausto, Inc.
P.O. Box 720651
Atlanta, GA 30358
800-255-2923
404-394-3156
Chimney exhaust fans

Fan America, Inc.
1748 Independence Blvd.
Suite F-5
Sarasota, FL 34234
813-359-3616
Quiet fans and multi-port ventilators

Fantech, Inc.
1712 Northgate Blvd. Suite B
Sarasota, FL 34234
800-747-1762
813-351-2947
Central-exhaust equipment, fans, wall vents, grilles, speed controls

Farr Co.
P.O. Box 92187
Los Angeles, CA 90009
800-333-7320
Full line of filtration equipment

Fasco Consumer Products, Inc.
P.O. Box 150
Fayetteville, NC 28302
800-334-4126
910-483-0421
Quiet fans, whole-house fans, general-purpose fans

Appendix E Equipment sources

First Alert
780 McClure Rd.
Aurora, IL 60504
800-323-9005
708-851-7330
Carbon-monoxide sensors

Foster Products Corp.
3210 LaBore Rd.
Vadnais Heights, MN 55110
612-481-9559
Duct-sealing mastic

E.L. Foust Co.
Box 105
Elmhurst, IL 60126
800-225-9549
708-834-4952
1" activated-carbon filters (mail-order supplier)

Furnace Tender
P.O. Box 890714
Houston, TX 77289
713-485-4421
Electrically operated dampers

Gas-Fired Products, Inc.
P.O. Box 36485
Charlotte, NC 28236
704-372-3485
Outdoor gas water heater

G-Controls, Inc.
10734 Lake City Way NE
P.O. Box 27354
Seattle, WA 98125
206-363-4863
Mixed-gas sensors

General Filters, Inc.
P.O. Box 8025
Novi, MI 48376-8025
810-476-5100
Medium-efficiency filters

Grainger, Inc.
333 Knightsbridge Pkwy.
Lincolnshire, IL 60069-3639
(over 300 local outlets in the U.S.)
Fans, dehumidistats, timers

Grasslin Controls Corp.
24 Park Way
Upper Saddle River, NJ 07458
201-825-9696
Timer controls

Greenheck
P.O. Box 410
Schofield, WI 54476-0410
715-359-6171
Various ventilating fans

Hardcast, Inc.
P.O. Box 1239
Wylie, TX 75098
800-527-7092
214-442-6545
Duct-sealing mastic

Heartland Products International
P.O. Box 777
Valley City, ND 58072
800-437-4780
701-845-1590
Dryer-vent closure

Home Equipment Mfg. Co.
P.O. Box 878
Westminster, CA 92684
800-854-6415
714-892-6681
Timer control

Honeywell, Inc.
1985 Douglas Dr. North
Golden Valley, MN 55422-3992
800-328-5111
612-951-1000
Filters, HRVs, TMAC control, misc. controls

Understanding Ventilation

Infiltec
P.O. Box 1533
Falls Church, VA 22041
703-820-7696
Blower doors, duct diagnostic fans

Intermatic Inc.
Intermatic Plaza
Spring Grove, IL 60081-9698
815-675-2321
Timer controls

Jenn-Aire Co.
3035 N. Shadeland Ave.
Indianapolis, IN 46226
317-545-2271
Range exhausts and dampered outlets

Johnson Controls, Inc.
P.O. Box 423
Milwaukee, WI 53201
414-274-5300
Carbon-dioxide controls

Kanaflakt, Inc.
1712 Northgate Blvd.
Sarasota, FL 34234
813-359-3267
Fans, grilles, controls, dampers

Kleen-Air Company, Inc.
269 West Caramel Dr.
Carmel, IN 46032
317-848-2757
Duct-mounted ozone generators

Lennox Industries
2100 Lake Park Blvd.
Richardson, TX 75080
214-497-5109
or
Lennox Industries
400 Norris Glen Rd.
Etobicoke, ON, Canada M9C 1H5
416-621-9321
HRVs

Lightolier
100 Lighting way
Secaucus, NJ 07096-1508
800-833-3664
201-864-3000
Motion sensors

Lomanco, Inc.
P.O. Box 519
Jacksonville, AR 72076
800-643-5596
501-982-6511
Whole-house cooling fans

Lymance International
P.O. Box 505
Jefferson, IN 47131
812-288-9953
Chimney dampers

Modus Instruments, Inc.
10 Bearfoot Rd.
Northboro, MA 01532
508-393-8991
Pressure transducers and gauges

Mon-Eco Industries, Inc.
5 Joanna Ct.
East Brunswick, NJ 08816
800-899-6326
908-257-7942
Duct-sealing mastic

Newtron Products Co.
P.O. Box 27175
Cincinnati, OH 45227
800-543-9149
513-561-7373
Electrostatic air filters

Nutone, Inc.
Madison & Red Bank Roads
Cincinnati, OH 45227-1599
800-543-8687
513-527-5100
HRVs, fans, controls

Appendix E Equipment sources

Nutech Energy Systems, Inc.
511 McCormick Blvd.
London, ON, Canada N5W 4C8
519–457–1904
HRVs, Scrubber air filters, grilles, controls, airflow balancing kit

Pacific Science & Technology Co.
64 NW Franklin Ave.
Bend, OR 97701
503–388–4774
Airflow measuring hoods

Panasonic
1 Panasonic Way
Secaucus, NJ 07094
201–348–5350
Quiet ceiling-mount fans

Penn Ventilator Co.
P.O. Box 52884
Philadelphia, PA 19115–7844
215–464–8900
Fans, speed controls, time-delay switches.

Permatron Corp.
11400 Melrose St.
Franklyn Park, IL 60131
800–882–8012
708–451–0999
Electrostatic air filters and thin carbon filters

Preston-Brock Mfg. Co., Inc.
1297 Industrial Rd.
Cambridge, ON, Canada N3H 4T8
519–653–7129
HRVs

Purafil, Inc.
P.O. Box 1188
Norcross, GA 30091
404–662–8545
Purafil adsorption material

Pure Air Systems, Inc.
P.O. Box 418
Plainfield, IN 46168
800–869–8025
317–839–9135
Central air filter systems

Radio Shack
Division of Tandy Corp.
P.O. Box 2625
Ft. Worth, TX 76113
817–390–3011
(Outlets in most major cities.)
Humidity meters

Raydot, Inc.
145 Jackson Ave.
Cokato, MN 55321
800–328–3813
612–286–2103
HRVs, filtration module

RCD Corp.
2310 Coolidge Ave.
Orlando, FL 32804
800–854–7494
407–422–0089
Duct-sealing mastic

RectorSeal Corp.
2830 Produce Row
Houston, TX 77023–5822
800–231–3345
713–928–6423
Duct-sealing mastic

Research Products Corp.
P.O. Box 1467
Madison, WI 53701–1467
800–334–6011
608–257–8801
HRVs, filters

Retrotec, Inc.
12–2200 Queen St.
Bellingham, WA 98226–4766
206–738–9835
Blower doors

Reversomatic Htg. & Mfg. Ltd.
790 Rowntree Dairy Rd.
Woodbridge, ON, Canada L4L 5V3
905–851–6701
Fans and multi-port ventilators

Riello Corp. of America
35 Pond Park Rd.
Hingham, MA 02043
617–749–8292
High-pressure oil burners

Rosenberg, Inc.
2500 W. Co. Rd. B, Suite 100
P.O. Box 130603
St. Paul, MN 55113
612–639–0846
Fans, speed controls, dampers

RSE, Inc.
P.O. Box 26
New Baltimore, MI 48047–0026
810–725–0192
Activated-carbon filter systems

Setra Systems, Inc.
45 Nagog Park
Acton, MA 01720
800–257–3872
508–263–1400
Pressure transducers

Shortridge Instruments, Inc.
7855 E. Redfield Rd.
Scottsdale, AZ 85260
602–991–6744
Airflow measurement devices

Skuttle Mfg. Co.
Route 10
Marietta, OH 45750–9990
800–848–9786
614–373–9169
Make-up air intakes

Snappy ADP
1011 11th St. SE
Detroit Lakes, MN 56501
218–847–9258
HRVs, ducts and fittings

Spacemaker Co.
1918 W. Chestnut St.
Santa Ana, CA 92703
714–542–4649
Outdoor water-heater enclosures

Staefa Control System, Inc.
8515 Miralani Dr.
San Diego, CA 92126
619–530–1000
Mixed-gas sensors

Stirling Technology Inc.
P.O. Box 2633
Athens, OH 45701
800–535–3448
614–594–2277
HRVs, central and window units

Tamarack Technologies, Inc.
15 Kendrick Rd.
P.O. Box 490
W. Wareham, MA 02576
800–222–5932
508–295–8103
Remote-mount bath fans, dehumidistats

Telaire Systems, Inc.
6489 Calle Real
Goleta, CA 93117
805–964–1699
Carbon-dioxide controller

Appendix E Equipment sources

The Energy Conservatory
5158 Bloomington Ave. South
Minneapolis, MN 55417
612–827–1117
Blower doors, duct diagnostic fans, digital micromanometers

Therma-Stor Products
Division of DEC International, Inc.
P.O. Box 8050
2001 S. Stoughton Rd.
Madison, WI 53708
800–533–7533
608–222–5301
Central air filters, heat-recovery water heaters, central-exhaust equipment, dehumidifiers, through-the-wall vents, etc.

Thermax
P.O. Box 300
Hopewell, NJ 08525–0300
800–929–0682
609–466–8800
HRVs, central and window units

The Watt Stopper, Inc.
2800 De La Cruz Blvd.
Santa Clara, CA 95050
800–323–9371
408–988–5331
Motion sensors

Thinking Vents, Inc.
P.O. Box 752
Mamaroneck, NY 10543–0752
914–698–7407
Passive ventilation system, through-the-wall vents

Titon Inc.
P.O. Box 6164
South Bend, IN 46660
219–271–9699
Window-mounted through-the-wall vents

Tjernlund Products, Inc.
1601 9th St.
White Bear Lake, MN 55110–6794
800–255–4208
612–426–2993
Fresh-air intake systems

Tork
1 Grove St.
Mount Vernon, NY 10550
914–664–3542
Controls

Tradewinds
270 Regency Ridge
Suite 210
Dayton, OH 45459
513–439–6676
HRVs

Trion, Inc.
P.O. Box 760
Sanford, NC 27331–0760
919–775–2201
Electrostatic precipitators

Trol-A-Temp
Division of Trolex Corp.
57 Bushes Ln.
Elmwood Park, NJ 07407–3204
800–828–8367
201–794–8004
TMAC ventilation control, motorized dampers & grilles

United McGill Corp.
P.O. Box 820
Columbus, OH 43216–0820
800–624–5535
614–443–5520
Duct-sealing mastic

Venmar Ventilation, Inc.
1715 Haggerty
Drummondville, PQ, Canada J2C 5P7
819–477–6226
HRVs, central-exhaust equipment, controls, grilles

Understanding Ventilation

Vent Air, Inc.
2801 SE Columbia Way
Bldg. 31, Suite 110
Vancouver, WA 98661
206–693–9112
Roof-mounted ventilation outlets

Vent-Axia, Inc.
230 Ballardvale St.
Building B
Wilmington, MA 01887
508–694–9336
Reversible fans, through-the-wall vents, controls

Ventex Systems Corp.
102–31234 Wheel Ave.
Abbotsford, BC, Canada V2S 1M3
604–852–2245
Odor-control toilet/fan combos

Viking Range Corp.
P.O. Drawer 956
Greenwood, MS 38930
601–455–1200
Range hoods

XeteX, Inc.
3530 E. 28th St.
Minneapolis, MN 55406
612–724–3101
HRVs

Ztech
2340 Gold River Rd.
Suite E
Gold River, CA 95670
916–635–7484
Fresh-air controller, motorized dampers

Index

A

Absolute humidity 103, 106–107, 115, 385. *See also* Moisture content of air; Relative humidity
Accessibility 271–272
Accidental pressures 21, 69, 385
Accidental ventilation 385. *See also* Ventilation: Accidental
Activated alumina 221–222, 385
Activated carbon 219–221, 385
Activated charcoal. *See* Activated carbon
Active 22, 37, 385
Active inlets 229
Active outlets 229
Adjusting airflows 262–265. *See also* Balancing
Adsorption 205, 217–222, 385
Affordable Comfort, Inc. 157
Air changes per hour (ACH) 161–162, 385
Air exchange 96–97
    Too little 82
    Too much 82–83
Air filtration. *See* Filtration
Air fresheners 43

Air handler 75, 386
Air Infiltration and Ventilation Centre (AIVC) 409
Air movement
    Active 37
    Interactions 37–39, 89–96, 170–172
    Layouts 247–248
    Passive 37
    Requirements for 35–37
Air noise 269. *See also* Noise
Air pollution. *See* Indoor air pollution; Outdoor air pollution
Air pressure 37, 62, 65, 73, 75–78, 78–80. *See also* Depressurization; Pressurization
    Controlling 52, 89, 125, 138–139
    Interactions 89–96, 231–232
        Rules 96
    Measuring 39–40
Air pressure and pollution 33–34
Air silencer 269, 386
Air-pressure controllers 201–202
Air-pressure difference 21, 35, 386

Understanding Ventilation

Air-to-air heat exchangers 87, 335, 386. *See also* Ventilation, Controlled: Heat-recovery systems
Airborne pollutants 32
Airflow grids 263, 295, 296, 321, 329, 336, 378, 379, 380, 385
Airtight construction 151, 386
Airtight drywall approach (ADA) 386
Airtight mechanical room 143
American Society of Heating, Refrigerating, and Air-Conditioning Engineers 53, 154, 162–163, 386, 409
Apparent sensible effectiveness (ASEF) 353–354, 386
Arsenic 30
Asbestos 31
Atmosphere 25
Atmospheric draft. *See* Natural draft
Atmospheric-spot-dust test 207–208, 386
Attached garages 143–144
Attic ventilation 43
Automatic timers 197
Axial fans 172, 386

B

Backdraft dampers 233, 252, 386
Backdrafting 131, 386
Backdrafting and spillage 57, 129–147, 282, 301, 315, 326, 339–340, 364
    Evaluating 145–147
    Solutions 137–145
    Summer 137
    Winter 136–137
Balanced ventilation 22, 386. *See also* Ventilation, Controlled: Balanced systems
Balancing 262–263, 330, 372, 386. *See also* Adjusting airflows
Basement dampness 54
Basement ventilating fan 286
Bath fans 308–309
Biological pollutants 33, 386. *See also* Moisture problems
Blower-door testing 152–156, 386
Bonneville Power Administration (BPA) 163, 177, 241, 294, 409
Breathing zone. *See* Headspace
British thermal units (Btus) 182, 386
Building envelope 386
Building materials 33
Bypasses 158–159, 386

C

Canada Mortgage and Housing Corp. (CMHC) 409
Canadian Standards Association (CSA) 352, 386, 409

Capacity 53, 161–174, 370, 374
    ASHRAE guidelines 163
    Effect of infiltration 170–172
    For comfort 162–163
    For formaldehyde removal 165
    For health 163–165
    Intermittent operation 166–169
    Limitations of recommendations 163–165
    Oversizing 166–169
    Overventilation 169–170
    To supply oxygen 162
Carbon dioxide 50, 53, 163–164, 386
    Due to exhaled breath 164
    In submarines 164
    Sensors 200–201, 386
Carbon monoxide 132–134, 386
    Detectors 146–147, 387
Carcinogens 30, 387
Carusorb 221
Category A rooms 244, 247–248, 387
Category B rooms 245, 247–248, 300, 318, 387
Ceiling fans 43
Central vacuum cleaners 72–73
Central ventilation 44, 387
Central-exhaust heat-pump water heater 290–291, 335, 357
Central-exhaust systems
    Freezing problems 283
Central-exhaust ventilation 387. *See also* Ventilation, Controlled: Central-exhaust systems
Central-supply ventilation 387. *See also* Ventilation, Controlled: Central-supply systems
Centrifugal fans 172, 387
Certification of HRVs 352–356
CFM_{50} 154
$CFM_{natural}$ 155
Chimney fans 144
Chimneys 73–74, 129–147, 387
    Airflow through 73–74, 320
    Dampers 74
    Safe draft pressures 146–147
Chlordane 121–122
Circulating air 22, 387
Cleaning products 33
Closed doors 78–80
Clothes dryers 69–72
    Exhaust fans 71
    Outlet closures 71
Combustion air 22, 142–144, 387
Combustion by-products 22, 29, 33, 52, 55, 132–134, 387
Combustion-air duct 142–144, 387
Combustion-air fan 144

Complexity 328–329, 341
Condensate drains 346–347
Condensation 103, 346–348, 387
Conditioned space 387
Continuous-duty motors 172, 387
Continuous-use motors 172, 387
Control, Lack of 364–365
Control of ventilation air 94–97, 326, 340
Controlled pressure 21, 81, 387
Controlled ventilation 387. *See also* Ventilation: Controlled
Controls 189–203, 232–233, 371
    Automatic 194–202
    Combined strategies 202
    Combining with furnace fan 202–203, 297
    Manual 190–194, 250
Convective air currents 65, 67, 387
Cost effectiveness 184–187
Costs 57, 175–187
    Capital 176–177, 280, 300, 314, 327, 340–341, 386
    Fan energy 174
    Filtration 227–228
    Health care 25
    Operating 177–184, 281, 301, 316, 327–328, 338, 341, 364, 391
    Tempering 179–184
Crawl-space ventilation 43, 54
Cubic feet per minute 161–162, 387

D

Dampers 232–233, 252, 296, 332, 387
Decay 113
Decibel 269, 387
Defrosting HRVs 347–348, 387
Degree days 181–182, 388
Dehumidification, Cost of 183–184
Dehumidifiers 56, 115
Dehumidistats 388. *See also* Humidity sensors
Delayed-off switches 192–193
Deliberate holes. *See* Inlets; Outlets
Demand-controlled ventilation (DCV) 195–196, 388
Depressurization 21, 37, 52–53, 71, 83, 117–127, 131, 277, 281, 299–300, 301, 373, 388
    Cause of backdrafting 135
    Measuring worst case 145–147
    Minimizing 282–283, 302
    Moisture problems 109–111
Depressurization-induced pollution 117–127, 281, 301, 315, 326, 339
Desiccants 344–345, 357, 358, 388
Design process 369–372

Dew point 103, 106, 388
Di-octyl phthalate (DOP) 208, 388
Differential-pressure transducers. *See* Air-pressure controllers
Diffusers 250, 388. *See also* Grilles
Diffusion 66–67, 107, 388
Dilution 22, 169, 388
Dilution air 22, 131, 135, 388
Displacement ventilation 248–249, 388
Distribution 243–265, 371, 388
DOP-smoke-penetration test 208, 388
Double-flow ventilation 85, 388
Downdraft kitchen range exhaust fans 306, 388
Draft 388. *See also* Natural draft
Draft free 151
Drying potential 111–112
Duct heaters
    Electric 239–240
    Hydronic 240
Duct insulation 253–255
Ductboard 254–255, 388
Ducts 252–258, 388
    Cleaning 252, 265
    Leaky 74–78, 155
    Locations 255–256
    Sealing 78, 145, 256–257, 294
    Sizing 258–262
    Testing for leaks 257
    Using heating/cooling ducts for ventilation 257–258
Dust mites 27
Dynamic wall 241

E

Earth tubes 241–242, 388
Effective Leakage Area (ELA) 154–155, 230, 388
Effectiveness 243, 244–249, 374, 388
Efficiency
    Fans 173–174
    Filters 206–209
    HRVs 350–356
Electrets 217, 388
Electric heating 141
Electrically-commutated motors 173, 194, 257, 258, 316, 321, 359, 388
Electricity costs 178–179
Electromagnetic radiation 31, 57, 389
Electronic air cleaners. *See* Electrostatic precipitators
Electrostatic air filters 216–217, 389
Electrostatic precipitators 215–216, 389
Energy Crafted Homes 141

Energy efficiency
    Duct insulation 253
    Fans 356
    Heat-recovery ventilators 350–356
Energy-Efficient Building Association (EEBA) 409
Energy-recovery ventilators 87, 335, 389. *See also* Ventilation, Controlled: Heat-recovery systems
Enthalpy 345, 354–356, 389
Environmental illness. *See* Multiple chemical sensitivity
Equipment costs 176
Equivalent Leakage Area (ELA) 154–155, 389
Equivalent length 259–260, 389
Evaporation 389
Exchange rate 22, 35, 96–97, 389. *See also* Capacity
Exfiltration 22, 45–46, 62, 389
Exhaust air 21, 389
Exhaust fans
    Multi-port exhaust 287–290
    Single-port exhaust 284–287
Exhaust-only ventilation 85, 277. *See also* Ventilation, Controlled: Central-exhaust systems
Extended-surface filters 213, 389

F

Fans, Ventilating 172–174
    Capacity 172–173
    Comparing 174
    Efficiency 173–174
    Wattage 356
Fiberglass 31. *See also* Insulation
Filtration 22, 43, 123, 205–228, 271–272, 327, 340, 389
    Costs 227–228
    Efficiency 206–209
    For hypersensitive people 228
    Limitations 205
    Location 209–210
    Miscellaneous strategies 222–225
    Packaged systems 225–227, 318
    Particulates 210–217
    Resistance to airflow 208–209
Flame rollout 133, 389
Flat-plate cores 343, 389
Flow-measurement hoods 263
Fluorescent lights 31
Forced-draft appliances 139–141, 389
Formaldehyde 29–30, 42, 170, 389
    And ventilation rates 165
Freezing in exhaust fans 283, 310
Freezing in HRVs 348
Fresh air 21, 389

Fresh-air ducts 296, 314, 318–321, 378, 389
Fuel utilization efficiency 183
Furnace damage due to cold air 142–143, 236, 296, 319
Future occupancy 168, 273

G

Gas 389
Gases as pollutants 29–30
General ventilation 21, 44, 389
Grilles 243–252, 261–262, 389, 393
    Adjusting 263–264
    Exhaust-grille locations 245–247
    Motorized 250–252
    Short circuiting 247
    Supply-grille locations 245
    Types 249–252

H

Headspace 245, 250, 389
Health effects, Negative 25, 27–28, 34, 48, 120, 122, 132–134, 223
Heat-pipe cores 344, 390
Heat-pump water heater 290–291, 335, 357
Heat-recovery ventilation 22, 87–88, 390. *See also* Ventilation, Controlled: Heat-recovery systems
HEPA filters 213–215, 390
Herbicides 121–122
High-pressure oil burners 141
Histoplasmosis 28
Holes, Deliberate 22, 229–242, 388. *See also* Inlets; Outlets
Holes, Random 22, 36–37, 45, 392
    vs. deliberate holes 230–231
Home Ventilating Institute (HVI) 163, 177, 259, 305, 308, 353, 390, 410
House plants 224–225
House tightening 158–159
House tightness 38, 62, 68, 124–125, 126, 134–135, 151–159, 231, 320
    Benefits 156–157, 158
    Measuring 152–156, 156
    Surveys 156
Humidifiers 31
Humidity sensors 199–200
Hygrometers 103, 390
Hypersensitive people 34, 47, 123, 167–168, 172, 214–215, 223, 253, 270. *See also* Multiple chemical sensitivity (MCS)
    Filtration for 228

Index

I

In-line fans 284, 390
Incidental costs 176–177
Indoor air pollution 26
    Sources of 27–32, 32–34, 118–124
    Strategies for reducing 41–46, 124–127
Induced-draft appliances 139–141, 390
Infiltration 22, 45–46, 62, 390
    Estimating 152, 154–155, 155–156, 170–172
    Shortcomings 156–157
Inlets, Fresh-air 229–242, 390. *See also* Through-the-wall vents
    Locations for 234–235
    Tempering incoming air 236–242
Insect screening 235–236
Installation costs 177
Installation process 369–372
Instruction plates 273
Insulation 33, 123–124, 159, 254–255, 270, 390
Invisible 267–268
Ionizers 222, 358, 390
Irritants 30, 390

J

Jump ducts 80, 390

K

Kerosene space heaters 138
Kilowatt hour (kWh) 178, 390
KISS principle 271, 390
Kitchen ranges 138
Kitchen-range exhaust fans 305–308

L

Labor costs 177
Latent heat 350–351, 354–356, 390
Lawrence Berkeley Laboratory (LBL) 154, 345, 410
Lead 30, 43, 57
Legionnaire's disease 27
Liters per second 162, 390
Living space 390. *See also* Occupied space
Local-exhaust systems
    Freezing problems 310
Local-exhaust ventilation 21, 84–85, 390. *See also* Ventilation, Controlled: Local-exhaust systems
Low-profile and microwave range hoods 306

M

Magnahelic gauges 40, 390
Maintenance 271–272, 280, 314, 342
    Costs 178
Make-up air 21, 83, 131, 390
Manometers 40, 390
Manual timers 192–193, 250–251
Mechanical filters 211
Mechanical ventilation 390
Media filters 211, 390
Medium-efficiency filters 212–213, 390
Mercury 30
Metabolism, By-products of 30, 34, 53
Metals as pollutants 30
Microns 207, 390
Military Standard 282 test 208
Minerals as pollutants 31
Minimizing the number of exhaust fans in a house 304
Mixed-gas sensors 201, 390
Modulating dampers 252, 390
Moisture content
    Of air 102–103
    Of wood 113, 391
Moisture control 53–56, 344–348. *See also* Ventilation, Reasons for
Moisture problems 53–56, 101–116, 122, 169, 326. *See also* Decay; Depressurization-induced pollution; Relative humidity
    From building materials 55
    Hidden 107–113, 339
        Mixed climates 111–113
        Summer 109–111, 281–282, 301, 315, 345
        Winter 108–109, 280–281, 300, 316, 346
Mold 27–28. *See also* Moisture problems
Monday-morning effect 168
Motion sensors 197–199, 391
Motorized dampers 233, 252, 296–297, 302–303, 348
Multi-port exhaust fans 279, 391
Multiple chemical sensitivity (MCS) 26, 48–49, 391. *See also* Hypersensitive people
Mutagens 30, 391

N

National Aeronautics and Space Administration (NASA) 224
National Coalition Against the Misuse of Pesticides (NCAMP) 28, 391
National Pesticides Telecommunications Network 122

Understanding Ventilation

Natural aspiration. *See* Natural draft
Natural draft 129–132, 391. *See also* Chimneys
Natural pressures 21, 61–67, 391
Natural ventilation 391. *See also* Ventilation, Natural
Negative pressure 21, 391. *See also* Depressurization
Negative-ion generators 222, 358, 391
Net free area 235, 391
Neurotoxins 30, 391
Neutral pressure 22, 85, 125, 391
Neutral-pressure plane 64, 136–137, 391
Noise 269, 305, 307, 309–310

O

Occupancy sensors. *See* Motion sensors
Occupant education 272–273
Occupied space 42, 52, 238, 391
Odor free 270
Olf 50, 391
On/off switches 191–192
Operating costs 177–184
Outdoor air pollution 25–26, 123
Outdoor boilers 144–145
Outdoor furnaces 144–145
Outdoor water heaters 144–145
Outgassing 30, 391
Outlets, Stale-air 229–242, 391. *See also* Through-the-wall vents
    Locations for 235
Overhead range hoods 305–306
Overventilation 114, 169–170
Ozone 26, 215, 222, 391
Ozone generators 223–224, 391

P

Partial-bypass filters 218, 391
Particulate filters 210–217, 391
Particulates 205, 391
Parts per million 391
Pascal 40, 391
Pass throughs 79, 392
Passive 22, 37, 392
Passive inlets 229
Passive outlets 229
Passive ventilation 392. *See also* Ventilation, Controlled: Passive systems
Perfect mixing 169
Permanent-split-capacitor motors 172, 173, 392
Permeable walls 392
Pesticides 28–29, 121–122, 392

PicoCuries per liter 120, 392
Planning ahead 265, 273, 369–370
Pop-up downdraft kitchen range exhaust fans 306–307
Porosity 67, 392
Positive pressure 21. *See also* Pressurization
Predictability, Lack of 365
Pressurization 21, 37, 85, 125, 315, 392. *See also* Air pressure
    Moisture problems 108–109
Primary ventilation 44
Psychometric chart 104–107, 392
Purafil 221

Q

Quiet. *See* Noise
Quiet fans 309–310

R

R-2000 program 141, 151, 163, 263
Radiation as pollution 31
Radon 26, 31, 52, 120–121, 124, 392
    And smoking 120
    From building materials 120
Random holes. *See* Holes, Random; House tightness
Recirculating range hoods 305, 392
Recirculation 22, 392
Relative humidity 102–107, 392. *See also* Moisture problems
    Effect of ventilation 113–115
    Measuring 103
    Predicting 104–107
    Recommended level 115–116, 316
Remote-mounted kitchen fans 307–308
Remote-mounted local-exhaust fans 310
Return air 74–78, 392
Reversible fans 286–287
Roof-turbine ventilators 366
Room-to-room fans 168
Rotary cores 343–344, 392

S

Sealed-combustion appliances 141, 392
Secondary ventilation 45
Sensible heat 350–355, 392
Sensible recovery efficiency (SRE) 354, 392
Sensitive people. *See* Hypersensitive people
Separation 22, 42–43, 47, 49, 56, 114, 393
Shaded-pole motors 172, 173, 393

Short circuiting 247, 393. *See also* Distribution
Simplicity 271, 363–364
Single-port ventilators 393
Smoke detectors 31
Smoking 31–33, 120, 271
Soil gases 118–123
Solar preheaters 241
Sone 269, 393
Sound attenuator 269, 393
Source control 22, 41–42, 47, 49, 56, 114, 270, 393
Special-purpose exhaust fans 311
Speed controls 194
Spillage 131, 393. *See also* Backdrafting and spillage
Spot ventilation 45, 393. *See also* Ventilation, Controlled: Local-exhaust systems
Squirrel-cage fans 172, 393
Stack effect 63–66, 393
Stale air 21
Static pressure 173, 393
Static-pressure drop 209, 393
Strain hygrometer 103, 393
Stuffiness 164
Supply air 21, 393
Supply fan coupled to local-exhaust fan 303–304
Supply fans 317–318
Supply-only ventilation 85, 313, 393. *See also* Ventilation, Controlled: Central-supply systems
Surface-mounted downdraft range exhaust fans 306
Surrogate 164, 393
Sweating ducts 253–254
Sweating of HRVs 347
Swimming pools, Ventilation of 358
Synergism 30, 393

T

Taking responsibility 270–271, 272
Temperature difference 181, 393
Tempering 179, 236–242, 338–339, 393
    Costs 179–184
Teratogens 30, 393
Termiticides 33, 52, 121–122
Through-the-wall vents 83–84, 237–239, 279–280, 291–294, 298, 302–303, 318, 374, 393
    In bedrooms 230
Tight houses. *See* House tightness
Toilet ventilating fan 310–311
Total recovery efficiency (TRE) 354, 393
Tracer-gas testing 152, 393
Transfer grilles 79

Transparency 267–270, 393
Treated lumber 30, 124
Trisodium phosphate (TSP) 393
Tube fans 172

U

U.S. Environmental Protection Agency (USEPA) 394, 410
Undercutting doors 244, 278
Upgraded bath fan 375
Urea-formaldehyde foam insulation (UFFI) 394

V

Vapor 394
Vapor pressure 394
Vent-free gas fireplaces 138
Ventilating chimney 362, 394
Ventilation, Accidental 37, 43, 69–80
Ventilation, Controlled 37, 43–44, 81–97, 394
    Balanced systems 85–87, 126, 294–297, 377, 379, 380
        Advantages 138–139, 326–327
        Coupling to furnace/AC 329–333
        Disadvantages 327–329
        Equipment 329
        Strategies 324–326
    Central-exhaust systems 85, 125, 277–298, 299–311, 323–333, 375, 376
        Advantages 280–281
        Coupling to furnace/AC 294–297
        Disadvantages 138–139, 281–283
        Equipment 283–294
        For cooling 277–278
        Strategies 278–280
    Central-supply systems 85, 126, 313–322, 378
        Advantages 138, 314
        Coupling to furnace/AC 318–321
        Disadvantages 316
        Equipment 317–318, 321
        Strategies 313–314
    Heat-recovery systems 87–88, 335–359, 379, 380
        Advantages 114–115, 239, 338–340
        Core types 342–344
        Cost effectiveness 184–186
        Coupling to furnace/AC 358–359
        Disadvantages 340–341
        Energy efficiency 184–186, 350–356
        Equipment 356–358
        Moisture control 344–348
        Strategies 338

Understanding Ventilation

Ventilation, Controlled, *continued*
    In loose houses 157–158
    Local-exhaust systems 44–45, 84–85, 375
        Advantages 300
        Avoiding excess depressurization 302
        Disadvantages 138–139, 300–301
        Equipment 304–311
        Minimizing number of fans 304
    Passive systems 88–89, 361–367
        Advantages 363–364
        Disadvantages 364
        Equipment 365–366
        Strategies 362–363
    Strategies 44–45
Ventilation controls. *See* Controls
Ventilation costs. *See* Costs
Ventilation effectiveness. *See* Effectiveness
Ventilation, Natural 37, 43, 61–68
    Limitations 156–157
Ventilation rate. *See* Capacity
Ventilation, Reasons for 47–57, 49–57
    For moisture control 113–115. *See also* Moisture control
        External sources 54–55
        Internal sources 55–57

Ventilation, Controlled, *continued*
    To dilute indoor pollutants 51–52
    To dilute metabolic pollutants 53
    To dilute non-indoor pollutants 52
Ventilation, Uncontrolled 82–83
Ventilation-induced pollution. *See* Backdrafting and spillage; Depressurization-induced pollution; Moisture problems
Vibrations 268–269, 269
Volatile organic compounds (VOCs) 29–30, 170, 394
    Health effects 30
    Outgassing rates 30

W

Water gauge (w.g.) 39, 394
Weight-arrestance test 207, 394
Wetting potential 111–112, 394
Whole-house fans 277–278, 394
Wind 62–63, 394
Window-mounted HRVs 357
Windows, Open 303, 362

Also available from The Healthy House Institute

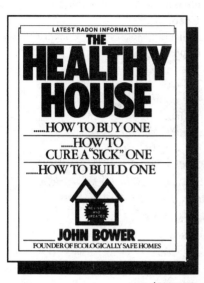

$17.95

Since its release in 1989, **The Healthy House** has been praised by builders, architects, and homeowners as an essential reference for selecting less-toxic construction materials. An encyclopedic resource, it discusses the potential negative health effects of poor indoor air quality and offers many practical solutions. Illustrated and fully referenced, it also contains a listing of 200 suppliers of healthier materials. Published by Carol Communications, Inc. Trade paperback edition, $8^1/_2$" x 11", 392 pages.

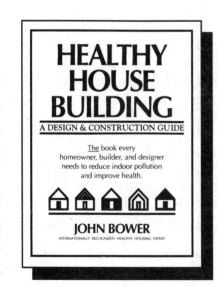

$21.95

How a house is put together is as important as what materials are used. With over 200 photos and illustrations, as well as a complete set of house plans, **Healthy House Building** takes you step by step through the construction of a Model Healthy House. With discussions of alternative materials and techniques, this is the most complete, in-depth guide to healthy construction and remodeling available today. Published in 1993 by The Healthy House Institute. Trade paperback, $8^1/_2$" x 11", 384 pages.

Order today from The Healthy House Institute, 430 N. Sewell Road, Bloomington, IN 47408. Or call 812–332–5073. MasterCard and Visa orders accepted.

Shipping: $3.00 for the first book or video plus $2.00 for each additional book or video. Indiana residents, please add 5% sales tax.

Also available from The Healthy House Institute

$19.95

Your House, Your Health is the companion video to *Healthy House Building*. Author John Bower discusses the basic causes of indoor air pollution and walks you through a Model Healthy House, pointing out all the important features. This excellent visual introduction takes much of the mystery out of what goes into a healthy house. Produced by The Healthy House Institute in 1992. VHS video, 27.5 minutes.

$17.95

The Healthy Household offers hundreds of practical suggestions for improving your indoor environment. With chapters devoted to cleaning products, personal-care items, clothing, linens, interior decorating, lifestyles, housekeeping, air and water quality, and electromagnetic radiation, no other book matches its completeness. It is an essential reference for anyone interested in their family's health—especially those who are chemically sensitive. Published in 1995 by The Healthy House Institute. Trade paperback, 6" x 9", 480 pages.

Order today from The Healthy House Institute, 430 N. Sewell Road, Bloomington, IN 47408. Or call 812-332-5073. MasterCard and Visa orders accepted.

Shipping: $3.00 for the first book or video plus $2.00 for each additional book or video. Indiana residents, please add 5% sales tax.

Also available from The Healthy House Institute

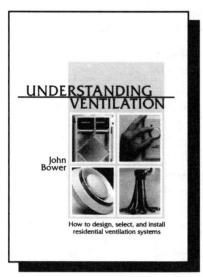

$31.95

Understanding Ventilation covers all aspects of exchanging the air in houses: infiltration, equipment selection, sizing, costs, controls, filters, distribution, and possible problems that a ventilation system can cause—and all in easy-to-understand language. Any architect, builder, or homeowner interested in maintaining a healthy, comfortable indoor environment will find this manual indispensable. Fully illustrated. Published in 1995 by The Healthy House Institute. Hardcover, $8\frac{1}{2}$" x 11", 432 pages.

Personal consultation services

John Bower offers personal consultation services either by telephone or on-site, to discuss your specific situation: a new construction project, an existing problem, general questions about healthy-house design, healthier lifestyles, house plan review, etc. A brief consultation can prevent mistakes, saving you thousands of dollars, and improve your indoor environment. This service is available at $50.00 per hour, with a $25.00 minimum.

Order today from The Healthy House Institute, 430 N. Sewell Road, Bloomington, IN 47408. Or call 812–332–5073. MasterCard and Visa orders accepted.

Shipping: $3.00 for the first book or video plus $2.00 for each additional book or video. Indiana residents, please add 5% sales tax.

ORDER FORM

▌▌▌ THE HEALTHY HOUSE INSTITUTE
430 N. Sewell Road   Bloomington, IN 47408   (812) 332–5073

NAME

ADDRESS

CITY, STATE, ZIP

PHONE (   )   DATE

Qty.	Description	Price Each	Amount

PAYMENT METHOD

☐ Check or Money Order   ☐ Master Card   ☐ Visa

Expiration date _____

Signature _____

	Subtotal	
	Indiana residents add 5% sales tax	
	Shipping: $3.00 for the first book or video, and $2.00 for each additional book or video.	
	TOTAL	

Payment must be in U.S. funds